ELECTRIC POWER AND ENERGY IN CHINA

Zhenya Liu
State Grid Corporation of China, China

WILEY

Library of Congress Cataloging-in-Publication Data

Liu, Zhenya, 1952-
  Electric power and energy in China / Zhenya Liu.
      pages cm
  Includes bibliographical references and index.
  ISBN 978-1-118-71635-9 (cloth)
  1. Power resources—China. 2. Energy development—China. 3. Energy consumption—China.
4. Electrification—China. 5. Electric power—China. I. Title.
  TJ163.25.C6L58 2013
  333.790951—dc23

                                                          2013016714

ISBN: 9781118716359

Set in 10/12pt Times by Laserwords Private Limited, Chennai, India

ELECTRIC POWER
AND ENERGY IN CHINA

Contents

About the Author

Zhenya Liu is the Chairman of the State Grid Corporation of China. As a senior engineer, he is entitled to the special governmental allowances offered by the State Council of the Chinese Government and is one of the most renowned experts in the global economy and energy sector.

The project 'UHV AC Transmission Key Technology, Equipment and Engineering Application' led by Mr Liu was granted the 2012 National Award for Science and Technology Progress (Special Prize). Mr Liu also proposed an innovative theoretical and technical roadmap for UHV transmission tailored to China's needs. Under his leadership, a series of major challenges have been well-addressed, such as the UHV AC voltage control, test technology, system integration and UHV DC large-capacity converter valve, DC engineering standardisation, etc.

Mr Liu has also led the world's first UHV AC transmission project and state-of-the-art UHV DC transmission project to success, through which China has fully mastered the UHV core technology with independent intellectual property rights as well as the manufacturing capacity of full-spectrum equipments. In addition, Mr Liu's publications, including *Electric Power and Energy in China*, and *Ultra High Voltage Power Grid*, have been well-received in China and are bestsellers in their field.

Preface

Energy is an important material foundation for economic growth and social development. The use of energy has brought about revolutionary changes to human society, making production and human life more convenient and comfortable. A historical survey of human social development reveals that every major leap in the progress of civilisation was accompanied by the advancement and transformation of energy sources. At the same time, the development and use of energy have also transformed the natural environment on which humanity depends upon for its survival. Environmental destruction, climate changes and depletion of resources have posed huge challenges for human development. It is humanity's shared desire to build up an energy supply framework that is safe, efficient, clean and economical.

The sustainable development of the energy industry is a key strategic issue that affects China's overall economic and social development. Since China began its reforms and opened up its economy, its energy industry has seen rapid growth, providing robust support for the country's development. However, what has gradually emerged from this process is the tension between economic growth and energy development on one side and resources and the environment on the other. China is rich in energy resources, but its per capita consumption of these resources is lower than the global average. In China's energy structure, coal has long been dominant. High quality and clean energy sources like oil and natural gas, in contrast, are used in smaller proportions, and the level of electrification is low. In recent years, China's demand for energy has continued to grow rapidly. The pollutants and greenhouse gases produced by the massive consumption of fossil fuels have had severe effects on the environment. Given China's limited resources endowments and increasing dependence on foreign oil, there is an inherent risk in the country's energy security. Unscientific development has led to repeated shortages and inadequacies in coal, electricity, oil, gases and transportation. China is expected to continue its economic and social development, and its demand for energy will keep growing. This, together with the increasing complexities of the international energy situation and the climate change issue, will result in greater limitations and conflicts in China's energy development. The pressure to safeguard the country's energy security will also intensify.

To resolve China's energy problems, we must survey the world with our feet planted in China, so as to acquire an overall understanding of the situation. This book begins by analysing the state of energy in the world and in China, and studying the problems faced by the country's energy development and their underlying causes. Recognising the complexity of the energy issue and adopting a grand energy vision, this book proposes a fundamental approach to resolving China's energy problems. The basic idea involves planning for changes in the mode of economic development, the mode of energy development and the dynamics of global competition; pursuing new industrialisation and modernisation of the energy sector with Chinese characteristics; and creating a relatively comfortable and favourable international milieu. The process of changing the mode of energy development is the process of transforming China's energy strategy. The transformation of energy strategy comprises various dimensions: the transformation of the energy structure from high-carbon to low-carbon mode, the transformation of the energy use pattern from one being extensive to one of intensiveness and efficiency, the transformation of energy allocation from ensuring local balance to an optimised distribution network covering large areas, the transformation of energy supply from ensuring domestic supplies to coordinated utilisation of both overseas and domestic resources, and the transformation of energy services from a one-way supply to smart interactive supply.

The year 2003 was the milestone year marking the strategic transformation of China's energy development. The proposal of the Scientific Outlook on Development brought the Theoretical System of Socialism with Chinese Characteristics closer to perfection. With the Scientific Outlook on Development as a guide, China's energy policies underwent new adjustments and changes, which facilitated the changing of the mode of energy development and the transformation of China's energy strategy. In that year, China's GDP per capita exceeded US$1 000, economic and social development entered a new phase, and the pace of industrialisation and urbanisation accelerated. Thus, the divergence between the supply and demand of energy became more acute, placing critical pressure on changing the mode of development. With the combined efforts of various sectors, there have been some changes in the mode of energy development in China, but there is still a long way to go.

This book analyses the basic premise in resolving China's energy problem and the transformation of the mode of its energy development. It argues that the core of the energy strategy should be electrical power, and that the core mission should be the implementation of the 'One Ultra Four Large' (1U4L) strategy. On this basis, the following issues are systematically discussed: the development and use of energy, the transmission and distribution of energy, the end consumption of energy, the energy market, energy warnings and contingency actions, innovation in energy technology, and ensuring the sustainable development of the energy sector.

The book is divided into nine chapters:

Chapter 1 analyses the world energy situation and gives a summary of its characteristics. It also gives an overview of China's energy situation and the major problems the country faces, and analyses the causes that affect China's energy development.

Chapter 2 proposes the basic idea to resolve China's energy problems and the basic path in implementing China's energy strategy. It analyses the central link in the energy strategy and argues that its core should be electrical power and the implementation of the 1U4L strategy.

Chapter 3 gives a general idea of the development and use of energy in China. It explores the major issues involved in the development and use of coal, oil, natural gas, hydropower, nuclear power, as well as new and renewable energy sources like wind and solar power. The chapter also proposes international energy cooperation and the active exploitation of overseas resources.

Chapter 4 analyses the implication and principles of constructing a modern and comprehensive power transmission framework. It explores strategic issues such as improving the mode of energy distribution by simultaneously developing coal and electricity transmission but speeding up the latter, building up a powerful and smart ultra high voltage (UHV) grid, and refining the oil and gas distribution networks.

Chapter 5 analyses the challenges of China's energy consumption and the structural problems of the green energy consumption model. It focuses on the three strategic measures of implementing energy conservation as a strategic priority, increasing electrification and developing electric vehicles.

Chapter 6 explores the issue of developing energy markets, analysing key points in the building up of the coal market, electricity market and pricing framework for oil and gas. It also discusses the issue of regulating the new energy market.

Chapter 7 is a study of energy warnings and contingency actions. It analyses the implications of boosting the energy warning system and contingency plans. It also describes the thoughts and emphasises on improving the energy warning and contingency systems and the build-up of energy reserves.

Chapter 8 analyses the challenges in the innovation of China's energy technology. It explains its basic principles, the important areas and the goals, and explores effective measures in refining the system for energy technology innovation.

Chapter 9 describes the measures to facilitate the transformation of the mode of energy development from the perspectives of laws, regulations and policies, system standardisation and corporate development. It analyses important issues like being policy-oriented, being more prominent in the international discourse on standardisation and supporting the development of large-scale power corporations.

Looking ahead, the future for China's energy development is grim, with many challenges and difficult tasks. To transcend these difficulties and provide the energy security for economic and social development, a spirit of reform and innovation, led by Scientific Outlook on Development must be harnessed to bring about changes in the mode of energy development, adjustments in the development and use of energy and the optimisation of energy transport and allocation. At the same time, a green energy consumption model must be constructed, the energy market framework fine-tuned and the energy warning system and contingency actions improved. The supportive and leading role of energy technology, the standardization and guidance of laws, policies and standards, and the driving force of the main energy market participants should also be fully harnessed. All these tasks should be tirelessly pursued with a down-to-earth approach.

The ideas and perspectives expressed in this book on the changing mode of energy development are the results of my 40-year career in the energy industry and my thoughts on China's energy strategy. I have also referred to the writings of experts and scholars. As the opinions put forth in this book are my own, there will naturally be inadequacies. By publishing this book, I hope to contribute to the study of China's energy strategy.

Zhenya Liu
May 2013

1

Energy: An Overview

The constant changes in global economy and politics, and the development of a new revolution in energy technology, are bringing about a profound transformation in the world's energy structure. Globalisation of the world economy means that this transformation will have a significant impact on China's energy development. Even as the energy industry in China is achieving great successes, it is facing a slew of challenges. A thorough understanding of the energy situation within and outside China, and in-depth analyses of the major problems in China's power industry and their causes are prerequisites to formulating a scientific energy strategy for the country.

1.1 An Overview of the World's Energy Situation

In recent years, economic and social factors, resources and the environment, science and technology, as well as the inherent laws governing energy development, have helped shape new trends and characteristics in the global energy situation.

1.1.1 The Global Energy Situation

As the world today becomes more and more dependent on energy due to economic and social development, energy development is increasingly constrained by ecosystems and the environment. The issue of security is also a matter of concern among many in the international community. It is against this backdrop that a new revolution in energy technology is gestating and developing in the whole world.

1.1.1.1 Unprecedented Level of Dependence on Energy due to Economic and Social Development

Energy development and exploitation is a hallmark of human progress. From a historical perspective, advancements in human civilisations were always accompanied by revolutions in energy technologies. In an agricultural economy, energy is required to satisfy only general human needs, and its main sources were biomass fuels like firewood. In the

---

second half of the 18th century, the proliferation of the steam engine heralded the first industrial revolution in the history of mankind. Mechanised production began to replace manual labour and the demand for energy increased radically. Coal gradually replaced firewood and so on to become the world's main source of fuel. Energy began to play an important role in the economic and social development of mankind; it became the key element of modernised production and the foundation on which modern material civilisation was built. Between the 19th and the early 20th centuries, the births of two revolutionary technologies—the internal combustion engine and electricity—further consolidated the status of energy as the bedrock of the modern economy and society. The demand for fossil fuels began its meteoric rise. Following the end of the Second World War, with the growth of the automobile industry and the emergence of multinational oil companies, oil consumption rocketed, gradually overtaking coal in the share of the energy consumption structure. The discovery of large numbers of natural gas fields propelled the exploration and exploitation of gas. By the early 21st century, the three fossil fuels of oil, coal and natural gas became the main sources of the world's energy supply, accounting for over 80% of the world's total energy consumption. They provide almost 100% of the energy used for transportation and over 65% of the primary energy for power generation. The exploration and use of energy resources facilitated humanity's progress and the development of the world economy. Modern agriculture, industry and services need the support of energy to stay in operation. It can be said that without energy, modern civilisation will vanish completely.

With human society's increasing dependence on energy, energy has become the key element of sustainable economic and social development. The adequacy and cost of energy supply is having a greater affect on development. The two oil crises in the 1970s had a devastating effect on a world economy that was heavily dependent on oil. In the early years of 21st century, energy prices as represented by oil have seen a large increase, which added greatly to the cost of social and economic development. The era of cheap energy is over. For developing nations still at the industrialisation stage, the ability of energy to put a brake on development is much more pronounced. Developed countries had earlier on completed their industrialisation process by using large quantities of cheap energy resources to fuel the rapid growth of their economies. As developing countries enter their stages of rapid growth, they can no longer obtain large quantities of high quality but cheap energy resources from the global market so easily. They cannot simply replicate the developmental models of developed countries; they must rely on innovation to transform their modes of development and achieve sustained growth.

1.1.1.2 Energy Production and Consumption Patterns are Undergoing Profound Transformation

Since the first Industrial Revolution, fossil fuels have occupied a leading position in human energy production and consumption. Prior to any breakthrough in energy use and before the discovery of new energy sources that are plentiful enough to replace fossil fuels, fossil fuels remain the basic energy sources for the development of the world economy. Human consumption patterns based on fossil fuels will not see a fundamental change for quite some time. In 2009, the world's primary energy consumption was approximately 17.33 billion tonnes of standard coal, with the fossil fuels of coal, oil and natural gas accounting for 80.9%.

Table 1.1 The level and composition of the world's primary energy consumption.

Type of energy	1971	1980	1990	2000	2005	2009
Coal	26.1%	24.8%	25.3%	22.9%	25.3%	27.2%
Oil	44.0%	43.0%	36.7%	36.5%	35.1%	32.9%
Natural gas	16.2%	17.1%	19.1%	20.8%	20.7%	20.9%
Nuclear power	0.5%	2.6%	6.0%	6.7%	6.3%	5.8%
Hydropower	1.9%	2.0%	2.1%	2.2%	2.2%	2.3%
Others	11.3%	10.5%	10.8%	10.9%	10.4%	10.9%
Total (billion tonnes of standard coal)	7.904	10.327	12.516	14.312	16.322	17.331

Source: International Energy Agency (IEA).

Even as fossil fuels have long taken the lead as the main suppliers of energy, the world's energy production and consumption patterns are quietly undergoing profound transformation. Informed by their concern over energy supply security and global climate changes, many countries are beginning to look for alternatives to traditional fossil fuels, and decreasing the share of fossil fuels, especially oil, in energy consumption. The share of oil in the world's primary energy consumption reached a peak in 1973, after which it began to fall steadily. The percentage had dropped to 32.9% by 2009, a drop of 13 percentage points from 1973. In terms of energy end-use, the share of oil also dropped from a peak of 48.1% (in 1972) to 41.6% in 2009. In the same period, the total share of natural gas and nonfossil fuels in global primary energy consumption has shown an increase. Between 1971 and 2009, the share of natural gas went up by 4.7 percentage points, while the share of nonfossil fuels grew by 5.8 percentage points. Table 1.1 shows that quantity and composition of the world's primary energy consumption.

The fall in the share of oil and the rise in nonfossil fuels are especially evident in developed countries. In 1973 the share of oil in primary energy consumption by Organisation for Economic Cooperation and Development (OECD) member countries was 52.5%. Nonfossil fuels accounted for only 5.9%. By 2009, the share of oil had dropped to 37.4%, but nonfossil fuels had risen to 19.0%.

In terms of energy end-use, the share of fossil fuels continued to decrease, while the share of electricity saw a significant increase. Increasing quantities of fossil fuels like coal and natural gas were converted into electricity. Between 1971 and 2009, the share of fossil fuels like coal, oil and natural gas in the world's energy end-use fell by 9 percentage points, while the share of electricity almost doubled, hitting 17.3% in 2009. The level and composition of the world's energy end-use is shown in Table 1.2.

Industry and transportation are the upmost major energy end-users. In the decade after 1998, the transport sector's energy end-use surpassed that of industry. However, with the rise of global oil prices and the industrialisation of developing countries, industry's energy end-use once again exceeded transport in 2008, becoming the biggest end-user sector. In 2009 the total energy end-use of both industry and transport was almost equal, accounting for 27.4% of the world's energy end-use.

The geographical spread of the world's energy use was also undergoing profound changes. The rapid economic growth of developing countries increased the demand for energy. In 2004 energy consumption by developing non-OECD countries surpassed

Table 1.2 The level and composition of the world's energy end-use.

Type of energy	1971	1980	1990	2000	2005	2009
Coal	14.6%	13.0%	12.1%	7.6%	8.4%	9.8%
Oil	46.8%	45.3%	41.4%	44.2%	43.5%	41.6%
Natural gas	14.2%	15.4%	15.2%	16.1%	15.6%	15.2%
Electricity	8.8%	10.9%	13.3%	15.4%	16.4%	17.3%
Thermal power	1.6%	2.2%	5.3%	3.5%	3.4%	3.0%
Others	14.0%	13.2%	12.7%	13.2%	12.7%	13.1%
Total (billion tonnes of standard coal)	6.080	7.688	8.990	10.053	11.255	11.899

Source: International Energy Agency (IEA).

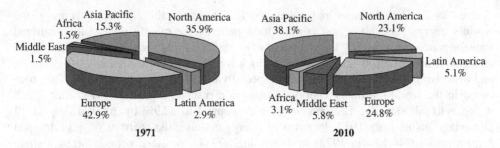

Figure 1.1 Distribution of the world's primary energy consumption.
Source: British Petroleum (BP), *Statistical Review of World Energy*, 2011.

developed OECD countries for the first time. The weight of global energy use began to shift towards developing countries. Among them, the percentage of world energy consumption by China, India, Russia, Brazil and South Africa rose from 26.7% in 1990 to 32.8% in 2009. The annual growth in their energy consumption is 2.1 times the world average.

In terms of energy use, the Asia Pacific is the fastest growing region in the last 40 years. Between 1971 and 2010, it accounted for 55.2% of newly added primary energy use in the world. The region's share in world energy consumption increased from 15.3% to 38.1% (see Figure 1.1). In 2003 the Asia Pacific overtook Europe as the region that consumed the most energy in the world.

In terms of energy consumption per capita, developed countries were still well ahead of developing countries. In 2009 the energy use per capita in OECD countries was 6.11 tonnes of standard coal, but in non-OECD countries it was 1.70 tonnes, a mere 27.8% of the former.

1.1.1.3 Resources and the Environment Exerting More Restrictions on Energy Development

The world's resources of fossil fuels are relatively abundant on the whole. At the end of 2010, the remaining proven reserves of coal stood at 860.9 billion tonnes. The

Table 1.3 Countries with the biggest remaining proven reserves of fossil fuels.

Rank	Coal		Oil		Natural Gas	
	Country	Reserves (billion tonnes)	Country	Reserves (billion barrels)	Country	Reserves (trillion cubic metres)
1	United States	237.3	Saudi Arabia	264.5	Russia	44.8
2	Russia	157.0	Venezuela	211.2	Iran	29.6
3	China	114.5	Iran	137.0	Qatar	25.3
4	Australia	76.4	Iraq	115.0	Turkmenistan	8.0
5	India	60.6	Kuwait	101.5	Saudi Arabia	8.0

Source: British Petroleum (BP), *Statistical Review of World Energy*, 2011 (2010 data).

reserve-production ratio (RPR)[1] is 118 years. For oil, there were 188.8 billion tonnes of remaining proven reserves with an RPR of 46.2 years. For natural gas, the figures are 187.1 trillion cubic metres and 58.6 years respectively. Table 1.3 lists the top five countries with the biggest remaining proven reserves of fossil fuels.

The sustainable supply of resources has become an important factor in restricting the world's energy development. Although the remaining reserves of fossil fuels have yet to impose a substantive limitation on global energy supply, the nonrenewable nature of fossil fuels and the increasing costs in exploration have highlighted the problem of ensuring the sustainable supply of energy. The early 21st century has witnessed the increasing scarcity of overall supply of fossil fuels, especially oil. Given factors like geopolitics, local wars, decreased investments, aged oil fields and increased demand, the demand and supply balance of the world's oil tends to be fragile. The stability of oil supply in the future has become a widespread concern in the international community. For most countries, especially developing countries in the industrialisation phase, the limitations that energy supply sustainability imposes on their socioeconomic development are gradually becoming more obvious, given that new and renewable energy sources have yet to sufficiently replace fossil fuels.

The issue of greenhouse gas emissions produced by burning fossil fuels is also becoming a cause of concern for many people. On a global scale, the carbon dioxide produced by burning fossil fuels accounts for 56.6% of all greenhouse gases emitted by human activity and 73.8% of carbon dioxide emissions (Figure 1.2). In 2009 the amount of carbon dioxide produced by burning fossil fuels reached 29 billion tonnes, 2.1 times the amount in 1971. Emissions per capital also increased from 3.74 tonnes in 1971 to 4.29 tonnes in 2009.

Greenhouse gas emissions produced by the burning of fossil fuels is a major factor that contributes to global climate change. As the issue of climate change becomes better understood, the whole world has recognised the need for the international community to work together to tackle its challenges. Climate change has become a focal issue in international politics, economics and diplomacy. Like the Charter of the United Nations and World Trade Organisation Rules that came before, various accords on climate change are

---

[1] Reserve-production ratio (RPR) is the ratio of the remaining recoverable reserves to annual production. RPR represents the number of years the existing remaining reserves can support current production rate.

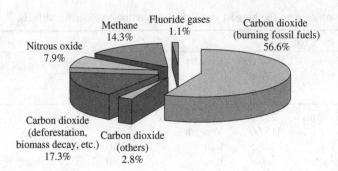

Figure 1.2 Composition of the world's greenhouse gas emissions produced by human activity in 2004.
Source: Intergovernmental Panel on Climate Change (IPCC), *Climate Change 2007: Synthesis Report*.

becoming the most important regulatory mechanisms that will affect world development and world energy development in the future.

In fighting climate change, developing countries are under dual pressures. Developed regions and nations like the European Union (EU), Japan and the United States (USA) have seen improvements in their environmental qualities. For them, the climate change issue has replaced traditional environmental concerns. Most developing countries, however, are still in the industrialisation stage. On the one hand, they face traditional ecological and environmental pressures that come with energy exploration and exploitation; on the other hand, they face the pressure of reducing greenhouse emissions. They have far bigger challenges compared to developed countries. Given the lag in technology, capital and discourse power, developing countries are at an overall disadvantage in the negotiations on carbon emissions with developed countries.

1.1.1.4 Energy Security is a Widespread Concern in the International Community

Energy security refers to the uninterrupted availability of energy supply that can meet the needs of economic and social development at a price that is affordable, while respecting ecological and environmental concerns. With the increasing interconnection and mutual dependence of the world's economies, facilitating the global supply and demand balance of energy and safeguarding global energy security have become urgent tasks that every country in the world must undertake together.[2] Energy security is a widespread concern in the international community and has become a major international issue. The underlying causes for the global community's concern about energy security include the following: (1) The nonrenewable nature of fossil fuels generates a worry about their future shortage. (2) The imbalance of regional supply and consumption of the world's energy. The remaining proven reserves and production of fossil fuels, especially oil, are concentrated in a small number of regions and countries like the Middle East, Russia and so on, but the growth in energy consumption is concentrated in the newly emerging economies

---

[2] Jinping Xi in a speech delivered at the International Energy Conference in June 2008.

in the Asia Pacific. The imbalanced distribution of resources and user areas causes a worry about the sustainability and stability of energy supply. (3) Many uncertainties exist in energy-producing regions and transport channels. The world's major oil- and gas-producing areas and transport bottlenecks are plagued by many problems such as a complex tangle of interests, political instability, regional tensions, etc. There is a risk of the energy supply being interrupted.

Due to the extreme importance of energy security and the enormous challenges that it poses to the world, it is high on many countries' agendas. Ensuring their own energy security has become a core component in the energy strategies of these countries. Through these measures, the USA can achieve energy independence and be a leader in the green energy industry of the future. To guarantee a stable, reliable and reasonably priced energy supply in the long run, the EU is formulating a shared energy policy. Key measures to ensure the EU's energy security include conserving energy, building up a common energy market, developing renewable energy and smart power grids, and strengthening international cooperation. Energy is at the core of Japan's economic policy. It is proposed that energy security can be attained through promoting vehicles that run on new energy, popularising energy saving technology, making greater use of solar and wind power, developing a new generation of transmission grids. Japan plans to increase the share of self-supplied energy (including traditional domestic energy resources and overseas obtainable energy resources Japan has invested in) from the current 38% to 70% by 2030.

1.1.1.5 A New Revolution in Energy Technology is Brewing

In response to the complexities of the energy security situation, and to resolve the ecological and environmental problems, including climate change, caused by energy use, many countries are attaching a high level of importance in energy technology innovation in the beginning of the 21st century. They are working to develop a new generation of energy technology in the hope that this technological breakthrough will facilitate the use of new and renewable energy, the increase in the efficiency of energy use and the creation of an energy framework with a sustainable supply.

The global financial crisis that began in 2008 has left the world economy badly battered. To free themselves from the effects of the crisis and to revive their economic strength, certain countries are expanding their investments in energy. At the same time as they are ensuring energy supply and stimulating economic growth, they are putting in place their new competitive edge in the postcrisis world. This has encouraged innovation in energy technology on a global scale. The reality demonstrates that 'the global financial crisis has given rise to a new technological revolution, and the world may enter a period that will witness an innovation boom and the rapid growth of new industries, where green development will become the major trend'.[3] A breakthrough in energy technology is an important prerequisite and the key to making green development a reality. A new technological revolution is inevitably a new revolution in energy technology.

---

[3] Keqiang Li (2010) A deep understanding of the main line of the theme in *Jianyi (The Communist Party of China (CPC) Central Committee's Proposal on Formulating the 12th Five-year Program (2011–2015) on National Economic and Social Development)* to facilitate comprehensive coordination of the sustainable development of the economy and society, *People's Daily*, 5th edition, 15 November 2010.

A survey of global energy and technology reveals that a new revolution in energy technology is brewing; the signs are already evident.

In the area of energy production, countries are actively pushing for the low-carbon and efficient exploration and use of traditional fossil fuels, as well as seeking new energy sources to replace fossil fuels. New innovations in energy production that have garnered considerable interest include the eco-friendly and clean use of coal, large-scale exploration and exploitation of renewable energy sources like wind and solar power, a new generation of nuclear power and unconventional exploration and use of oil and gas.

In the area of energy transport, countries are optimising the allocation of resources and improving the overall coordination and interaction of energy transport systems to ensure the secure transport of energy. At the core of the revolution in energy transport technology are large capacity transmissions using ultra high voltage (UHV), safe and stable operations of large grids, power generation with renewable energy sources and their integration, and smart grids.

In the area of energy consumption, various countries are increasing their energy efficiencies, and gradually bringing about the replacement of oil with electricity in areas like transport. Vehicles with power-saving technology and run on new energy, as represented by the electric car, have a huge market potential.

The new revolution in energy technology will bring about a profound change. The revolution will be led by technological innovation and centred on electric power. With the development of a smart energy system as its direction, the goals of the revolution are to enhance the energy structure, increase energy efficiency, decrease energy consumption, share social resources and achieve sustainable development. On the whole, new energy and the development of smart power grids will become the major engines of the new revolution in energy technology.

1.1.2 Characteristics of the Global Energy Situation

Economic and social development, changes in the international structure and advancements in science and technology have resulted in an unprecedented and profound transformation of the energy industry. Against the backdrop of pursuing the sustainability of energy supply and dealing with the problem of climate change, six major characteristics of the global energy situation have emerged: structural diversity, clean development, long-distance allocation, consumer-level electrification, smart systems and financialisation of resources.

1.1.2.1 Diversity of Energy Structure

With the technological advancements in developing and utilising energy, the growing shortage of fossil fuels and the concern among the world's countries for their own energy security, achieving diversity in energy structure and reducing the dependence on a single type of energy have become a shared strategic alternative among various countries. In the early 1970s oil was the main type of energy consumed in the world, accounting for over 46% of the world's energy consumption at one point in time. Following the two oil

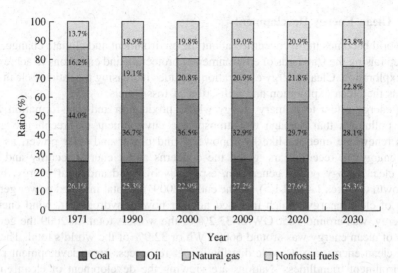

Figure 1.3 Changes in the structure of global primary energy consumption.
Source: IEA, *World Energy Outlook 2011*.

crises, the major developed countries began to push for the development of electricity generated by nuclear power and natural gas to reduce their consumption of oil. This caused a rapid fall in the share of oil in the world's energy consumption, and an apparent rise in the percentages of natural gas and nuclear power. After 1990 new energy and renewable energy have become the focus of energy development in the world. The share of nonfossil fuels keeps increasing and the world's energy structure keeps diversifying (Figure 1.3). According to the IEA's forecast of the world's energy development, the share of coal in the world's primary energy consumption will fall from 27.2% in 2009 to 25.3% in 2030. Over the same period, oil will fall from 32.9% to 28.1%. Natural gas, however, will rise from 20.9% to 22.8%, as will nonfossil fuels, which will rise from 19.0% to 23.8%. The global energy structure will be a tetrarchy of coal, oil, natural gas and nonfossil fuels.

Diversity of the energy structure provides more alternatives to meet energy needs and plays an important role in ensuring energy security. Following the oil crises of the 1970s, it was precisely this diversification strategy that developed countries in the west pursued that successfully reduced the share of oil in their energy consumption and their dependence on imported, especially Middle East oil. Currently, the US level of dependence on Middle East oil is only at 20%, while Europe's dependence is only at around 40%.

Diversification of the energy structure is helpful in overcoming the limitations of resources and the environment, and encourages sustainable development. The global trend in energy development is the transition from fossil fuels to new and renewable energy, and energy diversification is the inevitable choice for a stable transition. In the structure of energy consumption, the falling share of traditional fossil fuels like coal and oil and the increasing use of new and renewable energy is a landmark feature of the diversification trend.

1.1.2.2 Clean Energy Development

As the world becomes more concerned about the environment and climate change, nations must keep raising the standards of environmental protection and emissions to achieve clean energy exploration. Clean energy exploration includes increasing the ratio of clean energy, as well as the clean exploration and utilisation of fossil fuels.

Clean energy refers to primary energy whose production and transformation will not produce pollutants that damage the atmospheric environment or greenhouse gases. It includes renewable energies like hydropower, wind power and solar power, as well as nuclear energy. In recent years, given the concerns about energy security and climate change, clean energy power generation, especially by wind and solar energy, has seen rapid growth (Figures 1.4 and 1.5). At the end of 2009, the total installed power generation capacity of clean energy, which included nuclear power, hydropower, wind energy and solar energy, was around 1650 GW or 33.2% of the world's total. In 2009 the generating capacity of clean energy was around 6600 TWh or 32.9% of the world's total. The advantages of clean energy are its wide distribution of resources, great development potential and environment-friendliness. Nations are viewing the development of clean energy as an important alternative in their push for the sustainable development of energy. Renewable energy like wind energy has become indispensable components of the energy supply systems of certain developed countries. According to IEA forecasts, the share of clean energy in the global installed generation capacity will be raised to 44.9% by 2030, and its share of generation will be 41.5% (Figure 1.6).

The focus of clean development and utilisation of fossil fuels is on coal. The technology involved in the clean development and utilisation of coal include coal washing and processing, clean and efficient combustion, combined utilisation of resources and pollution control. As coal is widely distributed with abundant reserves all over the world, developed

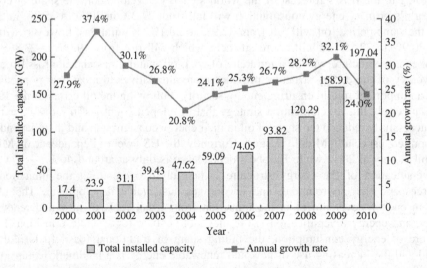

Figure 1.4 Total installed capacity and growth of global wind energy.
Source: Global Wind Energy Council (GWEC), *Global Wind Report 2010*.

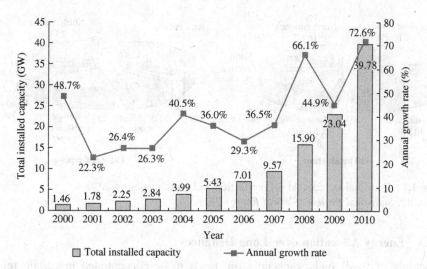

Figure 1.5   Total installed capacity and growth of global solar energy.
Source: BP, *Statistical Review of World Energy 2011*.

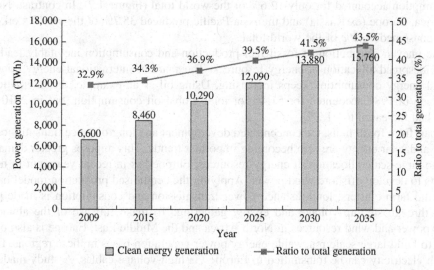

Figure 1.6   Power generation of clean energy and its ratio to the total generation in the future.
Source: IEA, *World Energy Outlook 2011*.

countries and regions like the USA, the EU and Japan attach a great deal of importance on developing and applying the technology for the clean development and use of coal. They see it as an important measure in reducing the pressure of energy supply and protecting the environment. After years of development, technologies like ultra supercritical power generation, integrated gasification combined cycle (IGCC) and circulating fluidised-bed are gradually maturing, whereas demonstrations of carbon capture and sequestration (CCS) are already being applied in industry.

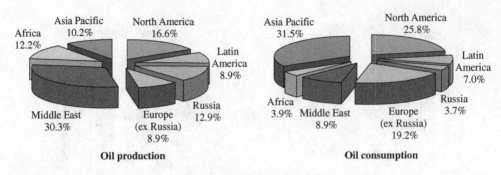

Figure 1.7   Distribution of world oil production and consumption in 2010.
Source: BP, *Statistical Review of World Energy 2011*.

1.1.2.3   Energy Allocation over Long Distances

The supply of fossil fuels, especially oil, tends to be concentrated in certain regions, but the geographic distribution between producers and consumers is uneven. In 2010 the Middle East and Russia produced 43.2% of the world's oil, but their combined oil consumption accounted for only 12.6% of the world total (Figure 1.7). In contrast, North America, Europe (ex Russia) and the Asia Pacific produced 35.7% of the world's oil, but they consumed 76.5% of the world total.

The uneven distribution of fossil fuel production and consumption inevitably leads to the widespread allocation of energy resources. The share of international energy trade in global energy consumption keeps increasing. Using oil as an example, the international oil trade in 1980 accounted for 51.4% of total global oil consumption, but in 2010 the ratio had increased to 61.2%.

Apart from fossil fuels, the concentrated development and long-distance transmission of wind and solar energy are also becoming important trends. This imposes greater demands on the large-scale allocation of energy resources. Europe has in recent years put in more efforts to develop offshore wind power. Applying the centralised production model based on wind farm clusters, long-distance power transmission and consumption is made possible through step-up voltage and energy gathering. To make full use of the abundant solar power and wind resources in North Africa and the Middle East, Europe is also planning to build large-scale renewable energy power generation bases in these regions, from which electricity can be transmitted to Europe via high-voltage cables. A study made by the US Department of Energy reveals that high quality onshore wind energy resources in the USA are mostly concentrated in the Midwestern Plains, but the electricity load is mainly concentrated in the Eastern and Western Seaboards. To make wind-generated electricity a reality, high-capacity, long-distance power transmissions must run between the wind energy resource areas and the power load areas.

Power grids are important carriers in the major enhancement of energy distribution and there is much focus on their construction. The interconnection of large grids has become a trend in the development of grids worldwide. In places like North America, Europe, southern Africa and the Middle East, the scale of grid interconnection between nations is getting bigger. The North America power grid is an interconnected grid that covers the USA, most parts of Canada and some parts of Mexico. Its total installed capacity is over

1000 GW. The European grid, which covers over 30 countries and has a total installed capacity of almost 900 GW, is gradually expanding into Eastern Europe, Russia, Central Europe and North Africa.

1.1.2.4 Electrification-oriented Energy Consumption

Electric power has the advantages of being clean, efficient and convenient. All primary energy can be converted into electricity. Electricity can be conveniently transformed into mechanical energy, thermal energy and other forms of energy, and precise control can be achieved. These features of electricity have resulted in its widespread utilisation in modern society. In fact, electrification has become one of the important indicators of modernisation.

The histories of developed countries indicate that socioeconomic growth is always accompanied by a constant rise in the level of electrification. The level of electrification is closely connected to a country's level of economic development: the higher the economic development, the higher the level of electrification. The changes in the relationship between the gross domestic product (GDP) per capita of developed countries and the power use per capita in those countries are shown in Figure 1.8. Although there are differences between these countries in resources, climate and living habits, and that the process of electricity development for each country is different, the graph shows basically a positive correlation between GDP per capita and electricity use per capita.

The level of a country's electrification is usually measured using two indices: (1) the ratio of power generation capacity to primary energy consumption and (2) the ratio of electric power to energy end-use consumption. Table 1.4 gives the historical changes in the ratios of electric power to energy end-use consumption in major countries. Figures

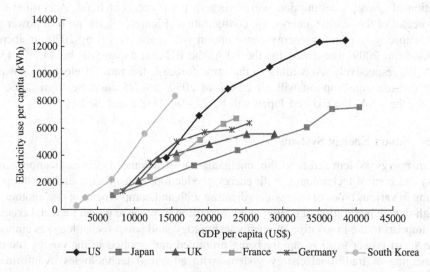

Figure 1.8 Graph showing the trends in the relationship between GDP per capita and electricity use per capita in developed countries.[a]
[a]GDP in constant 2000 US$.
Source: Calculated using IEA statistical data.

Table 1.4 Ratios of electric power to energy end-use consumption in
major countries.

Unit: % Country	1971	1980	1990	2000	2005	2009
US	10.1	13.3	17.5	19.5	20.3	21.4
Japan	14.7	19.0	21.5	23.6	24.3	25.5
UK	13.4	15.4	17.0	18.8	19.9	21.0
France	8.6	12.7	18.1	20.2	21.4	22.7
Germany	10.7	13.6	16.2	17.9	18.5	19.0
India	3.1	4.3	7.3	10.0	11.5	13.8
Brazil	5.8	10.6	16.2	17.8	17.9	18.3
Russia	–	–	11.4	12.4	13.4	13.9
China	2.9	4.3	6.3	11.7	15.8	18.5

Source: The numbers in the table are calculated using IEA statistical data,
which diverge slightly from China's statistical data.

in Table 1.4 show that in both developed and developing countries, the ratios are on an
upward trend. By 2009, the ratios of most developed countries are above 20%. Globally,
the ratio of power generation capacity to primary energy consumption is also on an
upward trend. In 1990 this ratio was 34.0% globally, but by 2009 it reached 37.7%, an
increase of 3.7%.

With the new revolution in energy technology, renewable energy will see greater utilisation. New electric products like electric vehicles will play a more significant role in
national economies. The global level of electrification will further increase and the electrification of energy consumption will become a more apparent trend. According to an
IEA forecast of the world's energy use configuration (Figure 1.9), the ratio of power generation capacity to primary energy consumption will reach 41.4% by 2030, an increase
of 3.7% from 2009. The ratios for the USA, the EU and Japan will be 44.7%, 43.0%
and 49.5% respectively. According to the same forecast, the ratio of electrical power to
energy end-use consumption will hit 22.0% in 2030, a 4.6% increase from 2009. The
ratios for the USA, the EU and Japan will be 25.4%, 23.0% and 30.2%.

1.1.2.5 Smart Energy Systems

A smart energy system refers to the amalgamation of advanced communications, information and control technologies with energy production, transmission and consumption,
resulting in various types of energy coordinating with and complementing one another, and
the high-degree integration of energy and information flows. The global financial crisis has
given impetus to the innovation of energy technology, and smart technology, as embodied
by the smart power grid, is rapidly being promoted and applied in the energy industry.
The merging of traditional energy systems with advanced technologies in information
communications, networks and controls, as well as progressive management concepts, is
pushing the energy industry towards smart systems.

Developed nations are aggressively building up smart energy systems based on smart
grids. The USA promulgated the Energy Independence and Security Act in 2007, legally

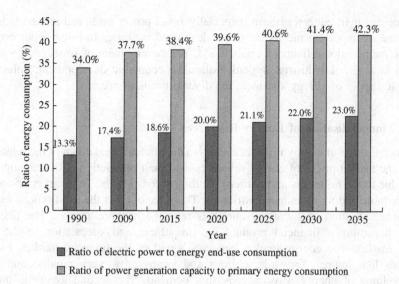

Figure 1.9 Electrification trend of global energy consumption.
Source: IEA, *World Energy Outlook 2011.*

affirming the position of smart power grids in its national strategic framework. The Act identifies smart grids as one of the core components of US energy strategy, and stipulates a higher level of technological and capital input. Documents like *Vision and Strategy for Europe's Electricity Networks of the Future* spell out the EU's strategic framework in developing smart power grids. Under this framework, member states develop their own power grid experiments, such as Smart City in the United Kingdom (UK), Germany's E-Energy demonstration project and Denmark's EDISON demonstration project. Japan has incorporated smart power grids into its Development Strategy and Economic Growth Plans, and has set up smart grid demonstration projects on ten islands in the Kyushu and Okinawa areas. South Korea's Strategy of Green Energy Industry outlines the country's conception of a smart power grid network. It has increased investments in enhancing the commercialisation of smart grids and a demonstration project has been set up on Jeju Island.

China places great importance on smart power grids, fuelling rapid development in this area. The country sees smart grids as a strategic emerging industry and provides ample support. The 12th Five-Year Programme on the National Economic and Social Development of the People's Republic of China, promulgated in March 2011, lists smart power grids as one of the important tasks of energy infrastructure. China's power grid industry has conducted experimentation on smart grids with good results. It has initiated many pilot projects and important technical results have been obtained in the areas of large-scale power transmission, smart metering, high capacity cell storage, grid connection control of new energy power generation. These have propelled China as the world leader in smart grid development. In September 2011 an international forum on smart power grids was held in Beijing, amply demonstrating China's achievements in the research and development of smart power grids, and enhancing the country's influence in the international smart power grid industry.

Developing smart energy systems, especially smart power grids and related industries, has become an important means for major developed countries to bring about economic growth in the postglobal financial crisis world. The development of smart energy systems will lead to a global industrial upgrade, stimulate economic development, enhance the sustainable supply of energy and meet the diverse needs of users.

1.1.2.6 Financialisation of Energy Resources

The financialisation of energy resources refers to the gradual breakdown of the relationship between the market prices of energy resources, and their production costs and supply and demand due to the finiteness and nonrenewability of fossil fuels. The market prices begin to acquire more and more financial attributes. The price of oil is the most typical example.

The financialisation of energy resources is reflected in three aspects. The first is the resource becoming a financial product for investment and speculation in the global financial market. The energy market becomes a part of the financial market. Financial derivatives like futures, forwards, options and swaps grow very rapidly, and futures trading volume of energy way exceeds spot demand. The second aspect is the price of energy futures becoming the main basis for pricing in the energy resources market. The energy futures market appeared in the late 1970s, providing an effective tool for price discovery and risk avoidance due to its high levels of transparency and market liquidity. As a result, pricing decisions on oil and other fuels have been effectively shifted from the spot market to the futures market. The third aspect is the evident permeation and integration of the energy and financial systems. The energy industry issues shares and bonds through the financial and capital markets to raise money, but it also makes use of financial derivatives to avoid market risks.

With the financialisation of energy resources, speculative behaviour is having a bigger influence on world energy prices, especially the price of oil. Figure 1.10 shows the changes in the average annual prices of crude oil since 1970. Between 1970 and 2010, the average annual growth of global crude oil consumption was only 1.9%, but prices increased 6.9 times over the period. The price increase far exceeds the growth in demand. Before the 1990s, fluctuations in oil prices were mainly the result of oil-producing countries controlling oil production. Since the 1990s, the financial and capital markets have had a more palpable influence on prices. That international oil prices skyrocketed after 2003 has everything to do with the manipulation of the oil market by international financial and capital markets.

A survey of the changes in international oil prices shows that though supply and demand is still the basic factor affecting price changes, the effects of nonsupply and demand factors are becoming more apparent. Speculators in oil and financial markets often take advantage of unexpected world events and make a big deal out of their psychological effects on consumers, which results in the volatile world oil prices. To a certain extent, the fluctuations of oil prices have gradually deviated from the market value of oil itself, demonstrating the characteristics of financial instruments.

The financialisation of energy resources is good for the functioning of the energy market, but it also brings in the risks of the financial market, and adds to the uncertainties of the world's energy development. Currently, the level of financialisation of energy resources in China is still relatively low, and the country has no influence on the international

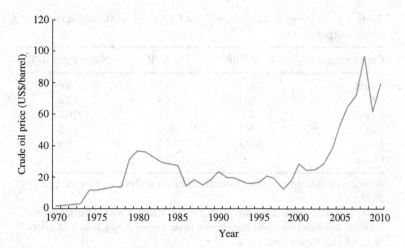

Figure 1.10 Trend of international oil prices.
Source: BP, *Statistical Review of World Energy 2011.*
Note: Price for the indicated year.

pricing of energy resources. China can only passively accept the decision reached by the world financial markets and it is vulnerable to fluctuations in world energy prices. To counter the risks brought about by the financialisation of energy resources, China must begin from a strategic perspective to build up a risk warning and prevention system, and take preventative measures like refining the energy financial system, building up a risk prevention framework and enhancing the country's status in the world energy market.

1.2 An Overview of China's Energy Situation

In absolute terms, China possesses abundant energy resources. However, the per capita share of these resources and the volume of high quality supplies are on the low side. Since the advent of China's economic reform and market liberalisation, the development of the energy industry has been rapid, with substantial growth in the level of production and consumption. At a basic level, the development has met the needs of economic and social development. Since becoming a net importer of energy in the 1990s, the country, to better safeguard energy supplies, has maintained significant progress in international energy cooperation.

1.2.1 Energy Endowment

China's resources of traditional fossil fuels comprise mainly of coal; high quality fossil fuels like oil and natural gas are comparatively insufficient. In terms of remaining economically recoverable reserves as at 2009, China's coal reached 163.69 billion tonnes, oil 2.16 billion tonnes and natural gas 2.9 trillion cubic metres. The remaining economically recoverable reserves of all three fossil fuels totalled 123.86 billion tonnes of standard coal, with coal accounting for 94.4%, oil 2.5% and natural gas 3.1%. China's oil and gas explorations are at the early to middle stages, and the reserves have the potential to

Table 1.5 Reserve-production ratios of China, the world and several countries (per year).

Country/region	Coal RPR	Oil RPR	Natural gas RPR
World	118	46.2	58.6
US	241	11.3	12.6
Russia	495	20.6	76.0
India	106	30.0	28.5
Brazil	>500	18.3	28.9
China	35	9.9	29.0

Source: All data are obtained from BP, *Statistical Review of World Energy 2011*.
Note: The data in the table are 2010 figures. There is a slight discrepancy between the data contained therein and those released by China's Ministry of Land and Resources.

increase in quantity in the future, but in overall terms, the total amount of oil and gas resources is at great variance to China's economic and social development needs.

Compared to other resources-rich countries, the total quantity of China's main fossil fuels is abundant, but its RPR is quite low and supply sustainability of resources is insufficient. Table 1.5 is a comparison of the fossil fuel RPR of China, the world and several countries. Although China has the world's third most abundant remaining reserves of coal, the huge quantities being extracted means that the RPR is only 35 years, equivalent to 29.7% of the world average. The RPR of oil is 21.4% of the world average and natural gas 49.5%.

Given China's large population, its per capita share of fossil fuel resources is lower than the world average. In 2009 the country's per capita shares of remaining proven recoverable reserves of coal, oil and natural gas were 85.9 tonnes, 1.5 tonnes and 1840 cubic metres respectively. These were only 70.0%, 5.6% and 6.6% respectively of the world average (Table 1.6).

China has the huge potential to develop renewable energy resources. Its available hydro-power resources are 540 GW, the highest in the world, and the annual utilisable resources

Table 1.6 Remaining proven recoverable reserves of fossil fuels per capita in China and the world.

Type	China	World	China/world
Coal (tonnes)	85.9	122.7	70.0%
Oil (tonnes)	1.5	27.0	5.6%
Natural gas (cubic metres)	1 840	27 843	6.6%

Source: The data in the table are obtained from BP, *Statistical Review of World Energy 2011*, International Monetary Fund (IMF), *World Economic Outlook Database*.

of biomass are around 899 million tonnes of standard coal. The potential available resources of Grade 3 winds at an altitude of 50 metres (equivalent to wind power density ≥300 W/square metre) are 2580 GW. With onshore wind resources of 2380 GW and off-shore wind resources of 200 GW, the annual utilisable resources are 634 million tonnes of standard coal at 2000 hours a year. According to estimates based on installation of solar power systems on 5% of desertified land, China's total installed generation capacity for solar energy can reach 3460 GW, with annual utilisable resources of 596 million tonnes of standard coal at 1400 hours a year.

1.2.2 Energy Production

Since the implementation of its reform programme, China's energy industry has seen rapid development. Its energy production capacity has increased significantly to support its relatively quick and sustained economic and social growth. In 2010 China's total energy production was 2.97 tonnes of standard coal, four times that of 1980. At an annual growth of 5.3%, China is the world's top energy-producing country. The country has insisted on guaranteeing its energy supply by domestic resources, and its energy self-sufficiency rate has for a long time been maintained at a high level of 85% and above.

An energy production and supply framework has taken shape in China, one that is based on coal and centred in electricity, with the all-round development of oil, natural gas, new energy and renewable energy. Coal accounted for 76.6% of China's total energy production in 2010, while oil and natural gas accounted for 9.8% and 4.2% respectively. Primary power sources like hydropower, nuclear power, wind power and so on took up 9.4% (Figure 1.11).

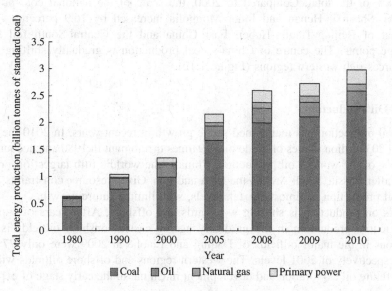

Figure 1.11 China's total energy production and composition.
Source: National Bureau of Statistics of China, *China Energy Statistical Yearbook 2011*.

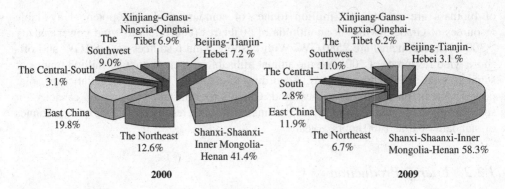

Figure 1.12   Distribution of coal production in China.
Source: National Bureau of Statistics of China, *China Energy Statistical Yearbook*.

1.2.2.1   Coal Production

Coal is China's basic energy resource, and the country has high concentrations of production areas. China's coal production in 2010 reached 3.24 billion tonnes, 5.2 times the production in 1980 and half of the world's total production. It was the world's top coal-producing country. Coal production is mainly concentrated in the provinces of Shanxi, Inner Mongolia, Shaanxi and Henan. In 2009 the quantity of coal produced in the three provinces and one autonomous region accounted for around 58.3% of the country's total output. In contrast, the coal production of the more economically developed regions of Beijing-Tianjin-Hebei, East China[4] and the Central-South[5] was only 17.8% of the total. Compared to 2000, the share of the national coal production of Shanxi, Shaanxi, Henan and Inner Mongolia increased by 16.9 percentage points, while that of Beijing-Tianjin-Hebei, East China and the Central-South fell by 12.3 percentage points. The centre of China's coal production is gradually shifting towards the resources-rich western regions (Figure 1.12).

1.2.2.2   Oil Production

China's oil production has maintained stable growth in recent years. In 2010 the country produced 203 million tonnes of crude oil, 1.9 times the amount in 1980 and accounting for around 5% of the world's oil production. China is the world's fifth largest oil-producing country, after Russia, Saudi Arabia, the USA and Iran. Given resource constraints, China's annual oil production is almost at peak levels, with limited future growth.

China's oil production is shifting westwards and offshore. After years of exploration and extraction, production in the gradually ageing eastern onshore oilfields is falling. Productions in the major oilfields of Daqing and Liaohe in 2009 were only 77.7% and 72.2% respectively of 2001 levels. The western regions and offshore oilfields will be the focus of future oil exploration and development in China. In the early stage of exploration

---

[4] The East China (*Huadong*) region includes the municipality of Shanghai, and the provinces of Jiangsu, Zhejiang, Anhui, Fujian, Jiangxi and Shandong.

[5] The Central-South (*Zhongnan*) region includes the provinces of Hubei, Hunan, Guangdong, Guangxi and Hainan.

and a period of rising production, they are the key regions that can ensure the stability of China's oil production.

1.2.2.3 Natural Gas Production

China's natural gas production is entering its rapid development stage, with a high production growth rate. In 2010 the country produced 94.85 billion cubic metres of natural gas, 6.6 times the volume in 1980 and accounting for around 3% of the world's natural gas production. China is the world's seventh largest natural gas-producing country, after the USA, Russia, Canada, Iran, Qatar and Norway. China's natural gas-producing regions are concentrated in the eight key areas of Sichuan-Chongqing, the Tarim Basin, Ordos, the Qaidam Basin, the Songliao Plain, East China Sea, Bohai Bay and the Ying-Qiong Basin. These areas account for 95% of the annual nationwide production. In recent years, China has expanded the natural gas pipe network and built more liquefied natural gas (LNG) receiving terminals along its coast. The natural gas supply situation in China is one where gas from the west is transported to the east, gas from the sea is transported onshore and gas from the north is transported to the south.

China is rich in alternative natural gas resources like coal seam gas (CSG) and shale gas, but their developments are still at the initial stage. The volume of shallow CSG resources below 2000 metres in China is about 36.81 trillion cubic metres, which is the world's third largest reserve of its kind. The potential volume of onshore shale gas resources is 134.42 trillion cubic metres, of which 25.08 trillion cubic metres are recoverable. In 2010 China's CSG production reached 1.5 billion cubic metres, of which 1.2 billion cubic metres were commoditised. The extensive exploration and use of CSG and shale gas can increase the energy supply and reduce environmental pollution. With technological advancements in exploration techniques, alternative natural gases like CSG and shale gas will become important in the future of natural gas supply to ensure the continuous growth of China's natural gas production.

1.2.2.4 Electricity Production

Since the implementation of its reform programme, China's installed capacity and generation have seen rapid development, with annual growth exceeding the growth of energy production. At the end of 2010, China's installed generation capacity was 966 GW, with an annual generating capacity of 4230 TWh. China is the world's second largest electricity producer.

Thermal power accounts for the bulk of China's electricity production (Figures 1.13 and 1.14). At the end of 2010, the thermal power generation installed capacity was 710 GW, with an annual generating capacity of 3420 TWh. Most of it was electricity produced by coal; the ratio of electricity produced by oil and natural gas was very low. Hydropower generation installed capacity was 216 GW, with an annual generating capacity of 686.7 TWh. China is the world's biggest producer of hydropower. Nuclear power is slowly developing, with 13 nuclear power units in operation, having an installed generation capacity of 10.82 GW and an annual generating capacity of 74.7 TWh. Wind power generation is also growing at a rapid rate, with 29.58 GW connected to grid. During the period of the 11th Five-Year Plan, the scale of the development of wind power doubled

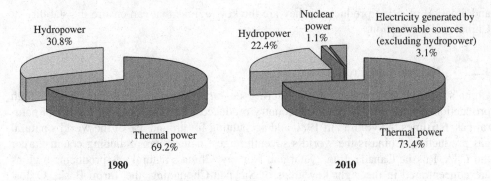

Figure 1.13 China's installed capacity structure.
Source: China Electricity Council, *Statistical Data on the Electricity Industry*.

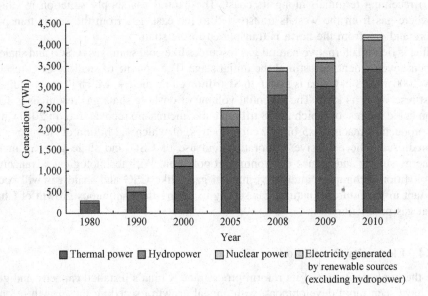

Figure 1.14 Changes in China's power generating capacity.
Source: China Electricity Council, *Statistical Data on the Electricity Industry*.

every year for five consecutive years. Solar power generation is entering the large-scale development stage, with the installed capacity of connected photovoltaic grids at 260 MW.

China's electricity production is very different in structure from that in the rest of the world, in particular some developed countries. In developed countries, the greater part of their electricity is generated by natural gas and nuclear power. In China, the overwhelming part of its electricity is generated by coal. The share of coal-fired electricity in the total electricity produced in China is higher than the world average by almost 40% (Table 1.7).

Table 1.7 China's power generation compared to the world and some developed countries (%).

Power generation mode	World	US	Japan	Germany	France	China
Coal	40.5	45.5	26.8	43.9	5.3	78.7
Oil	5.1	1.2	8.8	1.6	1.2	0.5
Natural gas	21.5	22.9	27.4	13.4	4.0	1.6
Nuclear power	13.5	19.9	26.9	23.0	76.2	1.9
Hydropower	16.2	6.6	7.2	3.2	10.6	16.5
Renewable sources (excluding hydropower)	3.2	3.9	2.9	14.9	2.7	0.8

Source: The data in the table are obtained from IEA, which are slightly different from the data released by China Electricity Council.

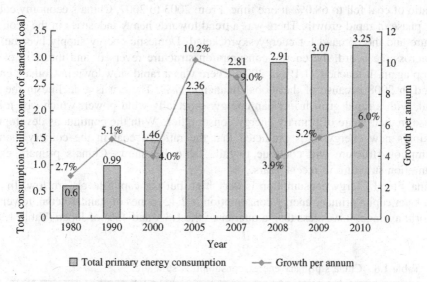

Figure 1.15 China's total primary energy consumption.
Source: National Bureau of Statistics of China.

1.2.3 Energy Consumption

1.2.3.1 Primary Energy Consumption

China's rapid economic and social growth has meant a sustained growth in energy demands. In 2010 China's total primary energy consumption reached 3.25 billion tonnes of standard coal (Figure 1.15), 5.4 times the 1980 level. China has become the world's biggest consumer of energy.

China's energy resource endowments and its energy policy of meeting demands by domestic supplies have dictated the domination by coal of China's long-term primary energy consumption structure. Compared to developed countries, China's use of coal is on the high side, whereas the use of oil, gas and clean energy is rather low. In 2010 the share of coal in China's primary energy consumption structure was 68.0%, or 45 percentage points more than the USA and 44.6 percentage points more than Japan. The share of oil and gas was 23.4%, or 38 percentage points less than the USA and 34.8 percentage points less than Japan. The share of clean energy was 8.6%, or 7 percentage points less than the USA and 9.8 percentage points less than Japan. Since the implementation of its economic reform, China's primary energy consumption structure has on the whole been moving towards consuming more high quality energy resources. Before the 1990s, China met its constant growing energy demands mainly by increasing coal production, resulting in the continuous rise in the share of coal in the country's primary energy consumption. Subsequently, the ratio of coal began to decrease, especially between 1995 and 2002. During the period, China's tight energy supply and demand situation eased up and there was a more rapid shift towards high quality energy resources in the consumption structure. The ratio of coal fell to 68.0% at one time. From 2003 to 2007, China's economy entered a new phase of rapid growth. There was a trend towards heavy industries in the economic structure and the demand for energy skyrocketed. Domestic energy supply became very tight across the board, the energy consumption structure reversed, and the ratio of coal went up again. It reached 71.1% in 2007. There was a rapid slowdown in China's energy demand in 2008 because of the global financial crisis. There was a decline in the ratio of coal, with a rapid growth in clean energy, especially wind power, which saw a swift increase in their share of primary energy consumption. With the continuous development of China's new energy, it is predicted that the ratio of coal in the country's energy consumption structure will continue to fall. Table 1.8 shows China's primary energy consumption structure in recent years.

China's total energy consumption is very high but per capita levels are low. In 2010 China's per capita primary energy consumption is 2.42 tonnes of standard coal, lower than the world average. It was less than a quarter of the USA and 43% of Japan (Table 1.9). In

Table 1.8 China's primary energy consumption structure.

Year	Total energy consumption (billion tonnes of standard coal)	Ratio (%)			
		Coal	Oil	Natural gas	Hydropower, nuclear power, wind power
1980	0.6	72.2	20.7	3.1	4.0
1990	0.9	76.2	16.6	2.1	5.1
2000	1.46	69.2	22.2	2.2	6.4
2005	2.36	70.8	19.8	2.6	6.8
2007	2.81	71.1	18.8	3.3	6.8
2008	2.91	70.3	18.3	3.7	7.7
2009	3.07	70.4	17.9	3.9	7.8
2010	3.25	68.0	19.0	4.4	8.6

Source: National Bureau of Statistics of China, *China Energy Statistical Yearbook 2011*. The data are calculated based on the consumption of coal for power generation.

Table 1.9 China's per capita energy consumption compared with the world and several countries. Unit: tonnes of standard coal/person.

Country/region	1980	1990	2000	2008	2009	2010
US	11.31	10.93	11.50	10.71	10.05	10.28
Russia	–	8.49	6.03	6.91	6.51	–
France	4.97	5.50	5.93	5.94	5.68	5.84
Germany	6.51	6.33	5.86	5.83	5.56	5.79
Japan	4.20	5.07	5.84	5.54	5.30	5.57
South Korea	1.54	3.10	5.64	6.67	6.72	7.22
India	0.43	0.54	0.64	0.77	0.84	–
Brazil	1.34	1.43	1.56	1.84	1.77	–
China	0.61	0.86	1.15	2.19	2.30	2.42
World	2.33	2.39	2.36	2.61	2.57	–

Source: The data for China are calculated based on the data from National Bureau of Statistics of China; the rest are from IEA.

line with economic and social development, China's energy consumption has the potential for greater growth.

1.2.3.2 End-use Energy Consumption

In the area of end-use energy consumption, China's demand for high quality energy has grown very rapidly in recent years, with an increasing share of the total. The share of coal in the end-use energy consumption structure continues to decline, especially in the last decade of the 20th century, which saw a sizeable decrease. At the same time, given the high growth of the transport sector and the rise in the living standards of Chinese citizens, the shares of oil and natural gas in end-use energy consumption saw a considerable increase. In the early years of the 21st century, coal consumption decreased slightly. However, the increase in the world oil price dampened oil consumption, and the share of oil also went down. End-use electricity continued its rapid growth, accounting for almost 20%.

In 2010 China's end-use energy consumption totalled 2.28 billion tonnes of standard coal. The ratios of coal, oil, natural gas, electricity and thermal power were 44.0%, 25.5%, 4.8%, 21.3% and 4.4% respectively. The share of coal fell by about 25 percentage points from 1990, whereas electricity increased by around 12 percentage points. There is a marked increase in the use of high quality energy in end-use energy consumption (Figure 1.16).

Industry has always been the sector accounting for a large share of China's end-use energy consumption. In 2010 industry accounted for 68.5% of all end-use energy consumption, while domestic use was 10.8% and transport took up 10.6% (Figure 1.17).

End-use coal consumption was gradually moving towards industry. Between 1990 and 2010, the share of industrial-use coal in end-use coal consumption grew by 27.9 percentage points to reach 87.3%. Domestic-use coal, in contrast, fell by 20.3 percentage points to 7.4%.

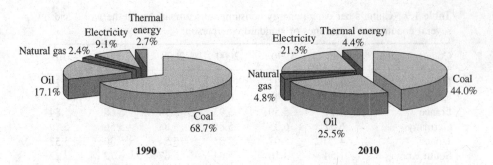

Figure 1.16 China's end-use energy consumption structure according to energy type.
Source: National Bureau of Statistics of China, *China Energy Statistical Yearbook*.

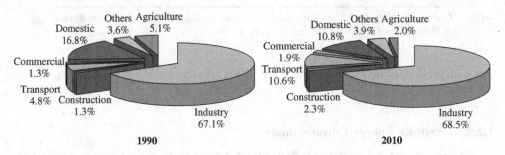

Figure 1.17 China's end-use energy consumption structure by sector.
Source: National Bureau of Statistics of China, *China Energy Statistical Yearbook*.

The transport sector is the biggest end-user of oil. Since 2008, transport has surpassed industry in end-use oil consumption. In 2010 the transport sector used 37% of oil, while industry used 35.3%.

There is higher growth in the domestic use of natural gas; its share in end-use gas consumption has already passed the one-quarter mark. The use of gas by industry continues to fall but the sector still consumes over 50%. The share of the transport sector in gas consumption has also risen by 10%.

There is an obvious shift in the industrial use of electricity towards heavy industry. Despite the declining share of industrial-use electricity in end-use electricity consumption, the share of heavy industry in industrial-use electricity continues to increase, reaching 82.8% in 2010.

1.2.4 International Energy Cooperation

With the growth in domestic energy demand, China is no longer a net energy exporter but a net importer of energy. Before 1993, apart from having self-sufficiency in resources like coal and oil, China had a certain amount of surplus for export. In 1993 China became a net oil importer, with massive growth in imports. The effects of the international energy market on China's energy security became very obvious. China's coal supply was

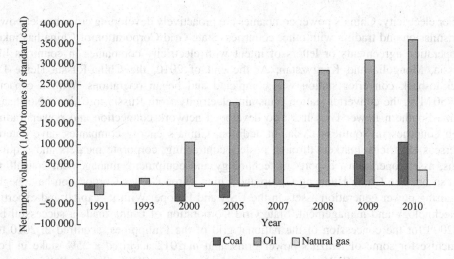

Figure 1.18 China's net imports of fossil fuels.
Source: National Bureau of Statistics of China, *China Energy Statistical Yearbook*.

insufficient in 2007 and demand for imported coal increased. This, together with the adjustment in the tariff policy for coal exports, caused China's net coal import numbers to scale up to positive territory for the first time. In 2010 China's net coal imports reached 146 million tonnes (Figure 1.18).

The continuous increase in energy imports has gradually changed China's energy security policy from one of self-sufficiency to one where there is equal focus on domestic supplies and using overseas resources. The role of international energy cooperation in China's energy development is becoming more apparent, and its status more important. The country is taking more proactive steps to participate in international energy cooperation and dialogue. It has built up a framework for bilateral dialogue with the world's major energy producers and consumers, and increased cooperation with various countries on energy issues. China is actively pursuing the 'go global' (*zou chuqu*) strategy, encouraging its energy corporations to venture outside its borders to acquire overseas resources.

China's power companies are quickening their pace in their overseas ventures in oil and natural gas through three major modes of investment. The first is directly acquiring exploration and development rights through agreements or bids. The second mode is obtaining overseas oil and gas resources by corporate mergers and acquisitions, e.g. Sinopec's acquisition of Swiss company Addax, PetroChina's acquisition of Australian energy firm Arrow. The final mode of investment is through providing loans or other assistance to resource-rich countries in exchange for promises of long-term supplies of oil and gas, e.g. the oil-for-loans agreements with countries like Russia, Kazakhstan and Brazil. Through these investment methods, China's overseas oil and gas interests have increased by a large margin, thus relieving the pressures caused by domestic shortage.

China's energy firms obtain overseas coal resources mainly by investing in coal mines. Companies like the Shenhua Group, ChinaCoal and Yanzhou Coal have invested in coal mining projects in Australia, and the Shenhua Group has invested in a coal-fired power generation project in Indonesia.

For electricity, China's power companies are proactively developing cross-border power transmission and trading with other countries. State Grid Corporation of China has inked cooperation agreements or letters of intent with electricity companies in countries like Russia, Mongolia and Kyrgyzstan. At the end of 2010, the China-Russia Heihe DC back-to-back converter station was completed and began operations. With a capacity of 750 MW, the converter station transmits electricity from Russia to China's Northeast. China Southern Power Grid has also developed network connection and power trading with countries in Southeast Asia. In addition, China's energy companies have entered overseas electricity markets through project contracting, corporate mergers and acquisitions, assets operations, exports of technology and equipment, management consulting, and so on. The China Huaneng Group and China Three Gorges Corporation have begun acquiring power generation assets in the USA and Europe. With its capital and expertise in technology and management, State Grid Corporation of China made a successful bid in 2008 for the concession of the national grid of the Philippines, acquired in 2010 the franchise for some of Brazil's power grids, and in 2012 acquired a 25% stake in Portuguese energy group REN—Redes Energéticas Nacionais, SGPS, S.A. At the same time, State Grid Corporation of China has also leveraged on its technical expertise in UHV power transmission and large power grid security and stability, and signed cooperation agreements with Russian national grid operators and US power companies to provide services like technical consultancy.

1.3 Major Energy Problems that China Faces

The development of China's energy sector has been remarkable, but at the same time there are many problems and challenges. China's socioeconomic growth is fuelling the sustained increase in energy demand, but given the growing constraints imposed by resources and the environment, and the growing dependence on foreign energy sources, the country is facing greater pressure in ensuring the stable supply of energy. The real imbalance between the distribution of energy resources and demand within China provides an objective need to further enhance long-distance energy transport and distribution over large areas. China's energy exploration and utilisation efficiency is on the low side, and improvements can be made to its energy structure. Energy developments in the cities and the countryside are uncoordinated, and they need to be improved.

1.3.1 The Problem of Sustained Supply

1.3.1.1 Energy Demand Will Grow at a High Rate in the Long-term

Since the implementation of its reform programme, China's economy continued to grow rapidly. From 1978 to 2010, the country's GDP grew at an annual rate of 9.9%. In 2010, China's GDP reached 40 trillion yuan to become the world's second largest economy after the USA.

According to the 'New Three-Step' strategy, China will achieve the target of becoming a moderately well-off nation by 2020. The country will be modernised by 2050, becoming a moderately developed, socialist and modern nation that will be affluent, democratic, refined and harmonious. Looking forward, China's economy still has the potential to sustain stable and high growth. Industrialisation and urbanisation will quicken, and consumption patterns will rapidly upgrade from the previous focus on clothing and food to residential and transport consumption. The transformation of the social structure will speed up, with the continuous enhancement of the socialist market economy, and the structural change from the urban–rural dichotomy to urban–rural integration. These medium- to long-term transformations will continue in the next 20–30 years.

Outside China, the world economy is slowing down under the impact of the global financial crisis. The development pattern has undergone a certain degree of change and various types of protectionism are rearing their ugly heads. Despite these negative developments, the multipolarisation of the world and economic globalisation are still going strong, and a new technological revolution is taking shape. Peace, development and cooperation are still the watchwords in today's world; the global situation is on the whole conducive to China's peaceful development. The important strategic window of opportunity for China's development, which had begun in the early 21st century, is still extant and will even be extended.

Within China, the industrial structure is irrational, the developments of the cities and the countryside are uncoordinated, the relationship between investment and consumption is imbalanced, and the gap in income distribution has widened. Despite these factors that are detrimental to economic development, the potential of the domestic market remains huge, capital is plentiful, the quality of the labour force has improved and the infrastructure continues to be enhanced. There is still ample space for economic and social development.

Based on an analysis of the situation both at home and abroad, China's economy is forecast to see stable and rapid growth in the long-term future. Table 1.10 shows the forecasts of China's economic growth by Chinese and international research organisations. Based on a comprehensive study of the figures, China's GDP in 2010–2020 will be between 7.9% and 8.1%, between 5.4% and 5.6% in 2020–2030, and between 4.4% and 4.6% in 2030–2050.

Table 1.10 Forecasts of China's economic development by Chinese and international research organisations (%).

IEA (2010)	2008–2020	2010–2015	2020–2035	2008–2035
	7.9	9.5	3.9	5.7
EIA (2010)	2007–2020	2015–2020	2020–2030	2007–2030
	7.4	6.4	4.7	6.3
IEEJ (2010)	2007–2020	2020–2030	2030–2035	2007–2030
	7.1	4.7	3.8	5.6
Chinese Academy Of Engineering (CAE)	2010–2020	2020–2030	2030–2040	2040–2050
	8.4	7.1	5.0	3.6

Source: IEA, *World Energy Outlook 2010, 2010*; Energy Information Administration (EIA), *International Energy Outlook 2010, 2010*; Institute of Energy Economics, Japan (IEEJ), *Handbook of Energy & Economic Statistics in Japan 2010, 2010*; China Energy Medium- and Long-term Development Strategy Research Project Group, *China Energy Medium- and Long-term (2030, 2050) Development Strategy Research, Synthetic Volume*, Science Press, 2011.

China is now at the intermediate stage of industrialisation, where energy-intensive products form a large proportion of industrial products. As the world's most populous country, China's urbanisation has a huge demand for energy-intensive products, which is met by domestic production. Therefore, China's industrialisation and urbanisation will bring about fast-growing energy demand in the long-term future. Table 1.11 shows the forecast of China's future energy needs by Chinese and international research organisations.

Based on a comprehensive study, China will need 4.8–5.2 billion tonnes of standard coal of primary energy in 2020, 5.7–6.2 billion tonnes in 2030 and 7.0–7.7 billion tonnes in 2050. Table 1.12 shows the primary energy structure, in which total primary energy needs in 2020, 2030 and 2050 are respectively, 5.14 billion tonnes, 5.91 billion tonnes and 7.3 billion tonnes of standard coal.

It is predicted that in 2020, 2030 and 2050, China's total installed generation capacities will be 1860–2010 GW, 2270–2490 GW and 3480–3770 GW respectively. Tables

Table 1.11 Forecasts of China's future energy needs by Chinese and international research organisations.

Unit: billion tonnes of standard coal

Organisation	2015	2020	2030	2035	2050
IEA (2010)	4.12	4.51	5.10	5.34	–
EIA (2010)	3.65	4.37	5.86	6.55	–
IEEJ (2010)	–	3.63	4.52	4.93	–
CAE	–	4.07–4.35	4.55–4.95	–	5.19–5.79
Energy Research Institute, National Development and Reform Commission	–	3.85–4.77	–	4.60–5.85	5.02–6.69

Source: IEA, *World Energy Outlook 2010, 2010*; EIA, *International Energy Outlook 2010, 2010*; IEEJ, *Handbook of Energy & Economic Statistics in Japan 2010, 2010*; China Energy Medium- and Long-term Development Strategy Research Project Group, *China Energy Medium- and Long-term (2030, 2050) Development Strategy Research, Synthetic Volume*, Science Press, 2011; Task Group of the Energy Research Institute of the National Development And Reform Commission, *China's Path of Low-carbon Development 2050: An Analysis of Energy Needs and Carbon Emissions*, Science Press, 2010.

Table 1.12 China's future primary energy needs and structure.

Type	2020		2030		2050	
	Energy needs (billion tonnes of standard coal)	Ratio (%)	Energy needs (billion tonnes of standard coal)	Ratio (%)	Energy needs (billion tonnes of standard coal)	Ratio (%)
Coal	3.00	58.4	3.09	52.3	3.17	43.4
Oil	0.89	17.3	1.09	18.5	1.20	16.4
Natural gas	0.48	9.3	0.57	9.6	0.72	9.9
Nonfossil fuels	0.77	15.0	1.16	19.6	2.21	30.3
Total	5.14	100	5.91	100	7.30	100

Table 1.13 China's future installed generation capacities and structures.

Type	2020		2030		2050	
	Installed capacity (MW)	Ratio (%)	Installed capacity (MW)	Ratio (%)	Installed capacity (MW)	Ratio (%)
Total installed capacity	1 934 000	100	2 380 000	100	3 630 000	100
Coal	1 190 000	61.5	1 270 000	53.3	1 370 000	37.8
Gas	70 000	3.6	100 000	4.2	170 000	4.7
Hydropower	345 000	17.8	430 000	18.1	470 000	12.9
Nuclear power	80 000	4.1	160 000	6.7	430 000	11.8
Wind power	160 000	8.3	250 000	10.5	520 000	14.3
Solar power	24 000	1.3	75 000	3.2	340 000	9.4
Biomass, etc.	15 000	0.8	25 000	1.1	50 000	1.4
Pumped storage	50 000	2.6	70 000	2.9	280 000	7.7

Table 1.14 China's future generation and structures.

Type	2020		2030		2050	
	Generation (GWh)	Ratio (%)	Generation (GWh)	Ratio (%)	Generation (GWh)	Ratio (%)
Total generation	8 680 100	100	10 230 000	100	13 945 000	100
Coal	6 307 000	72.7	6 604 000	64.5	6 987 000	50.1
Gas	245 000	2.8	350 000	3.4	595 000	4.3
Hydropower	1 138 500	13.1	1 419 000	13.9	1 551 000	11.1
Nuclear power	576 000	6.6	1 152 000	11.3	3 096 000	22.2
Wind power	320 000	3.7	500 000	4.9	1 040 000	7.5
Solar power	33 600	0.4	105 000	1.0	476 000	3.4
Biomass	60 000	0.7	100 000	1.0	200 000	1.4

Note: Generation of hydropower does not include pumped storage.

1.13 and 1.14 show the installation structures and generating capacity structures of 2020, 2030 and 2050 at total installed capacities of 1934 GW, 2380 GW and 3630 GW respectively.

1.3.1.2 Greater Constraints on Improving Energy Supply Capacity Imposed by Resources, Environmental and Other Factors

Constrained by various factors such as environmental capacity, development conditions and technical capability, the potential for improving energy supply in China is somewhat limited. The country faces immense pressure in guaranteeing energy supply.

The increase in coal production is mainly hindered by environmental concerns. Coal mining causes ground subsidence and destroy water resources. The total area of ground subsidence caused by coal mining in China has reached 700 000 hectares, and the volume of groundwater ruined by coal mining amounts to 2.2 billion cubic metres every year.

Coal-burning is the main source of pollutants in China, such as sulphur dioxide, nitrogen oxides, smog, mercury and ultrafine particles. Given its high consumption of coal, China has one of the worst air pollution problems in the world. The country's sulphur dioxide emission is the world's highest, and acid rain has occurred. The mercury emission problem caused by coal-burning has garnered greater concern in recent years. It is estimated that 45% of worldwide anthropogenic mercury emission comes from coal combustion. As the mercury in the atmosphere will be carried by air currents, the management of mercury emission is becoming an international issue. There are already international or regional agreements on mercury control, and the restriction of mercury emission is becoming a global trend. To reduce mercury emission, China added emission limits of mercury and its compounds in the 2011 amendment of the Emission Standard of Air Pollutants for Thermal Power Plants, and requested that prevention of mercury pollution be incorporated into the planning for large-scale power producers in the 12th Five-Year Plan. In consideration of the constraints imposed by environmental concerns and water resources, China's coal production capacities in 2020 and 2030 are capped at 4 billion tonnes and 4.2 billion tonnes respectively. There is a certain gap between these figures and coal demands.

Increases in oil and gas productions are mainly hampered by resource constraints. China has rich resources of oil and gas, but its per capita amount of resources is very much lower than the global average. Although there is a high potential for future growth in China's recoverable oil and gas resources, the conditions for resources extraction are poor, resulting in high exploration and production costs. Currently, China's main oilfields are already in the middle and late stages of development, it is very difficult to increase oil production by any great amount in the future. Research has shown that China's oil production peak is around 200 million tonnes, whereas for natural gas it is around 300 billion cubic metres. There is a considerable gap between these figures and the demand for oil and gas to fuel economic and social development.

The development of nonfossil fuels is mainly hindered by environmental concerns, technical capabilities and costs. China has some potential in developing hydropower resources, but the main difficulties are resettlement and the environment. The main obstacles to the development of nuclear power lie in the site selection for nuclear power plants, environmental safety and the supply of nuclear fuel. The development of other nonfossil fuels like wind power, solar power and biomass is mainly hampered by development costs and technical capabilities. In the short and medium term, a host of obstacles must be overcome before production capacities of nonfossil fuels can be increased significantly.

With the escalating problem of global climate change, carbon emission has also become an important constraint restricting the increase of energy supply capacity. Where the amounts of energy provided are equal, coal gives off 30% more carbon dioxide than oil and 70% more than natural gas. Mine gases released during coal mining are also an important source of greenhouse emissions. China's coal-oriented energy consumption structure is at a very disadvantaged position in terms of carbon emission when compared to developed nations that consume mainly oil and gas. According to IEA figures, China produced 24% of the world's carbon dioxide emissions from fossil fuel combustion in 2009. Its per capita carbon dioxide emission has already surpassed the world average (Figure 1.19). Given the continuous rise in China's total energy consumption, it can be foreseen that the country's carbon emission will increase significantly. As the international community

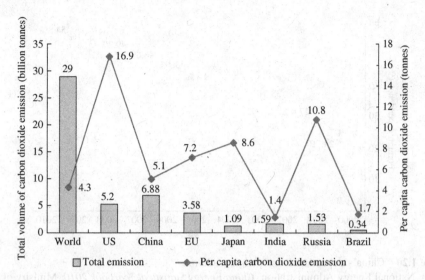

Figure 1.19 Carbon dioxide emissions from fossil fuel combustion produced by the world and major countries in 2009.
Source: IEA, *CO₂ Emissions from Fuel Combustion 2011.*

becomes more vocal in asking China to control its carbon emission, the country will face tremendous pressure to reduce greenhouse emissions in the future.

With socioeconomic development and higher standards of living, atmospheric pollution, especially urban air quality, has become a matter of concern for many. Since the end of 2011, many people have zeroed in on the problem of particulate matter (PM) 2.5 pollutants. PM 2.5 refers to fine particles in the atmosphere that have diameters equal to or smaller than 2.5 micrometres, and they are particularly damaging to human health and atmospheric air quality. In China, the regions that are most heavily polluted by PM 2.5 are Beijing-Tianjin-Hebei, the Yangtze River Delta region and the Pearl River Delta Region. The main sources of PM 2.5 in these regions are the residues from combustion in processes such as coal-burning, industrial production and vehicle exhaust emissions. To reduce PM 2.5 emissions in these severely affected areas, there must be strict limits imposed on new coal-fired power plants in East and Central China, soot pollution in the cities must be controlled, and electric vehicles should be actively developed to replace vehicles that run on petrol.

1.3.1.3 Dependence on Energy Imports Continues to Grow

China's energy supply capabilities, especially in oil and gas, are way too insufficient to meet future demands. To guarantee the energy needs, especially for oil and gas, for China's future economic and social growth, the country must rely on both domestic and international markets and two different resources.

China's dependence on imported energy will reach a relatively high level in the future. Since 2009, the country's dependence of foreign crude oil has surpassed the 50% mark, reaching 54.8% in 2010 (Figure 1.20). Judging by the current development trends, China's

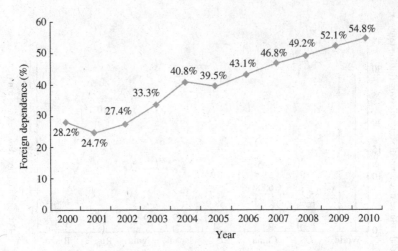

Figure 1.20 China's dependence on foreign crude oil.
Source: National Energy Administration, *China Energy Statistical Yearbook 2010*; Ministry of Land and Resources, *Communiqué on Land and Resources of China 2010*.

demand for crude oil is forecast to reach around 650 million tonnes in 2020 and around 790 million tonnes in 2030. Based on peak production of domestic crude oil at 200 million tonnes, China's dependence on foreign crude oil will reach 69% in 2020 and 75% in 2030. In 2010, the country's dependence on imported natural gas was around 15%. With increasing demands for natural gas, the figure is expected to reach 21% in 2020 and 27% in 2030.

China's increasing dependence on imported energy, especially crude oil, is going to pose huge threats to the country's energy security. The first is the risk to China's energy supply stability caused by social unrest in energy-producing countries. Most of China's imported crude oil originates from the Middle East and Africa. These are areas plagued by intense ethnic and religious conflicts and political instability, which severely threaten the stability of the oil market. The outbreak of conflict will not only damage China's energy infrastructure in those areas, it may also lead to the nonperformance of energy agreements or contracts due to factors like regime change. The second threat is the risk to the safety of overseas energy transport channels. Most of the crude oil imported from the Middle East must pass the Strait of Hormuz, across the Indian Ocean and then sail along the Strait of Malacca. Crude oil from northern Africa must pass through the Suez Canal and sail along the Strait of Malacca. Crude oil imported from western Africa must pass the Cape of Good Hope, across the Indian Ocean and then sail along the Strait of Malacca. Crude oil from Latin America must sail across the Pacific Ocean. Therefore, the Strait of Hormuz, the Indian Ocean shipping lines and the Strait of Malacca are of vital importance to China's oil imports. To ensure the safety of oil transport shipping channels, China must further strengthen its political and military influence. This is an enormous task. The third threat is the risk to China's economic development caused by fluctuations in world energy prices. The increase in world oil price will raise the cost of China's economic operations. As estimated based on crude oil imports in 2009, each time the world oil price goes up by US$1.00 per barrel, China must pay US$1.4 billion more for the purchase of imported crude oil. The increase in the price of crude oil will lead to rising costs in industries like transport, petrochemicals and light industries. It will also push up the prices of basic industries like coal, electricity and construction, as well as the prices of many consumer goods. Inflationary pressures will increase, company profits will be depressed and international competitiveness of products will suffer.

1.3.2   The Problem of Transport and Allocation

1.3.2.1   Spatial Mismatch between Energy Production and Energy Demand

Given the spatial mismatch between China's energy production and energy demand, there is an objective need to bring about a major improvement in the allocation of energy resources. The eastern region's economy is relatively developed with a greater demand for energy, but it has poor energy resources. In contrast, the economies of the central and western regions are smaller with less demand for energy, but they have rich energy resources. In 2010 China's eastern region accounted for 61.7% of China's GDP, 53.5% of primary energy consumption and 57.1% of electricity consumption, but it only produced 10.5% and 7.3% of the country's coal and hydropower resources respectively (Table 1.15). With China's economic development, the energy demands of economically advanced regions will further increase, but the energy production centres will gradually move west with more intense resource development. Therefore, the spatial mismatch between energy

Table 1.15   The distribution of China's energy resources, electricity consumption and GDP (%)*.

Unit: % Item	Eastern Region	Central Region	Western Region
Share of hydropower (technically exploitable amount)	7.3	11.2	81.5
Share of coal basic reserves	10.5	38.2	51.3
Share of power installed capacity	46.0	22.7	31.3
Share of thermal power installed capacity	52.4	23.1	24.5
Share of electricity consumption	57.1	19.3	23.6
Share of primary energy consumption	53.5	21.7	24.8
Share of GDP	61.7	19.7	18.6

Source: The data in the table are calculated and organised according to information contained in *The 2003 Review of National Hydropower Resources, China Statistical Yearbook 2011, China Energy Statistical Yearbook 2011*, etc.
*The technically exploitable amounts of hydropower are 2003 figures; the rest are 2010 data. The eastern region includes the municipalities of Beijing, Tianjin and Shanghai, and the provinces of Heilongjiang, Liaoning, Jilin, Hebei, Shandong, Jiangsu, Zhejiang, Fujian, Guangdong and Hainan. The central region includes the provinces of Shanxi, Henan, Hubei, Anhui, Hunan and Jiangxi. The western region includes the provinces of Qinghai, Gansu, Shaanxi, Sichuan, Yunnan and Guizhou, the autonomous regions of Xinjiang, Tibet, Inner Mongolia, Ningxia and Guangxi, and the municipality of Chongqing.

production and energy consumption regions will become even more obvious. This makes the large-scale and long-distance deployment of energy resources across regions inevitable. In the future, the overall flow of energy in China will see coal and oil from the north moving south, and natural gas and electricity from the west moving east. This will place greater demands on energy resource allocation capabilities.

1.3.2.2   Insufficient Capability in Improvement of Extensive Energy Allocation

The current energy transport system in China still cannot meet the demands of large-scale and long-distance energy allocation, resulting in repeated crises in coal, electricity and transport. The railway is the main mode of coal transportation. The volume of coal transported by rail accounts for around 60% of China's total coal production, and transport demands by coal on the national railways account for 50% of nationwide cargo transport demands (Figure 1.21). China's railway industry has seen tremendous development since 1949, but it is still insufficient to meet the rapid increase in coal transport demand. The inadequate transport capability of the railways has an increasingly serious bottleneck effect on coal transport. Coal resources in the eastern region will gradually deplete but consumption will continue to rise. With the westward shift in coal production, the role of the railways in coal transportation will get bigger.

China's energy investments have for a long time been focused on production, with inadequate attention paid to transportation. Take electricity for example. Since the implementation of its economic reform programme, the country's electricity supply has been tight most of the time. The long-term electricity shortage has brought about a serious situation where power generation is emphasised at the expense of power transmission, without any management of power usage. Investments in electricity lean towards the source, resulting in the imbalance between investments in the power source and the power

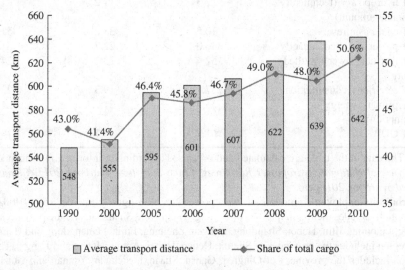

Figure 1.21   The rail-based transport of coal in China.
Source: National Bureau of Statistics of China, *China Statistical Yearbook*.

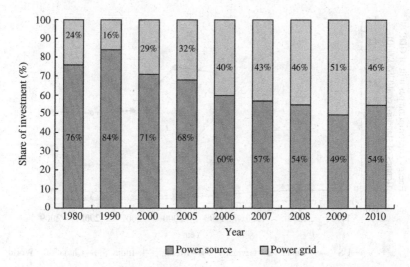

Figure 1.22   The shares of power source and power grid investments in China's investments in electricity.
Source: *Statistical Data on the Electricity Industry*, China Electricity Council.

grid. The development of power grids is much delayed, causing suboptimal improvement in resource allocation. Figure 1.22 shows the shares of power source and power grid investments in China's total investments in electricity. As a major way of transporting energy, inadequacies in the long-distance power transmission by power grids also increase the pressures on coal transport. In recent years, insufficient capabilities in rail transport and inadequate cross-regional resource allocation capacities of power grids have meant occasional power shortages in certain regions.

1.3.3   The Quality Problem of Development

1.3.3.1   Comparatively Low Energy Development and Utilisation Efficiency

The energy consumption indicator of China's energy industry is comparatively higher than the international advanced level. In 2010 the overall energy consumption of coal mining and washing in China was 20.8 kg of standard coal/tonne, while the overall energy consumption of crude oil was 160 kg of standard coal/tonne. These figures are higher than international advanced levels by 3.8 kg of standard coal/tonne and 45 kg of standard coal/tonne respectively. Coal-fired power generation in thermal power plants consumed 333 g of standard coal/(kWh), higher than the international advanced level by 23 g of standard coal/(kWh). The power consumption rate by power plants was 5.43% and the line loss rate of power grids was 6.53%. These figures are higher than international advanced levels by 1.63 percentage points and 1.53 percentage points respectively.

The energy consumption levels of China's main energy-consuming products are also generally higher than international advanced levels. The comparable energy consumption per tonne of steel (large and medium-sized enterprises) in China was 697 kg of standard coal/tonne in 2010, the overall energy consumption of cement was 139 kg of standard

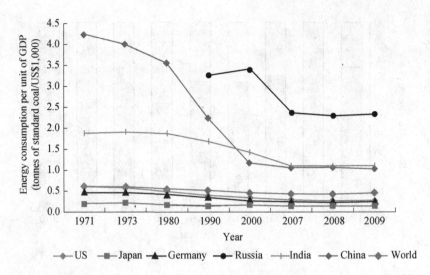

Figure 1.23 Comparison of energy consumption per unit of GDP between major countries tonnes/ per thousand US dollar.
Source: Calculations based on IEA statistical data.

coal/tonne and the overall energy consumption of ethylene was 976 kg of standard coal/tonne. These numbers are higher than international advanced levels by 87, 21 and 347 kg of standard coal/tonne respectively. Energy consumption of calcium carbide was 3,395 kWh/tonne, higher than the international advanced level by 395 kWh/tonne, while energy consumption of synthetic ammonia was 1,591 kg of standard coal/tonne, higher than the international advanced level by 591 kg of standard coal/tonne.

The economic output efficiency per unit of energy in China lags behind that of advanced countries. The energy consumption per unit of GDP reflects to a certain extent the level of a country's energy efficiency. On an internationally comparable basis, China's energy consumption per unit of GDP in 2009 was 1.03 tonnes of standard coal/US$1,000.[6] This is 2.3 times higher than the world average, 3.8 times higher than the USA and 7.4 times higher than Japan (Figure 1.23). China's GDP accounts for around 8% of the world's GDP, but its energy consumption takes up around 19% of the world total.

1.3.3.2 Urgent Need to Improve Energy Structure

Compared to developed countries, China consumes a higher volume of coal and uses less oil, natural gas and clean energy. On a worldwide comparable basis, the share of coal in China's primary energy consumption structure in 2009 was around 45 percentage points higher than the USA, 46 percentage points higher than Japan and 40 percentage points higher than the world average. The share of oil and gas was around 42 percentage points lower than the USA, 39 percentage points lower than Japan and 34

---

[6] GDP in constant 2000 US dollars.

Table 1.16 Comparison of primary energy consumption structures between China and major countries and regions (%).

Unit: %

Country/region	Coal	Oil	Natural gas	Hydropower, nuclear power, wind power, etc.
US	22.5	37.1	24.7	15.7
EU	16.4	34.9	25.0	23.7
Japan	21.4	42.4	17.1	19.1
Russia	14.7	21.3	54.0	10.0
India	41.9	23.8	7.3	27.0
China	67.2	16.9	3.4	12.5
World	27.2	32.9	20.9	19.0

Source: The data in the table are obtained from IEA, which are slightly different from the data released by the National Bureau of Statistic of China. The data are from 2009.

Table 1.17 Comparison of end-use energy consumption structures between China and major countries and regions

Unit: %

Country/region	Coal	Oil	Natural gas	Electricity	Thermal power and others
US	1.6	50.7	21.3	21.4	5.0
EU	3.9	44.1	21.2	20.9	9.9
Japan	8.3	54.8	10.2	25.5	1.2
Russia	4.3	25.1	30.5	13.9	26.2
India	13.8	29.7	4.8	13.8	37.9
China	36.0	23.6	3.5	18.5	18.4
World	9.8	41.6	15.2	17.3	16.1

Source: The data in the table are obtained from IEA, which are slightly different from the data released by the National Bureau of Statistic of China. The data are from 2009.

percentage points lower than the world average (Table 1.16). Clean energy consumption was around 3 percentage points lower than the USA, 7 percentage points lower than Japan and 7 percentage points lower than the world average. The share of coal in China's end-use energy consumption structure was around 34 percentage points higher than the USA, 28 percentage points higher than Japan and 26 percentage points higher than the world average (Table 1.17).

An energy consumption structure dominated by coal has brought about problems such as environmental pressures. Adjusting and improving the energy structure to lower the ratio of coal and raise the ratios of low-carbon and carbon-free energy like natural gas, nuclear power and renewable energy, thereby diversifying and cleaning up the structure, is a major task for China's energy industry.

1.3.3.3 Arduous Task of Coordinating Urban and Rural Energy Developments

Given the effects of various factors like history and social organisation, China's socioeco-nomic development exhibits a dual structure divided between urban and rural. Agriculture and the countryside are at comparatively weaker positions compared to industry and urban areas. The urban disposable income per capita in 2010 was 3.23 times of the rural net income per capita. If various forms of welfare protection are included, the actual difference may be greater.

Due to the dual urban–rural structure, energy development in the countryside lags far behind urban areas. In terms of everyday consumption of energy in the countryside, commercial energy accounts for a relatively small portion. In many areas, firewood and straw remain the major energy source, and large quantities of wood are directly burnt. Not only does it cause pollution, it also results in the destruction of forests and soil erosion. Energy infrastructure in the rural areas is backward, and with weak power grid structures there is a high incidence of transmission loss. Indices like the penetration rate of electric appliances, personal energy consumption per capita and electricity utilisation still lag far behind urban levels (Table 1.18). The lack of high-grade energy and the low level of energy utilisation directly affect the economic and social development of the countryside, which results in many modern production methods and lifestyles not being able to be actualised. Employment opportunities remain depressed and people remain in poverty for prolonged periods of time. A host of negative effects are also seen in the environment and people's health.

The objective reality of China's dual urban–rural structure makes the task of energy development very difficult. Energy infrastructure like power grids in the countryside lack

Table 1.18 Standard of living and energy consumption of urban and rural residents in China.

Index			1990	2000	2005	2009	2010
Income per capita[*] (Yuan/person)		Urban	1 510	6 280	10 493	17 175	19 109
		Rural	686	2 253	3 255	5 153	5 919
Penetration rate of main energy-consuming appliances (units/100 households)	Refrigerator	Urban	42.3	80.1	90.7	95.4	96.6
		Rural	1.2	12.3	20.1	37.1	45.2
	Colour TV	Urban	59.0	116.6	134.8	135.7	137.4
		Rural	4.7	48.7	84.1	108.9	111.8
	Washing machine	Urban	78.4	90.5	95.5	96.0	96.9
		Rural	9.1	28.6	40.2	53.1	57.3
	Air-conditioner	Urban	0.3	30.8	80.7	106.8	112.1
		Rural	–	1.3	6.4	12.2	16.0
Domestic energy consumption per capita (1000 g of standard coal/person)		Urban	298	210	279	325	315
		Rural	83	76	132	190	204
Domestic electricity consumption per capital (kWh/person)		Urban	90	202	319	439	442
		Rural	25	65	146	296	318

Source: The data in the table are obtained from National Bureau of Statistics of China, *China Statistical Yearbook 2011* and *China Energy Statistical Yearbook 2011*; China Electricity Council, *Statistical Data on the Electrical Power Industry 2010*.
[*]Income per capita of urban residents refers to disposable income per capita; income per capita of rural residents refers to net income per capita.

investments, the funding gap for upgrading is large and the returns on investments are hard to come by. A framework for sustainable development is lacking. There is an urgent need for increasing the financial support for infrastructure like power grids in the countryside and improving the coordination in the use of funds. Based on the development needs of urban and rural areas, residents in the cities and countryside should pay the same price for electricity when they use the same grid. Electricity should be universalised, and the levels of electricity supply capability and service should be raised.

1.4   Causes that Affect China's Energy Development

China's energy industry is a key basic industry that is connected to people's livelihood, and its development is closely linked to socioeconomic growth, as well as the combined effects of various domestic and global factors. The causes that affect the development of the country's energy industry can be analysed from three perspectives: the economic development model, the energy development model and the global competitive environment.

1.4.1   The Economic Development Model

A survey of the history of modernisation shows that modernisation of the economy is closely linked to industrialisation. One of the features in the traditional path of industrialisation taken by developed countries involved the massive consumption of resources (especially nonrenewable resources) to support rapid economic growth. Heavy industrialisation was an inevitable phase that major developed nations went through in the industrialisation process. This development model places excessive importance on achieving growth at all costs and making the maximum profits, and ignores the depletion of natural resources. Their stance on the natural environment had been to 'pollute first and fix it later'. In the course of industrialisation, major developed countries had suffered environmental pollution of varying degrees of seriousness, which had severely affected the health of their citizens. Examples include the 'Los Angeles smog', London's 'pea souper' and the Minamata diseases in Japan. It was only in the later stages of their industrialisation that developed countries, with their superiority in technology and capital, shifted a great deal of their high energy-consuming and high polluting industries to developing countries. They began to rely on technological innovation to fuel their own economic growth and reduced the damage to the environment.

Compared to the developed countries, China had a late start in industrialisation and it started from a low base. To reduce the gap between itself and the developed nations, China followed the same rule that had guided traditional industrialisation, which was to fuel economic growth through increasing its investments in production factors like capital, labour and natural resources. China's adherence to the traditional path of industrialisation has resulted in its industrial structure showing three main characteristics: energy-consuming, labour intensive and export-oriented. These characteristics had taken form not only because of the restrictions of the general rules of industrialisation, but also because of China's own conditions. China has the world's biggest population. It has rich labour and natural resources, but its level of technology and capital stock is low. For this reason, it must make use of its superiority in manpower and resources in the early stages of its industrialisation to manage the shift of industries worldwide. China must develop low-tech manufacturing and processing industries to produce large quantities of low value-added products, and make use of its cost advantage to seize the international market. In the process, the country has had to endure the damage done to its environment. For a period of time, China's economic development model has been an extensive form of economic growth, characterised by high investments, high depletion, high pollution and low efficiency. The extensive form of economic growth has determined China's current situation of continual rapid growth in energy demand and low energy efficiency.

As China industrialises further, flaws that accompany the traditional path of industrialisation become more apparent. The domestic and international climates also dictate that China can no longer realise its goal of modernisation along this traditional route. From the international perspective, when the developed countries in the West were at their industrialisation stages, their level of knowledge about environmental issues was far lower compared to the present day. The environmental capacity then was also far greater than now. With the growing global consensus on environmental issues, indices like carbon emission are restricting the space for developing nations to grow. If developing nations continue to take the traditional route of industrialisation, they will face greater international pressures. In addition, developed countries have taken advantage of their early start to dominate and grab the better resources and markets in the world. This has made it more difficult and expensive for developing countries to acquire good and inexpensive resources. From the domestic angle, China's ownership of resources per capita is low and its ecological environment is fragile. The country cannot support a development model characterised by the high consumption of natural resources. Just after China implemented its economic reform, the main obstacles to its economic development were capital and technology. Currently, however, resources and environment factors are proving to be even more of a restriction to China's economic growth. To realise the sustainable development of China's energy industry, and to make use of limited energy resources to support the country's modernisation efforts, there is an objective need to adjust China's economic development strategy, and change its economic development model quickly.

1.4.2 The Energy Development Model

For a long time, affected by factors such as the county's stage of development and institutional mechanisms, China's energy industry development has lacked a unified strategic guideline. The different parts of the industry are developing on their own without any coordination, and there is no synchronisation between energy and socioeconomic development. The energy structure and composition are irrational, and there is insufficient accord in the pace and quality of development. The supply and demand of energy are balanced by individual regions, and there is not enough capacity for large-scale and better allocation of resources. Energy growth relies mainly on outward expansion, with little contribution from factors like technological progress, management innovation and improvements made in human resources. Energy investments are focused on developing energy resources, especially conventional energy resources. Insufficient investments are made in developing new energy, energy transport and energy allocation. Hampered by a late start in 'going global' and making use of foreign energy resources, China does not have much say in the international energy market. The energy price mechanism can still be improved, and the fundamental role of the market allocation of resources can be strengthened. The institutional mechanisms that are conducive to improving the energy structure, ensuring safe power supply, saving energy and reducing carbon emission, protecting the environment, and coordinating the energy developments between urban and rural areas have yet to come about.

In recent years, the structural contradictions of China's energy development are becoming more apparent. There are greater pressures on energy security and sustainable development. A major cause is the unscientific mode of energy development. It has become an urgent task to quicken the pace of changes in energy exploration and use, transport and allocation, and consumption patterns.

1.4.3   The Global Competitive Environment

The finite nature of fossil fuels and the dependence of modern socioeconomic development on energy have resulted in resources like oil and natural gas becoming strategic materials fought over by the world's nations. The greater control a country has over oil and gas resources, the greater its say and influence over world affairs. To control energy resources and obtain greater international competitiveness, the world's major nations have placed energy diplomacy at the cores of their diplomatic strategies. Certain countries have even used energy issues as bargaining chips in political negotiations. Daniel Yergin, Chairman of Cambridge Energy Research Associates, had said that oil was 10% economics and 90% politics.

Countries and regions like the USA, the EU, Japan and Russia have taken advantage of their early start and got involved in the world energy market at an earlier stage. They occupy a dominant position in global energy competition. With its mighty economic, political and military prowess, the USA continues to influence and control energy-rich regions and important energy transport channels. Through defending the US dollar's role as the settlement currency for global oil, the USA has strategic control over the world energy market. By strengthening its internal integration, the EU is implementing a multilayered and comprehensive foreign energy policy, at the same time as it places great importance on cooperation with the USA on global energy matters. Historically, Japan has always focused on energy diplomacy. By making good use of economic aid, technical support and joint development, the country ensures a steady energy supply. Russia is the world's biggest exporter of energy. Its two-pronged energy diplomacy features expanding energy export markets and transforming its resource advantage into a strategic one. Compared to these countries, China is a late starter in internationalising its energy strategy, and is relatively weak in obtaining resources and setting world energy prices in the global energy market. This increases the pressure on China's energy security.

In recent years, the issue of global climate change caused by greenhouse gas emission has become a worldwide concern. What emission reduction targets and roadmaps are laid down, and how individual countries' responsibilities and obligations to reduce greenhouse gas emission are determined have a tremendous effect on the energy and socioeconomic developments of each country. Therefore, they have also become bones of contention.

Given China's coal-dominant energy structure, its rapid industrialisation and urbanisation, and its position at the bottom of the industrial chain among the world's countries, it will face greater pressures in the improvement of its energy structure and its response to climate change. For China's energy development to be sustainable, it is imperative that a new energy security framework marked by mutually beneficial cooperation, diverse development, and coordination and assurance be adopted. At the same time, China must increase its negotiating powers with major countries over energy and related issues in order to occupy a more advantageous position in global competition.

2

Strategic Thinking on Energy

Any solution to China's energy problem must be formulated under the guidance of the Theoretical System of Socialism with Chinese Characteristics with the overall socioeconomic development in mind. The 'Grand Energy Vision' and scientific solutions must be established, and suitable means of implementation must be chosen. Energy modernisation with Chinese characteristics must be pursued by proactively changing the mode of energy development and pushing for a transformation of energy strategy. In changing the mode of energy development, electricity must be placed unwaveringly as the centre of the process. Implementing the 'One Ultra Four Large' (1U4L) strategy is an important and urgent task in the present time.

2.1 Basic Thinking Behind the Energy Solution

The energy problem is a very complex one that involves a host of factors. The basic solution to China's energy problem is to establish a 'Grand Energy Vision', and push for changes in the mode of economic development, the mode of energy development and the global competitive environment. The process of changing the mode of energy development is the process of transforming energy strategy.

2.1.1 Complexity of the Energy Problem

2.1.1.1 It is Comprehensive

The energy problem affects the nation's overall socioeconomic development. Energy development involves many aspects such as the country's science and technology capabilities, developmental stage, consumption patterns, ecological environment, as well as its diplomatic and military affairs. First of all, energy is intimately linked to technological progress, which is the fundamental force that fuels energy development. Every major change in energy development and utilisation could be attributed to breakthroughs in energy technology. Secondly, Energy is intimately linked to economic development. The energy sector is an important basic industry. With the flourishing of new energy, smart grids and so on, the energy sector in the new order is becoming the hotbed for

---

Electric Power and Energy in China, First Edition. Zhenya Liu.

the growth of strategic emerging industries. Thirdly, energy is intimately linked to social development. Energy is essential for meeting the basic needs and services of modern human society. The balanced distribution of energy supply and quality is an important reflection of social justice and harmony. Fourthly, energy is intimately linked to environmental protection. At the same time as people's lives have been made more convenient by the large-scale development and utilisation of energy, the natural environment has been affected. Environmental pollution, climate change and ecological deterioration are problems that refuse to go away. Finally, energy is intimately linked to national security. With China's increasing dependence on foreign energy sources, and the global energy situation and greenhouse gas issue exerting an increasingly prohibiting effect on China's energy development, energy diplomacy has become an important component of Chinese diplomacy. Ensuring the safety of strategic energy transport channels not only involves commercial interests and diplomatic relationships, but also entails military capabilities.

2.1.1.2   There Are Differences

The problems and conflicts of energy development that individual countries and regions face are different. From the time dimension, developed countries have, for the past century or so, made use of fossil fuels like coal, cheap oil and natural gas to complete their industrialisation. As a result, they occupy an advantageous position in the global competition for energy. With many developing countries now undergoing industrialisation, and the depletion of fossil fuels becoming increasingly serious, the international competition for energy is turning more intense. China's economic development does not enjoy the easy availability of energy that Europe and North America once had during their industrialisation. From the space dimension, the mainstay of China's energy structure will still be coal in the long run, given the country's energy endowments and energy policies. It will be so for a long time. This is different from the oil- and -gas-centred energy structures of OECD states. China's energy development has yet to experience what has been called the Oil Age. Besides, there are obvious differences in the level of development between regions within China, and between the urban and rural areas, which result in different tensions and problems in energy development. The solution to China's energy problems must be modified based on China's realities. The mode of development with Chinese characteristics must be comprehensive, coordinated and sustainable.

2.1.1.3   It Is Long-term

Since the dawn of human civilisation, the energy problem has produced profound influences on human lives and production as well as social progress. Important breakthroughs in energy technologies and the improvement and adjustment of energy structures were never achieved overnight. They often had to undergo a lengthy and gradual process. The acute energy problems facing China today are characterised by their own histories and realities. Some have come about because of China's energy endowment and stage of development, while others have been created by a combination of domestic and global factors. Some are the results of an accumulation of longstanding contradictions, while others are new challenges posed by the new order. There are no 'miracle drugs' to solve these problems instantly. What is needed is a tireless enquiry, with goals, planning and procedures, guided by a clear energy strategy.

2.1.1.4 There Are Uncertainties

History tells us that inherent in the process of energy development are many uncertainties. These uncertainties may be important breakthroughs or discoveries in the realm of science and technology, or they may be major amendments and changes in policies and laws. They may also be unplanned incidents or crises. It was the invention of the steam engine in the late 17th century that precipitated the large-scale development and use of coal, which replaced the leadership role of traditional renewable energy such as firewood in global energy use. The invention of the internal combustion engine in the late 19th century was the reason why oil was able to replace coal as the world's main energy source in the 1960s. The 2008 global financial crisis brought the world economy to its knees. After the crisis, many countries identified new energy as an important stimulus to economic growth, and the development and utilisation of new and renewable energies were given unprecedented attention. The radioactive fallout from the nuclear plants at Fukushima, Japan in 2011 prompted the international community to rethink the safety of nuclear power, which has significant repercussions on recent development of nuclear energy use around the world. The long-term effects of these repercussions are as yet unknown. History also tells us that even when the correct direction and suitable path of energy development were taken, it was often impossible to set a precise developmental process. Therefore, in the course of energy development, experience must be constantly evaluated, knowledge increased and trends identified, so as to adjust the energy strategy and ideas for development in tandem with the times.

2.1.2 Grand Energy Vision

Solving the energy problem requires guidance from a scientific worldview and methodology. The Theoretical System of Socialism with Chinese Characteristics provides powerful theoretical ammunition for the analysis of China's energy strategy. In using the Theoretical System of Socialism with Chinese Characteristics to guide and study China's energy strategy, the key step is the establishment of a Grand Energy Vision.

The Grand Energy Vision refers to the analysis and study of the energy problem using system theory, the concept of sustainable development, and a perspective that is comprehensive, all-inclusive, historical, open and universally connected. Seen in combination with China's realities, it means a coordinated view of how energy development is linked to the economy, society, the environment and diplomatic relations; how meeting energy demands and protecting the environment is linked to enhanced international competitiveness; how global energy resource development and utilisation is linked with that in China; how coal, hydropower, electricity, oil, gas and nuclear energy are related to each other; how fossil and nonfossil fuels are related; how traditional and new energies are related; and how the energy development, transport and consumption are connected.

2.1.3 Solutions to the Energy Problems

Based on the Grand Energy Vision, it is obvious that China's energy problem cannot be solved by adopting piecemeal solutions. What is needed is an overall consideration of the country's modernisation efforts. The stage of China's economic development, its economic structure and the mode of growth need to be carefully reviewed, as well as the country's energy endowment, energy structure and quality of energy development. Global issues like international relations, energy diplomacy and climate change are also a vital part of the equation. These issues must be thoroughly considered in order to go forward in changing the mode of economic development, the mode of energy development and the international competition framework.

2.1.3.1 Changing the Mode of Economic Development to Pursue New Industrialisation with Chinese Characteristics

The energy problem is a reflection of economic problems in the area of energy. Following sustained and rapid growth since China reformed and opened up its economy, the country has achieved stellar results in economic development, but the problems of imbalanced, uncoordinated and unsustainable development are gradually building up. Significant manifestations of these problems are chronic shortage of energy supplies and increasingly serious environmental issues. To solve China's pressing energy development problem at its roots, the country must quicken the strategic readjustment of its economy, change the traditional model of growth based on natural resources and capital investments, and diverge from the path towards industrialisation that is based on pursuing superficial fast growth and extensive development. The path of industrialisation that western developed nations had trodden, involving massive consumption of resources and sacrificing the environment, is not one that China can continue to take. Otherwise, the competitive edge that the country has enjoyed will be gradually eroded.

To change the model of economic development, the quickening of the change itself and the push for the optimisation and upgrade of the industrial structure must be established as an important strategic mission. Domestic demand, in particular consumption demand, must be increased to transform the main engines of economic growth from investments and exports to one marked by the coordination between consumption, investments and exports; from secondary industry to one marked by the collaboration between primary, secondary and tertiary industries; from the increased consumption of material resources to one marked by technological progress, better workers and innovative management.[1] Information technology (IT) must integrate with industrialisation so that IT and industrialisation can give each other a leg up and in so doing, forge a new path of industrialisation marked by high technological content, high economic efficiency, low consumption of resources, low environmental pollution and optimal use of human resources.[2]

China must ride on the new wave of developing a low-carbon economy. The country's strategic economic transformation and change in development model must be integrated with the development of a low-carbon economy and the fight against global climate change. This will bring about a model of development where the economy, society and the environment complement one another. Studies have shown that per capita emission of greenhouse gases has an inverted U-shaped curve relationship with the level of socioeconomic development and prosperity. This means that all human societies go through these stages of development: poor but clean, prosperous but polluted, prosperous

---

[1] Hu Jintao, in a report made at the 17th National Congress of the Communist Party of China in October 2007.
[2] Jiang Zemin, in a report made at the 16th National Congress of the Communist Party of China in November 2002.

and clean. China is on its way to reaching the prosperous but highly polluted stage. If the traditional model of development is not abandoned, China will face increasing international pressure over carbon reduction in the future.

By transforming its economic development model and forging a new path of industrialisation, China can break the 'treatment after pollution' mode. By enhancing the flexibility of energy supplies through increasing the efficiency of resource use and the level of environmental protection, and lowering energy consumption and energy intensity, it will be easier for China to adjust its energy structure. This is a necessary requirement to adjust to major changes in the global demand structure and increase China's ability to withstand risks in the international market. It is a necessary requirement to enhance the capacity for sustainable development. It is a necessary requirement to secure a vantage point and develop a new competitive edge in a postfinancial crisis world. It is a necessary requirement to rationalise the distribution of national income and promote social harmony and stability. It is a necessary requirement to meet China's target of becoming a moderately prosperous nation and meeting its people's expectations of leading better lives.[3]

Innovation is the key to making a fundamental change in the economic development model. Through innovation in institutional mechanisms and the removal of the obstacles inherent in them, scientific development can be gradually built up and perfected. This will accelerate the change in development mode at an institutional level. By promoting technological innovation and increasing the pace of building an innovative nation, the promoting, supporting and leading functions of technology in its role as the primary productive force can be fully harnessed to transform the economic development model.

2.1.3.2 Changing the Mode of Energy Development to Pursue Energy Modernisation with Chinese Characteristics

Whether one looks at it from the perspective of ensuring energy supplies, promoting energy conservation and emission reduction, and easing the pressures posed by energy security and environmental concerns, or from the point of view of enhancing energy service quality, and boosting the international competitiveness of the energy industry, the need to quicken the pace of changing the mode of energy development is an urgent task. To change the mode of energy development, the starting point must be the realities of China's energy industry. China must be fully cognisant of the features of international energy development and conform to the trend of low-carbon and smart development. Fuelled by innovation in institutional mechanisms and technology, reforms can be implemented in how energy is developed and utilised. Various forms of energy and domestic or global resource development and use can be promoted, and energy consumption models can be enhanced. In so doing, a modern energy framework marked by a clean, efficient, diverse and smart development, and a safe, efficient, clean and economic structure will emerge. The process of changing the mode of energy development is the process of transforming China's energy strategy. The transformation of energy strategy entails five aspects:

Firstly, transforming the energy structure from high-carbon to low-carbon. Following the global trend of low-carbon development, China should proactively develop zero- or

---

[3] Hu Jintao, in a speech given at a symposium for provincial- and ministerial-level leaders on thoroughly implementing the Scientific Outlook on Development and accelerating the transformation of the economic development mode in February 2010.

low-carbon energy such as hydropower, wind power, solar energy and natural gas, as well as pursue a steady development of nuclear and other clean energies. This will gradually lower the proportion of fossil fuels, especially coal, in total energy consumption, and achieve a low-carbon energy structure.

Secondly, transforming energy use from extensive mode to intensive and efficient mode. By reforming the mode of energy development and utilisation and energy consumption pattern, energy resources can be developed in a more intensified manner. By promoting the adjustment of total energy consumption and conserving energy use, by raising the efficiency of energy resource exploration, conversion, transmission and utilisation, and by increasing the economic output of energy consumption per unit, the comprehensive benefits of energy utilisation will be maximised.

Thirdly, transforming energy allocation from local balance to an optimal allocation in a larger scope. This means breaking the development thinking mode of balancing energy distribution locally. With the reverse distribution of China's energy resources and energy demands as a starting point, greater efforts must be put into building up energy transmission networks and channels and enhancing the energy transport system. The problem of over dependence on coal-handling must be resolved. In doing so, energy resources can be optimally allocated throughout the entire nation.

Fourthly, transforming energy supply from one that ensures domestic resources to one that ensures the coordination and utilisation of international and domestic resources. This means focusing on two markets and two resources: worldwide and domestic. At the same time as keeping a firm foothold within its borders, China should take more initiative in participating in global energy competition and cooperation. By increasing the import of various types of energy products, the country's deficiency in energy resources, especially oil and gas, can be topped up. This ensures adequate supply where domestic and worldwide resources complement each other.

Fifthly, transforming energy services from unidirectional supply to a smart and interactive system. Based on a rationally structured and solid energy system, China should focus on constructing a smart power grid. This will bring about a smart energy system that will effectively support various kinds of energy input and output, and result in a new energy model marked by convenient and quick turnaround between demand and supply and their mutual interaction.

2.1.3.3 Changing Global Competitive Structure to Create a Comparatively Comfortable and Favourable International Environment

Since the advent of the 21st century, the world has quickly entered a new phase of economic upheaval, paradigm shifts and systemic changes. An uncommon global financial crisis has had deep repercussions in global politics and economics. These have highlighted China's important role and position, and also increased international pressures on the country. These pressures do not only originate in the major developed countries, which have a say in international affairs, but also from other emerging nations in a similar position as China. At the same time as China's overall strength is enhanced with more opportunities to venture overseas, the country is entering a 'trouble-ridden' period. As China experiences more space for energy diplomacy, it is encountering more difficulties in policy manoeuvres. By keeping strictly within the bounds of peaceful development, by participating in international affairs in a more proactive, pragmatic and constructive manner, and by nudging the world political and economic framework towards a direction more favourable to China, the impact on sustainability in China's socioeconomic and energy developments will be significant.

China must put energy diplomacy in a more important position, have a clear understanding of the global energy situation, and pay attention to the geopolitics of global energy. It must insist on a new concept of energy security marked by win–win cooperation, diverse development and protection through collaborative efforts. It should actively seek bilateral and multilateral cooperation with countries or organisations, building closer strategic partnerships with major oil- and gas-producing countries and energy bodies like Organisation of the Petroleum Exporting Countries (OPEC). China should focus more on cooperation with high energy consumption countries to withstand energy supply risks together and maintain the stability of the global energy market. Based on the principle of mutual benefit for all parties, a global energy market governing body comprising energy-producers, energy-consumers and transit countries should be established.[4] The focus should be switched from simply ensuring the security of supply to a two-pronged goal of guaranteeing security of supply and stabilising international energy prices. China must push forward in the areas of energy diplomacy and environmental diplomacy, actively take part in the global 'carbon politics', and advocate the responsibility to reduce the emission of greenhouse gases, one that is shared by all but different from country to country. China should support its energy enterprises to venture out of China and engage in various kinds of public diplomacy to refute the talk of 'China energy threat' and 'China climate threat'. This will create the most flexible and favourable international environment possible for China's economic development and transformation of its energy strategy.

2.1.3.4 The Relationship between the Three Changes

The change in the economic development model leads the change in the energy development model, while the latter backs up the former. The change in the economic development and energy development models makes demands on the change in the global competition structure, while the change in the global competitive structure provides support for the change in the economic development and energy development models. The three changes affect and complement one another (Figure 2.1). Without the change in the economic development model and the change in the global competition structure, the change in the energy development model and strategy of China can never be truly successful.

2.2 The Way to Change the Mode of Energy Development

Beginning in 2003, China's energy development entered a strategic transformation phase. Changing the mode of energy development became an important strategic mission. Guided by state policies and supported by collective efforts, the change in the energy development mode and transformation of the country's energy strategy has in recent years

---

[4] Wen Jiabao, in a speech, 'China is Committed to Pursuing Green and Sustainable Development', given at the World Future Energy Summit in January 2012.

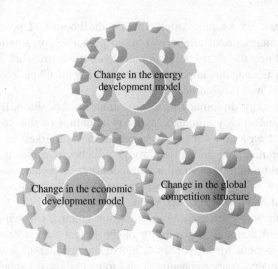

Figure 2.1 The relationship between the three changes.

achieved some results, but there is still much to be done. To bring about real and profound change and transformation, however, the following must be carried out in tandem: energy development and energy conservation, traditional energy development and new energy development, exploitation of domestic and foreign resources, improving energy distribution and transmission as well as innovation in technology and systems.

2.2.1 Transformation Phase of China's Energy Strategy

The development of the energy industry in China after 1949 can be divided into three phases.

The first phase was from 1949 to 1978, during the period of the planned economy. This development phase of the energy industry was marked by a production-led approach. The focus was on development, with coal as the mainstay and self-sufficiency as the goal. For a long time, coal accounted for more than 70% of end-use energy consumption, with oil and electricity totalling some 25%.

The second phase was from 1970 to 2000, from the start of the reform and opening up to the 16th National Congress of the Communist Party of China (NCCPC). This phase of energy industry development was marked by a focus on efficiency. Development and conservation were simultaneously pursued, and energy development was centred in electricity. Based on coal, there was diversity in development. The percentage of coal in end-use energy consumption had fallen to around 50%, with oil and electricity totalling some 40%. It was during this period, in1993, when China went from being a net exporter of oil to a net importer.

The third phase is from 2003 to the present. This development phase of the energy industry is guided by sustainable growth. The demand for energy conservation and emission reduction pervades the entire process of development, utilisation, distribution, transmission and consumption. The development of new energy and international energy cooperation was accorded top priority. The methods of energy development, utilisation, distribution, and consumption witnessed seismic changes, and the energy structure continues to be adjusted and enhanced. This phase is an important strategic transformation period for China's energy development, and it is expected to continue for quite a long period of time.

China's energy development entered its strategic transformation period after 2003, flagged up by the two milestones below.

2.2.1.1 Change in the Phase of Development

In 2003 China's GDP per capita surpassed US$1000 for the first time to reach US$1090. With this as a milestone, China's modernisation entered a new phase of development. This is a phase characterised by the gradual upgrading of the consumption structure, the continued adjustments and changes of investment and industrial structures, and the overall acceleration of the industrialisation and urbanisation process. The internal engines of economic growth are strong, and the national economy is expected to sustain rapid growth in the long run. The data in recent years have proven this point. China took only five years, from 2003 to 2008, to make the jump in GDP per capita from US$1000 to US$3000. Developed countries usually took 10–15 years to make the same leap, e.g. 15 years for Germany and 11 years for Japan. In 2010 China's GDP per capita was 29 941 yuan and the size of its economy surpassed Japan's to become the world's second largest economy. The rate of urbanisation reached 49.7%. The country's economic development was in the second half of the intermediate period of industrialisation.

The inevitable result of continued economic development is rapid growth in energy demand, and tight energy supply becomes a major constraint holding back socioeconomic development. The upgrading of socioeconomic development and consumption structure has resulted in persistent rapid growth in energy demand, which in turn intensifies the demand and supply imbalance, and the pressure on resources and the environment. In 2003 China's energy consumption growth reached 15.3%, a record high since 1980. The year after, it went up to 16.1%. In 2003 China's total oil consumption exceeded Japan's and became the world's second largest oil consumer next to the USA. There were shortages in the supplies of coal, electricity, oil and gas in the whole country, with brown-outs imposed in 19 provinces, autonomous regions and municipalities. People's work and lives were severely affected. China's energy demand and supply imbalance became even more pronounced in 2004, and brown-outs were expanded to cover 24 provinces. Problems related to energy development like resources and the environment were also becoming more evident. In 2003 China's total sulphur dioxide emission reached 21.587 million tonnes, a year-on-year increase of 12.0%. The serious situation has an impeding effect on the change in China's mode of energy development and transformation of its energy strategy.

Currently, there is no agreement among China's research institutions on the way to assess the level of industrialisation and the timetable for China to complete its industrialisation process. A report by the Chinese Academy of Social Sciences predicts that China will become basically industrialised by 2015–2018, and achieve total industrialisation by 2021. Other experts are of the view that going by the proportion of the employed population in nonagricultural industries, there is still a long road ahead for China's industrialisation. A comprehensive analysis of the progress of industrialisation in developed countries and China's development predicts that China will become industrialised before

2025, after which it will enter the postindustrialisation developmental phase. By 2030, China's urbanisation rate will reach around 65% (24.5% higher than the 40.5% in 2003). After that, the national economy and energy demand will enter a phase of low and stable growth.

2.2.1.2   Change in Energy Policy

It was clearly stated at the Third Plenary Session of the 16th Central Committee of the Communist Party of China in October 2003 that a people-oriented approach must be taken to establish a comprehensive, coordinated and sustainable Scientific Outlook on Development. In the 17th NCCPC convened in 2007, the Scientific Outlook on Development was officially written into the Party constitution. The Scientific Outlook on Development, which enriches and completes the Theoretical System of Socialism with Chinese Characteristics, is an important guiding principle for China's socioeconomic development. Under the guidance of the Theoretical System of Socialism with Chinese Characteristics, new adjustments and changes have appeared in China's energy policy.

Since 2003, China has introduced a series of legislation, policies and standards related to energy development. The development of new and renewable energies, the implementation of energy conservation and emission reduction, the improvement of the energy structure, and the strengthening of international energy cooperation, and so on were ideas that received much attention. Energy consumption per unit of GDP and emission targets of main pollutants became binding targets, and were incorporated into the planning of socioeconomic development. The share of nonfossil fuels in energy consumption will reach around 15% by 2020. This has become an important target in the adjustment of the energy structure. At the dialogue meeting between the leaders of the Group of Eight nations and developing nations in July 2006, President Hu Jintao proposed a new concept of energy security based on cooperation for mutual benefit, diversified development and protection through coordinated efforts. The White Paper, 'China's Energy Conditions and Policies', issued in December 2007, gave a systematic exposition of China's energy policies. These include giving priority to conservation, being domestic-oriented, diversifying development, relying on technology, protecting the environment and strengthening international cooperation for mutual benefit. China's energy policies also involve building up an energy supply system that is stable, economic, clean and safe. Sustainable energy development will support sustainable socioeconomic development. In November 2009, China made an official pledge to the international community that by 2020, its carbon dioxide emission per unit of GDP would be reduced by 40–45% compared to 2005.

Guided by state policies and supported by collective efforts in the energy and other sectors, the change in China's mode of energy development and transformation of its energy strategy has achieved certain results. There has been rapid growth in new and renewable energies, China is leading the world in the research and development in UHV transmission and smart power grid technology, and international energy cooperation, principally in oil and gas, continues to grow. In 2010 nonfossil fuels accounted for 8.6% of China's primary energy consumption, 2.1% higher than the ratio in 2003. Overall speaking, however, the change in China's mode of energy development and transformation of its energy strategy is still in the preliminary stage. Further adjustments and improvements can be made in its energy structure and distribution. The institutional mechanisms that affect and restrict sustainable energy development have yet to be dismantled. The scientific development, conversion, distribution and consumption of energy have yet to come about. So there is still a long way to go in changing China's mode of energy development and transforming its energy strategy.

2.2.2   The Way to Change the Mode of Energy Development

2.2.2.1   Simultaneous Emphasis on Energy Development and Energy Conservation to Raise the Efficiency of Energy Development and Utilisation

Development is the top priority. To meet the country's sustained and rapid growth in energy demand, China must boost its energy infrastructure and continue to explore and develop energy resources. In the long-term future, energy will be an important area of investment for China, especially in the development of new energy industries, which requires large amounts of capital investments and state-level policies to steer them. In terms of the pace of growth, ample consideration must be given to the socioeconomic development, i.e. the supply-side situation, as well as the cyclical periodicity of energy infrastructure, so as to prevent large-scale fluctuations. Focus must be on the entire industrial chain. Steered by unified planning, coordinated development must be sought in energy resources development, transmission and distribution using both market adjustments and macro controls. The demand and supply balance must be synergised with production, supply and distribution to bring about a genuine boost in energy supply capabilities.

At the same time as strengthening the country's energy development, China must continue to pursue its basic policy of resource conservation. The policy of saving energy must be given more importance and insisted upon as a strategic priority. To implement this strategic priority, China must on the one hand promote the intensive and efficient development of energy resources, by boosting the overall efficiency of the energy system, including development, transformation, transmission and storage. On the other hand, it must improve and adjust the industrial structure, mode of growth and consumption patterns, and raise the efficiency of energy end-use. Energy conservation and emission reduction will be the key initiatives in pushing forward the change in the mode of economic development, and through changing the economic development mode the energy intensity will keep on falling. Promote the concept of energy saving consumption. By using economic and legal means and other effective measures, strictly enforce energy conservation legislation, policies and standards to bring about energy saving in the whole country and in all sectors. This will lead China into a path of socioeconomic development that is marked by high efficiency and low consumption.

2.2.2.2   Simultaneous Development of Traditional and New Energies to Promote a Diverse and Low-Carbon Energy Structure

China's energy structure is mainly based on traditional energy; new energies like wind and solar powers account for less than 1% of the primary energy structure. Given factors such as economic structure, technological levels and resource conditions, traditional fossil fuels like coal, oil and natural gas will still be the mainstays of China's energy supplies in the long-term future. This is true especially for coal, which will account for no lower than 50%

of China's primary energy consumption structure before 2030. Compared to coal and oil, the unit calorific value of energy use for natural gas has a lower carbon dioxide emission volume, which means it is a low-carbon energy, and nuclear and renewable energies are clean, zero-carbon energies. Accelerating the development of electricity generation by natural gas, nuclear power and renewable energy meets China's need for an improved, low-carbon and diverse energy structure.

Currently, the developed nations have completed their fossil fuel improvement process and are developing low-carbon energies in earnest. They are moving towards a higher quality of energy. Also placing a high level of importance on upgrading the energy structure, China has set a strategic target to have nonfossil fuels accounting for around 15% of primary energy consumption by 2020. Although there is huge potential for developing China's natural gas, hydropower and conventional nuclear power, the potential scale of development is insufficient to support the massive energy demands in the country. To realise the transformation of China's energy strategy, to promote a change towards clean and low-carbon energies, and to bring about the coordinated development of China's economy, society and environment, new and renewable energies must be developed. At the moment, while China's renewable energy resources such as wind power, solar power, biomass, geothermal energy and tidal power may be abundant, development is stymied by problems in technology and costs. The next step will be to implement more effective measures to achieve breakthroughs in key technologies and step up development efforts. In particular, the development and utilisation of wind and solar powers should be stepped up by building many large-scale wind power and solar power generation bases, in conjunction with the distributed development. In addition, more attention should be given to the development and utilisation of unconventional oil and gas resources like dense oil, dense gases, coal bed methane, shale gases and combustible ice, as well as new energies like nuclear fusion energy and hydrogen energy.

2.2.2.3   Simultaneous Utilisation of Domestic and Overseas Resources to Build Up a Complementary Energy Supply System

Solving China's energy problem requires work on both the domestic and international fronts. The proper development and utilisation of domestic energy resources is the basic prerequisite of ensuring China's energy security. This is both an internal necessity to promote national economic development and safeguard China's economy and energy security, and a practical choice given the complex international political and economic situations. From the standpoint of energy security, 1 tonne of oil buried overseas is not the same as 1 tonne of oil buried within the country's borders. This is because only domestic energy resources can be controlled in terms of price and volume. At any point in time, there must be no waver in the energy policy of focusing on the domestic.

Seen from another perspective, the distribution of the world's energy resources is not balanced. China's per capita shares of coal, oil and gas resources are only 69.9%, 5.6% and 6.6% respectively of the global average. In the long term, with China's economic development and its active participation in the global division of labour, relying on domestic energy production alone is not going to ensure China's energy supply. The continued rise in the country's dependence on foreign energy is a predictable trend. At the same time as China boosts the development of its domestic energy resources and increases its domestic energy supplies, it must pursue the strategy of sustainable energy development by building up an energy structure centred on oil and supplemented by coal, electricity and natural uranium, increasing energy imports, and fostering international energy cooperation that benefits all. With large state-owned energy enterprises at its core, the energy strategy of 'going global' must be actively pursued. China's investments and interests in overseas energy production must be enhanced. The country must also take advantage of the international framework, and take active part in laying down rules for international trade, improving trade and developing a diverse and stable energy trading system.

By adhering to the new perspective of energy security, and making use of economic, diplomatic and political means, China must reinforce and develop relationships with the major energy-producing countries, and increase its cooperation with the main energy consumers. By enhancing China's capabilities in global energy development and its competitiveness in energy trade, the country can be assured of economic, stable and reliable energy resources from overseas. This will gradually change the country's energy supply framework from one that relies mainly on domestic resources to an energy and resources supply system that makes full use of both domestic and overseas resources. At the same time, China must have a thorough understanding of the complex nature of international energy competitiveness and geopolitical factors. By participating in global energy security mechanisms, strengthening the development of its energy warning system, and building up a system of energy reserves, China will be more capable of defending itself against global energy risks. At the same time as the country promotes international energy cooperation and strengthens its energy supply capabilities, China must contribute positively towards improving the world's energy supplies and ensuring global energy security.

2.2.2.4 Simultaneous Improvements in Energy Distribution and Energy Transmission for Wide-Area and High-efficiency Energy Allocation

Improving energy distribution and the mode of energy transmission is absolutely necessary in the transformation of China's energy strategy. Factors like China's energy endowments, energy imports, the distribution of the energy consumption market, ecological and environmental features of geographical locations, infrastructure of energy transport channels, and so on must first be considered. Also, the country's planning for its main functional areas must be referred to. The varying development levels of different regions and their long-term development trends must be combined for the scientific planning of China's energy overall development. The scale and focus of energy developments in the eastern, central and western regions must be determined rationally, as well as the distribution and development model of energy bases. The distribution of coal bases must be improved and large-scale coal-fired electricity bases must be built. By bringing about long-distance and large-scale electricity transmission, the development of a large number of coal-fired power plants in the land-scarce and ecologically-fragile eastern and central regions can be realised.

Based on the improvement in the distribution of energy bases, China should focus on increasing the distribution of energy resources over large areas. By boosting the energy transport infrastructure like railroads, roads, water transport, pipelines and power grids, China needs to make full use of the advantages of UHV transmission, and build smart power grids to clear energy transmission bottlenecks. In building a modern and comprehensive energy transmission system, and creating a rational, structured and systematic energy distribution platform, China can meet the demands of its national economic and social development and the need for optimising its resources distribution.

The key to improving China's energy distribution and mode of energy transmission is to break the development concept of achieving local or regional balance and the existing notion of transporting coal over long distances and transmitting electricity for short distances. The idea of an improved energy resources distribution framework across the whole country must be established, and the functions of modern grids like UHV in large-area distribution must be taken advantage of, to improve the capacity for long-distance, large-scale transmission. This will bring about a modern energy industry system with rational development and scientific distribution.

2.2.2.5 Simultaneous Promotion of Innovation in Science and Technology and Institutional Mechanism to Boost the Internal Dynamic for Sustainable Development

The root to solving China's energy problem lies in innovation. In accordance with the plans to foster an innovation-oriented country, a technologically innovative energy system that is sound and compatible with China's developmental needs and resources characteristics must be built. China must grasp the development opportunities offered by the switch between traditional fossil fuels and new and renewable energies, and take advantage of the role of innovation in bringing about important breakthroughs in energy technology. By achieving independence of core energy technology, China can quickly reduce the distance between itself and developed countries in terms of energy technology. By taking the commanding role in international energy technology and with technology as the primary productivity force, China's energy strategy transformation and sustainable development will be supported and assured.

Based on the goal of building a socialist market economy and the policy in which the relations of production must suit the development needs of productive forces, and going by its energy development stage and national conditions, China must actively learn from other countries' experience to improve and adjust the management and operational systems of its energy industry. Reforms in state-owned energy enterprises must be intensified, and legal frameworks, operational mechanisms and monitoring systems conducive to changing China's energy development mode and energy strategy transformation must be set up. Fostering an energy market with Chinese characteristics provides an excellent institutional environment for China's sustainable energy development and a boost for its energy industry's internal dynamics.

2.3 The Central Link in the Energy Strategy

The transformation of the energy development mode is a systematic project on a massive scale, and electricity is at the core of the whole process. Any energy strategy to bring about this transformation must be one where electricity is firmly in the centre. Focusing on electricity has important implications for the sustainability of China's energy development by ensuring energy supplies and reducing supply pressures. It also facilitates the improvement of the energy structure and takes the pressure off the environment while increasing energy efficiency and reducing energy intensity. Finally, it improves people's lives, enhances services and helps bring about social harmony.

2.3.1 The Position of Electricity in the Energy Strategy

The energy system is a complex one. It consists of different types of energy, including coal, hydropower, electricity, oil, natural gas and nuclear power, and involves various components like development, transformation, transmission, storage, consumption, and so on, as well as aspects like resources endowments, technology levels, industry structure, consumption patterns, international cooperation, and so on. Electricity is at the centre of the energy system (Figure 2.2). As early as 1985, 'the need to develop the energy industry with the focus on electricity' was mentioned in *The CPC Central Committee's Recommendations on the Formulation of the Seventh Five-Year Plan for National Economic and Social Development*. Promulgated in 1996, *The People's Republic of China's Ninth Five-Year Plan for National Economic and Social Development and Outlines for the Goals to be Achieved in 2010*, reaffirmed the policy of 'placing electricity at the centre of energy development'. In 2004 the State Council Executive Meeting passed *The Outlines for Planning Medium- and Long-term Energy Development (2004–2020) (Draft)*, which proposed a 'strategy committed to coal as the main body, electricity as the core, with the comprehensive development of oil, gas and new energies'.

The problems of global energy security and climate change are becoming more serious in recent years. Within China, there have been frequent shortages of coal, electricity, oil and gas, and recurring disruptions in energy transport. The environment has also been seriously affected. The new energy technology revolution, represented by new energies and smart power grids, is being fostered and developed, and electricity's position as the core of energy development is even more apparent. Electricity is the key to resolving the serious conflicts and problems of energy development, and pushing ahead the transformation of the mode of energy development. To focus on electricity is to have a firm hold on the crux of China's sustainable energy development.

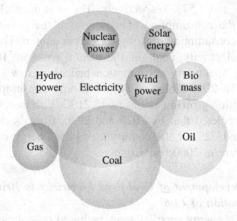

Figure 2.2 The relationship between electricity and other energy forms.

What an electricity-centred energy strategy means is that in implementing the strategy and bringing about a change in the mode of energy development, all measures must be based on China's national conditions concerning large coal reserves, rich renewable energy resources and insufficient oil and gas resources. In line with the global trend in energy development, the balance of electricity should be made an important pillar of energy balance, and electricity generation should be made the key mode of primary energy transformation and utilisation. Also, power grids should be the important basic platforms for energy distribution, and electrification should be the basic measure in the improvement of the energy structure and efficiency. The scientific development of the electricity industry will result in the clean and efficient development and a rational framework of primary energy resources. It will also bring about positive adjustments in the energy structure and distribution pattern, and relief the increasing pressures on energy supply and the environment. These will provide China with sustained energy security for the country's socioeconomic development.

2.3.2 The Significance of an Electricity-centred Energy Strategy

Objectively speaking, an electricity-centred energy strategy is determined by electricity's unique features, resources endowments and the patterns of energy development. Whether it is the relationship between electricity and other forms of energy, or ensuring energy security, improving the energy structure, conserving energy and reducing emissions, or building a harmonious society, electricity performs a very important function. The significance of electricity in China's sustainable energy development is four-fold.

2.3.2.1   Ensuring Energy Supply and Relieving Pressure on Energy Security

Need to Focus on Electricity Development to Meet Energy Demands
Whether it is within or outside China, electricity is the fastest growing form of energy in the last 20 years. Between 1990 and 2009, the total end-use energy consumption in the world grew 1.49% per annum, with end-use electricity consumption growing by 2.91% a year. In contrast, the per annum increases of end-use consumption of coal, oil and natural gas were 0.33%, 1.50% and 1.57% respectively. The growth of electricity consumption was much higher than the consumption of major fossil fuels. Within the same period, China's end-use energy consumption grew by 4.13% per annum. The increases per annum of end consumption of electricity, coal, oil and natural gas were 10.09%, 2.65%, 7.50% and 9.45% respectively. Electricity consumption had also grown much faster than the other types of energy. In 2010 China's end-use energy consumption grew by 4.18%, with electricity consumption growing as fast as 13.21%. With increasing urbanisation and industrialisation, it is predicted that by 2020 and 2030, China's electricity consumption will account for 28% and 32% of the country's end-use energy consumption, 12% and 16% higher than the figure in 2000 respectively.

Need to Improve the Development of Coal-fired Electricity to Bring About Efficient Development and Utilisation of Coal
The assurance of China's energy supplies must be based on policies guided by domestic conditions. In terms of resource endowment, coal will be China's basic energy in the

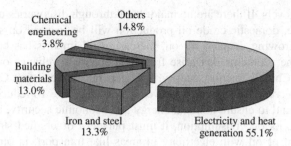

Figure 2.3   Breakdown of China's coal consumption in 2010.
Source: Coal Industry Research Center.

long-term future. Electricity generation is the most common way to utilise coal in China. In 2010 electricity and heat generation accounted for 55.1%, or more than half, of coal consumption that year (Figure 2.3). Therefore, the enhanced development of China's coal-fired electricity can result in coal resources being efficiently developed and utilised. This has significant implications for China's energy supply security.

From a global perspective, the main use for coal is also electricity generation. In 2009 electricity generation accounted for 65.1% of global coal consumption. In the USA, over 90% of all the coal consumed that year was used for generating electrical power, while in the EU the share was 78.7% and 72.5% in India (Figure 2.4). Compared to both the world average and developed economies like the EU and USA, China uses a significantly smaller percentage of its total coal consumption to generate electricity.

Need to Develop Electricity as Replacement to Relieve Pressure on Oil Supplies

The constantly increasing dependence on foreign oil is the biggest challenge to China's energy security. Given the country's limited resources, China's current oil production is

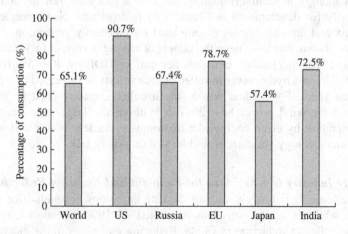

Figure 2.4   The percentage of coal used for electricity generation in the world and major countries (and regions) in 2009 (ex. China).
Source: International Energy Agency (IEA).

already near peak levels. If there are no major breakthrough discoveries in resource exploration in the future, domestic crude oil production will be focused on stable generation. To meet the fast-growing demand for oil, increasing oil imports has become an important alternative. The problems that arise from this are more reliance on foreign oil and increased risks in China's energy security. It is generally agreed that China's maximum reliance on foreign crude oil should be between 70% and 75%. To reduce this reliance on crude oil imports, and to ensure national energy and economic security, not only must the country economise its oil consumption, it must pursue a diversified strategy, especially for the replacement of oil with electricity in areas like transport, in order to reduce the pressure on oil supplies.

2.3.2.2 Improving Energy Structure and Relieving Pressure on the Environment

Clean Energy Needs to be Transformed into Electricity for Use

To resolve the environmental bottleneck facing its energy development and to tackle global climate change, China must develop clean and renewable energies. At the moment, clean energy like hydropower, whose development and utilisation technology is already very advanced and nuclear and wind power, whose technology is also relatively mature, need to be converted into electricity for convenient use. China's hydropower technology can produce over 540 GW power but its current utilisation rate is only 43%, very much lower than the 60–70% rate in developed countries. In the short and medium term, hydropower will be China's main focus in renewable energy development, and more developmental efforts should be put into hydropower. At the end of 2011, the installed capacity of China's hydropower was 230 GW, the world's highest. It is predicted that an extra installed capacity of 200 GW can be added in the next 20 years. Nuclear power has the benefits of low environmental pollution, ample reserves and low fuel costs. In the medium and long term, it is an important alternative to replace fossil fuels. In terms of technology, costs and market potential, wind power offers promising development prospects. The electricity generation technology is relatively mature and it is a renewable energy that should be accorded priority for development in China, after hydropower. New energies like solar power, biomass and tidal energy are mainly used for electricity generation.

Studies have shown that to achieve the targets of having nonfossil fuels taking up 15% of fuel use, and reducing carbon emission per unit of GDP by 40–45% (compared to 2005) by 2020, China's hydropower installed capacity must be increased by 120–140 GW in the next ten years. For nuclear power, the installed capacity must be increased by 60–70 GW and for wind power 80–120 GW. Without the large-scale development of electricity generation by clean energy like hydropower, nuclear power and wind power, improving China's energy structure would be just an empty talk.

The Electricity Industry is a Key Area for Pollutant and Emission Reduction

In 2010 the electricity industry's sulphur dioxide emissions accounted for 42% of the national total, while its smoke emissions accounted for 19%. The electricity industry is a key area for pollutant reduction in China. Reducing emissions in the electricity industry, reinforcing the centralised management of pollutants from coal-fired power plants, and developing and using clean coal technology have important implications for China's emission reduction targets. Not only does the large-scale use of a clean secondary energy like electricity and its replacement of other energies (including replacement of oil and coal by electricity) have significance for economic efficiency and energy security, it is also important in reducing the pollution of the atmosphere.

2.3.2.3 Increasing Energy Efficiency and Reducing Energy Intensity

Electricity is the Most Efficient Form of Energy

Different types of energy have different economic efficiencies. Compared to other forms, electricity has the highest economic efficiency. Based on Chinese data between 1978 and 2003, scholars have found that the economic efficiency of electricity is 3.22 times of oil and 17.27 times of coal. To put it another way, the economic value derived from electricity equivalent to 1 tonne of standard coal is equal to the economic value of oil equivalent to 3.22 tonnes of standard coal, and the economic value of coal equivalent to 17.27 tonnes of standard coal. This implies that for the same economic output, using more electricity helps conserve energy.

Raising the Level of Electrification Conducive to Reducing Energy Intensity

Overseas experience has shown that a country's level of electrification is closely linked to the level of economic development and energy intensity.[5] With greater economic development and higher levels of electrification, energy intensity will continue to fall. In other words, the level of electrification has an obvious negative correlation with energy intensity. According to the historical data from 1970 to 2000 in the five countries of the UK, the USA, Japan, Germany and France, for every 1% increase in the ratio of energy for electricity generation to the primary energy consumption, the energy intensity fell by 2.4% (Figure 2.5). According to the historical data from 1970 to 2004 in the same countries, for every 1% increase in the ratio of electricity consumption to the end-use energy consumption, the energy intensity fell by 3.7% (Figure 2.6). Therefore, in the course of industrialisation and urbanisation, promoting electrification at state level helps to increase China's energy efficiency and ease its energy supply pressures.

2.3.2.4 Improving Lives and Helping Build a Harmonious Society

Development of Electricity Affects All Aspects of the Economy and Society

Electricity is secondary energy that is clean, efficient and convenient, with a wide range of applications. It is an indispensable resource for production and living in the modern world. All primary energies can be transformed into electricity, and electricity in turn can be easily transformed into motion, light, heat, as well as electrophysical and electrochemical reactions. Electricity can be transmitted at the speed of light through conductors, and split to an unlimited extent in distribution systems. It can also be controlled over long distances with precision, and it makes automation in production possible. These features

---

[5] Energy intensity refers to the amount of energy input required for a specific amount of economic output. It is often indicated by the ratio of the country's total primary energy consumption to its GDP.

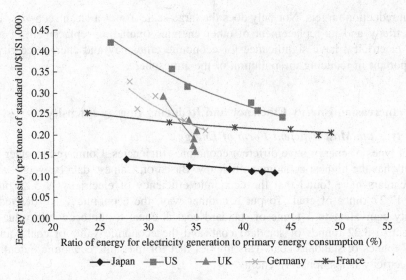

Figure 2.5 Graph showing relationship between energy intensity and the ratio of energy for electricity generation in major developed countries.
Source: International Energy Agency (IEA).

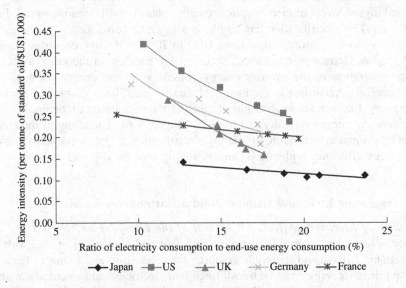

Figure 2.6 Graph showing relationship between energy intensity and the ratio of electricity to end-use energy in major developed countries.
Source: International Energy Agency (IEA).

have resulted in electricity becoming the most widely used energy in the modern world. Compared with other forms of end-use energy, electricity has greater network coverage and more users; electricity consumption is more evenly distributed among different industries; and the fundamental and public nature of electric power is more pronounced. The development of electricity does not only affect economic development, it also affects social harmony and progress.

Development of Electricity is the Key to Solving Energy Problems in the Rural Areas

The effect of electricity development on the building of a harmonious society is also reflected in the rural energy infrastructure. The short supply of high quality energy and commercial energy is an important factor restricting the socioeconomic development of rural China and also the chances of resolving the problems concerning 'Agriculture, Countryside and Farmers'. Currently, much of the energy consumption in rural China is derived from burning traditional biomass like firewood and straw. Not only is it inefficient, it results in massive deforestation, destruction of forest vegetation and even more soil erosion. Increasing the pace of electricity development and the push of electrification in rural areas is an important means of solving the energy supply problem in rural China. During the period of the 11th Five-Year Plan, the State Grid Corporation of China actively implemented the 'Power for All' project, and connected 1.34 million households in remote areas to power supply. Areas with access to electric power have seen new production activities and a renewed economic vitality.

Problems in Electricity Security have a Wider Reach and Greater Impact

The importance and wide impact of electricity development are demonstrated by negative examples such as major power incidents, especially power failures of major power grids. Given the modern world's increasing dependence on electricity, a sudden power failure that covers a large area will seriously affect social order, and may even lead to disastrous results. In recent years, major power failures across the world, like North America in 2003, Moscow in 2005 and Western Europe in 2006, had all resulted in massive losses. In comparison, in the last 20 years, the level of control over China's power grid operations has constantly increased. The operations themselves are well run and no major power failures covering large areas have occurred.

2.4   The 'One Ultra Four Large' (1U4L) Strategy

The implementation of the 'One Ultra Four Large' (1U4L) strategy[6] is China's core mission of electric power development. It is the strategic starting point of shaping a change in the way of energy development, with profound implications for energy supply security, intensive energy resource development, optimised energy transportation planning, improved energy allocation efficiency, environmental protection and responding to a new

---

[6] *Translator's note*: An energy development strategy focusing on the construction of UHV grids supported by the development of four massive bases of coal-fired, hydroelectric, nuclear and renewable power generation.

revolution in energy technology. The key to implementing 1U4L is the development of ultra-high voltage (UHV) grids, which are the prerequisite and bedrock of building up massive energy bases. The development of UHV grids will fully utilise UHV's superior performance in transmitting electricity over long distances, and create a structurally and functionally sound and intelligent platform for energy allocation. UHV grids will also provide technical support for energy cooperation between nations.

2.4.1 The Core Mission of Electric Power Development

To advance—with electric power as the main driving force—the transformation of the mode of China's energy development, the most important and urgent task at hand is to speed up the implementation of the 1U4L strategy. This strategy refers to the coordinated construction of powerful smart grids with UHV as the backbone (the 'One Ultra'), the accelerated development of overhead energy transportation, the realisation of coal transportation and electricity transmission, the intensive and efficient development of large coal-fired, hydroelectric, nuclear power and renewable energy bases (the 'Four Larges'). This will facilitate large-scale and long-distance power transmission and optimised energy resource allocations across the country to provide sustainable and dependable electricity supply for the benefit of China's economic and social development. UHV technology refers to power transmission at 1 000 kV AC and above, and ±800 kV DC and above. It is currently the highest voltage level and most advanced electricity transmission technology in the world.

The implementation of the 1U4L strategy can effectively resolve the Chinese energy sector's glaring problems in terms of supply security, utilisation efficiency, resource allocations, environmental constraints and technological innovation. Reflecting its important and far-reaching significance to sustainable energy growth, China sees this strategy as being central to its core mission to develop its power industry now and in the future.

2.4.2 The Need to Implement the 1U4L Strategy

Implementing the 1U4L strategy is the only way to support the scientific development of the China's power industry and consequently propel the change of energy development mode and the power sector's strategic transformation. Its significance is reflected in the following six areas:

2.4.2.1 The Need to Ensure Electric Power Supply

Based on an analysis of China's socioeconomic development, the nation's electric power demand for the next 20 years will continue to see sustainable growth. By 2020, electricity consumption nationwide is forecast to reach 8600 TWh and electricity load 1410 GW, or 2 and 2.1 times the levels of 2010 respectively. By 2030, China's social power consumption will reach 11 800 TWh and its electricity load 1940 GW, respectively 2.8 and 3 times the levels of 2010.

Therefore, expediting the growth of the electric power industry and implementing the 1U4L strategy have become urgent tasks in order to meet such colossal energy needs and ensure a sustainable supply of electricity. By the end of 2020 and 2030, the installed capacity of China's power generation industry is expected to reach 1934 GW and 2380 GW respectively. These numbers are respectively 2 and 2.5 times the levels of 2010. The installed capacity of coal-fired generation is expected to reach 1190 GW and 1270 GW respectively by then (1.8 and 2 times the 2010 levels). It's 345 GW in 2020 and 430 GW 2030 for hydropower (1.6 and 2 times the 2010 levels), 80 GW and 160 GW for nuclear power (7.4 and 14.8 times the levels of 2010), and 200 GW and 360 GW for renewable energy sources such as wind power (6.7 and 11.7 times the levels of 2010). Renewable and nuclear power generation will see relatively more rapid growth in installed capacity.

2.4.2.2   The Need for Intensive Development of Energy Resources

Taking into consideration the distribution of energy resources for power generation in China and the nation's distribution of productive forces, the practical option for bringing about the sustainable development of China's electric power industry is to construct massive bases and undertake intensive and efficient development. For the development of coal-fired generation, the focus will gradually shift westwards and northwards given the continued intensification in the opening up of coal resources. Building up large-scale coal-fired power generation bases in coal-rich regions and expanding the scale of in situ coal conversion will help improve the supply security of China's coal for generating electricity. By coordinating the use of environmental resources nationwide, efficient and clean coal conversion will be encouraged. In the area of hydropower development, China is blessed with rich hydropower resources, which are abundant in the great rivers like the Yangtze River, Jinsha River, Nu River and the Yellow River. With the great potential for intensive development and large-scale delivery, the construction of large hydropower bases will be the main avenue in exploiting the nation's water resources. For nuclear power development, the building of large-scale nuclear power bases is in line with the growing trend towards more powerful single units. Through the efficient use of valuable site resources, the development of nuclear power will benefit from economies of scale. For renewable energy development, China's concentration of wind and solar power resources makes the intensive development of large bases a natural choice. Based on the development concept of 'building a giant base and integrating it into the large power grid', building major wind and solar power bases in resource-rich areas is an important avenue in the large-scale development and exploitation of China's renewable energy sources. Comprehensive analyses indicate that in the next 20 years, two-thirds and above of China's additional installed capacity will come from large bases of coal-fired, hydroelectric, nuclear and renewable power generation.

2.4.2.3   The Need to Optimise Energy Transportation Plans

A glaring problem that China faces in power development is the inadequacy of long-distance transmission capacity of its large power grids and the over-reliance on transporting coal for energy development. In recent years, the structural irrationality of energy transportation has been the major cause of the continued serious under-supply of coal, electricity and transport capacity. By implementing the 1U4L strategy to develop UHV transmission through a powerful UHV grid backbone, the long-distance transmission capacity of large power grids will be given a significant boost. There will be an increase of in situ coal and electricity conversion, and coal transportation will be replaced by electricity transmission. This will gradually change the model of energy transportation from the one overly reliant on coal transport. According to the plans developed by the State Grid Corporation of China, the smooth implementation of 1U4L would mean that by 2020, the ratio of coal transport to electricity transmission in the 'Three West' region (including eastern Ningxia)[7] will be around 3.7:1. In Xinjiang, the ratio will be 0.8:1. By 2030, the ratio will be 2.2:1 in the 'Three West' region (including eastern Ningxia) and 1:2 in Xinjiang. The structure of power transmission will be greatly enhanced.

2.4.2.4 The Need for Higher Energy Allocation Efficiency

The implementation of 1U4L is conducive to greater grid interconnection for more networking benefits. By building a Strong and Smart Grid with synchronised UHV grids in East China, North China and Central China, the installed capacity can be reduced, thereby lowering the costs incurred in power development. With the interconnection between power grids, a large power grid can perform multiple functions. These include adjusting for surpluses and shortages, enabling exchanges between hydro and thermal power, realising cross-river basin compensation, reducing energy waste in over-production, achieving peak shifting and peak shaving, and effecting mutual backups and emergency rescue operations. The utilisation efficiency of energy will be increased as a result. It is projected that by 2020, the Strong and Smart Grid will result in networking benefits and reduce newly added installed capacity by over 30 GW and save over 100 billion yuan in investments. At the same time, 34.3 TWh and 37.2 TWh of energy that could otherwise be wasted from water and wind abandoning respectively will be saved, equivalent to some 31 million tonnes of coal. A Strong and Smart Grid will also contribute to the development and exploitation of demand-side resources. Interaction between the power grids and users can guide users towards proper utilisation of electricity to save energy and reduce consumption while facilitating load shifting, enhancing system operations and increasing energy efficiency. It is projected that by 2020, more intelligent and efficient power utilisation will save users some 400 TWh of electricity.

2.4.2.5 The Need to Protect Ecosystems

Developing and exploiting massive bases of hydroelectric, nuclear and renewable power generation for clean energy will directly enhance China's energy structure, drive the nation's shift to low-carbon energy and reap palpable benefits for the environment. Developing and exploiting hydropower, nuclear power and renewable energy like wind and solar power on a large scale also helps to tackle and resolve the effects the full lifecycle development and utilisation of these energy resources have on the environment. Building up large coal-fired power bases can, through centralised mitigation measures, effectively reduce the pollution and carbon emissions from coal-fired power generation. It can also promote the integrated development of coal and power resources and develop a circular economy for coordinated development between coal utilisation and socioeconomic and

---

[7] 'Three West' (*San Xi*) refers to Shanxi, Shaanxi and western Inner Mongolia (*Xi Neimeng*).

environmental progress. The development of UHV provides a pillar for the creation of big energy bases. It also ensures a higher level of electrification and replacement of other forms of energy by electricity, and greater reduction in urban air pollution in East and Central China. In recent years, land resources in East and Central China are becoming scarce and the environmental impact on these regions is increasing. It is no longer viable to continue building large-scale thermal power plants in East and Central China to meet energy demands. The 1U4L strategy of long-distance power transmission from the west and north of China to the load centres in the east and central region is the inevitable, and possibly the only choice.

2.4.2.6 The Need to Respond to a New Revolution in Energy Technology

In the face of a new global revolution in energy technology, many countries are increasing their investments in new energy and smart power grids. Some have even designated it as their national strategies with the intention to claim their new dominant positions in the global competition in energy, economic and technical know-how. For China, the implementation of 1U4L strategy is advantageous to the development of new energy and the Strong and Smart Grid. It also puts the country in a better position to meet the challenges posed by the new energy technology revolution, as represented by new energy and smart power grids. China is then able to seize the commanding heights in the international development of energy technology and maintain the initiative in future global competition. At present, China's new energy is basically at the same starting block as most developed countries, but it leads in smart power grid technology. It is imperative that China seizes this opportunity to accelerate its development. The 1U4L strategy and building up of a Strong and Smart Grid can also bring about energy conservation and environmental protection, as well as spur the rapid growth of strategic and new products like high-end equipment, new energy automobiles and new generations of information technology gadgets. It will improve China's technological expertise and global competitiveness in related industries like power equipment.

2.4.3 The Key to Implementing the 1U4L Strategy

The key to implementing 1U4L strategy is to expedite the construction of UHV grids as an intrinsic requirement in the scientific development of the energy sector. It also meets the objective requirement for utilising UHV's technical superiority and functions, ensuring the intensive development of big power-generating bases and the safe, efficient and economic power delivery as well as smooth accommodation, optimising energy allocations nationwide, and promoting energy cooperation between nations.

2.4.3.1 Development of UHV Grids is the Prerequisite and Bedrock of
        Building Large Energy Bases

Most of the massive coal-fired and hydroelectric power bases are far away from the load centres in East and Central China. To be cost-effective, power transmission across distances between 1000 km and 3000 km requires UHV technology. The big wind bases in the 'Three North' region[8] also need a UHV grid for large-scale power delivery and consumption. The development of large nuclear bases requires the same. Relatively speaking, a power grid comprised mainly of 500 kV AC and ±500 kV DC, being restricted in long-distance power transmission capacity and cross-regional network interconnection capacity and efficiency and due to its excess short circuit current, can no longer meet the actual needs of large-scale, intensive development of power sources and extensive optimisation of energy allocations. A UHV power grid with higher voltage levels and stronger resource allocation capability must be developed.

2.4.3.2 Development of UHV Grids Will Fully Utilise UHV's Superior Performance in Transmitting Electricity over Long Distances

Compared to the present 500 kV power transmission, UHV has clear technical and economic advantages in long-distance power transmission. In terms of AC transmission, 1000 kV transmission is four to five times as powerful as 500 kV transmission, and the economic transmission distance is about three times that of the latter. The line loss rate of 1000 kV transmission is only between one-quarter and one-third of that of 500 kV transmission, but the transmission capacity of the unit corridor width is between 2.5 and 3.1 times that of the latter. In terms of DC transmission, ±800 kV and ±1,000 kV transmissions are respectively 2.3 and 2.9 times as powerful as ±500 kV transmission, and the economic transmission distance of ±800 kV DC transmission is about 2.7 times that of the latter. The line loss rate of ±800 kV transmission is two-fifth of that of ±500 kV transmission, and the transmission capacity of the unit corridor width is 1.6 times that of the latter. UHV power grids can overcome the inadequacies of 500 kV power grids in long-distance power transmission, to facilitate the safe, economic and efficient development of the power industry.

2.4.3.3 Development of UHV Power Grids Will Create a Structurally and Functionally Sound and Intelligent Platform for Energy Allocations

For a long time, the power grid has been functionally confined to the transmission of electricity. With the development of UHV technology and its integrated application with smart technology, China's power grids are undergoing profound changes in function and form. A Strong and Smart Grid featuring UHV and smart technology is not only a carrier of electricity, but also an important component of a modern and integrated energy transport system. It is a basic platform with smart functions and a powerful capacity for energy and resource allocations. Without a robust network anchored by a powerful UHV grid backbone as the foundation for smart grid development, the improvement of energy and resource allocations will be greatly affected. Given the full integration with the Internet, television, radio and other networks, a Strong and Smart Grid will be functionally elevated to become a platform for comprehensive social services. The Strong and Smart Grid of the future will amalgamate the different functions of energy transportation, network marketing and public services into a single entity. It will play a vital role in ensuring energy security, energy conservation, emission reduction and socioeconomic development.

---

[8] 'Three North' (San Bei) refers to the Northeast (Dongbei), North China (Huabei) and the Northwest (Xibei).

2.4.3.4 Development of UHV Power Grids Will Provide Technical Support for Energy Cooperation between Nations

China's neighbours have rich energy resources like coal and hydropower with the potential for cross-border power transmission. However, long-distance transmission constitutes a major impediment and a practical solution lies in power transmission through a UHV network. Developing a UHV power grid and improving the technology for long-distance and large-capacity power transmission can strengthen the energy cooperation between China and its neighbours. The utilisation of the power-generating resources in neighbouring countries and the transmission of power across long distances into China can provide valuable technical support for the nation's power supply and energy security. In the long run, with the development of ultra long-distance, $\pm 1100\,kV$ UHV DC power transmission, and the breakthroughs in renewable energy technology and energy storage expertise, there exists, in theory, the feasibility of grid interconnection between Xinjiang and Qinghai in China and Central Asian and even European countries. When this comes to pass, the optimisation of resource allocations with UHV power grids will not only cover transmissions between different provinces, different regions, different countries, but also different continents to form electric power channels and networks that will traverse the Eurasian landmass.

3

Energy Exploration and Utilisation

Currently, the mode of energy exploration and utilisation in China is relatively extensive. The levels of efficiency in energy resource extraction and transformation can be raised. Optimisation of energy distribution and structure is urgently needed. The effects on the environment are increasingly evident and production safety is severely affected. To accelerate the future transformation of the mode of energy development and the energy strategy, and bring about sustainable development, the mode of energy exploration and utilisation must be changed. Based on the principles of safety, intensity, high efficiency, diversity and cleanliness, hydropower resources, nuclear power, and new and renewable energies must be proactively developed and utilised in safe and efficient ways; conventional and non-conventional oil and gas resources must be explored; and overseas energy resources must be tapped. In so doing, China's energy security and sustainable supply can be assured.

3.1 General Thinking Behind Energy Exploration and Utilisation

The transformation of the mode of energy development requires the transformation of the mode of energy exploration and utilisation. We must target the current problems in energy exploration and utilisation by establishing the relevant scientific principles, identifying priorities and clarifying thinking, so as to provide guidance for reshaping the way energy is explored and utilised.

3.1.1 Main Problems in Energy Exploration and Utilisation

3.1.1.1 Low Efficiency of Energy Exploration and Utilisation

The scale of energy resource exploration in China is low. As of 2010, there were as many as 10 000 coal production companies of various types across the country, but their average annual production was less than 300 000 tonnes. Production volumes of the top four and top eight coal companies accounted for just 22.0% and 31.9% respectively of the total industry production. Based on the theory of industry organisation, industry concentration is on the low side. Compared to the major coal-producing countries in the world, China's coal industry has a low concentration. In 2010 the production volume of China's top

Electric Power and Energy in China, First Edition. Zhenya Liu.

ten coal companies accounted for less than 36% of the national total. The share in the USA was 46% compared with 50% in Australia and 90% in India. The excessively low industry concentration stands in the way of building economies of scale and further improving production safety and capacity for sustainable development.

Inefficient energy resource exploration leads to the serious wastage of resources. As most are small and medium-sized coalmines, advanced production technologies and mining equipment are seldom utilised. During the extraction process, smaller pieces of coal are frequently discarded in favour of larger pieces. The resource recovery rate for key state-owned coalmines is generally around 50%. For state-owned local and rural mines, the rate is less than 30%, and some are even as low as 10–15%. This is very much lower than the global average of 60%. As for the exploration of renewable energy like wind power, the weak fundamental research in areas such as resource assessment, the low level of technological management, the lack of a large-scale and concentrated exploration mechanism, and the unsound coordination and operation mechanisms with power grids have resulted in low annual power generation utilisation hours. The exploration efficiency is lower than developed countries in North America and the EU.

Energy transformation and utilisation is not efficient. In China, the ratio of coal consumption to total end-use energy consumption is higher than the world average by almost 40%. A considerable proportion of coal is used by direct burning in medium- and small-sized boilers. Not only does it lower the energy utilisation efficiency, it disperses the emission of pollutants, and increases the difficulties and costs of processing them. The level of oil consumption by automobiles is higher than Japan by 20% and the USA by 10%.

3.1.1.2 Urgent Need to Optimise Energy Exploration and Utilisation Plan and Structure

There is an urgent need to optimise plans for coal exploration. For a long time, there has been super intensive exploitation in resource-poor regions like the eastern and central China. In the Beijing-Tianjin-Hebei-Shandong region, the Henan-Hubei-Hunan-Jiangxi region and the eastern China,[1] the coal resource development intensity coefficients[2] are 2.3, 3.8 and 2.3 respectively (Table 3.1), which are higher than the national average. The coal resource development intensity coefficients in coal-rich regions like Shaanxi, Inner Mongolia, Ningxia and Xinjiang are all below 0.6, with Xinjiang's as low as 0.1. After many years of super intensive mining, the better mines in the eastern and central China are depleting. Many large coalmines are in the middle and late stages of their life-cycles, and increased production will be difficult in the future.

There is an urgent need to improve the plans for electricity exploitation. For a long period of time, China has built a large number of coal-burning power plants in the eastern and central China, where there is a relative lack of energy resources for power generation and where the electricity load is high. This has led to stretched environmental capacity,

---

[1] Unless otherwise specified, 'The eastern China' in this book refers to the area covered by the Eastern China Power Grid, which includes Jiangsu, Shanghai, Anhui, Zhejiang and Fujian.

[2] Coal resource exploration intensity coefficient = amount of coal output as proportion of national total/amount of coal resource as proportion of national total.

Table 3.1 Exploration intensity of key areas of coal production and consumption in China in 2009.

Region	Amount of resources (billion tonnes)	Proportion of nation's total resources (%)	Coal resource development intensity coefficient
Shanxi–Shaanxi–Inner Mongolia–Ningxia–Xinjiang	857.066	77.7	0.7
Beijing–Tianjin–Hebei–Shandong	44.317	4.0	2.3
Henan–Hubei–Hunan–Jiangxi	31.128	2.8	3.8
The eastern China	31.131	2.8	2.3

Source: The data in the table are based on the individual regions' coal reserves and 2009 output numbers.

over reliance on coal handling in energy transport and other problems in the region. There have been frequent occurrences of electricity and coal shortages.

There is an urgent need to improve the structure of energy exploration and utilisation. A large proportion of China's energy structure is made up of high-carbon energy like coal, while low-carbon and clean energies make up a smaller share. In 2010 coal, oil and natural gas accounted for 91.4% of China's total energy output, while nonfossil fuels accounted for only 8.6%. The unreasonable energy structure imposes great pressures on China's emission reduction targets and the security of its energy supply.

3.1.1.3 Increasingly Evident Effects of Energy Exploration and Utilisation on the Environment

Coal mining and utilisation has had serious effects on China's environment. Not only does wastewater from coalmines contaminate surface water, it goes deep into the groundwater systems and pollutes the groundwater. It is calculated that China's annual gangue displacement is around 400 million tonnes, with Shanxi stockpiling over 800 million tonnes of gangue. In the stockpiling process, gangue self-immolates easily, producing large quantities of harmful gases. Even more seriously, when gangue gets wet in the rain, it will seep into and pollute groundwater sources. China's major coal producing regions suffer from serious soil erosion and land desertification with low vegetation coverage. The ecological environment is very fragile. The dispersal of coal dust during the coal storage, packaging and transport processes pollutes the environment of the coal producing areas and along the transport lines, seriously affecting the development of coal exploitation and regional economies. Pollutants like sulphur dioxide, nitrogen oxides, smog, and mercury that are produced during the coal-burning process also cause serious destruction to the environment. Heavy energy consumption and a coal-centred energy consumption structure have resulted in China being one of the countries with the worst air pollution in the world.

The exploration of oil and gas also affects the environment, e.g. volatile components polluting the atmosphere, oily wastewater polluting water, crude oil spillage affecting the natural vegetation, and etc. China's inland oil and gas fields are mostly located in areas with low ecological capacities. With more exploration and exploitation of oil and natural gas resources, the limits imposed by the environment on oil and gas exploitation are becoming more apparent. When accidents occur during the oil extraction process at sea, the pollutants will spread with the ocean currents and cause an ecological disaster. In June 2011 the spill from an oil field in Bohai Bay caused serious damage to the ecology of China's coastal areas.

3.1.1.4   The Situation of Energy Production Safety is Grim

In recent years, production safety in China's energy industry is, on the whole, stable with continued improvement. However, the total number of accidents is still very large, and serious accidents still occur from time to time. Production operations and construction practices that are illegal or in violation of rules and regulations are common despite repeated bans. The situation of energy production safety is still grim.

In 2010 the death rate per million tonnes of coal in China was 0.7, very much higher than the rates in other major coal producing countries. The death rate per million tonnes of coal in the USA in 2008 was 0.028. In South Africa it was 0.06; India was 0.15 and Australia has seen zero death since 2002.

With more intense extraction, China's oil and gas production safety is also facing a grim situation. In recent years, the rapid expansion of production scale, the inexperience in production, and poor supervision and management have resulted in many serious accidents such as explosions in oil and gas production facilities with human casualties.

China's electricity production safety is also not looking good. The effects of natural disasters, external damage and equipment quality on electricity safety are becoming more apparent. The leaks at the Fukushima nuclear plants in Japan in 2011 sounded alarm bells for China's nuclear energy development.

3.1.2   Principles of Energy Exploration and Utilisation

China's energy exploration and utilisation should be based on the principles of safety, intensity, high efficiency, diversity and cleanliness.

3.1.2.1   Safety

In the entire process of energy exploration and utilisation, the policy of safety first and prevention as priority should be strictly adhered to. Through various technological means and controls, major safety hazards can be prevented and resolved, and the levels of emergency response and rescue can be raised. This ensures the safe exploitation, utilisation and development of energy.

3.1.2.2   Intensity

The integrated and massive exploitation of energy resources should be proactively carried forward, and the construction of large-scale energy bases should be increased to bring about intensified exploitation of energy resources and improved allocation across the whole country. The convergence among different energy resources in exploitation should be strengthened; the plans for energy exploration should be improved; and the ratio of on-site processing and transformation should be increased. Extend the industrial chains of energy bases to push forward the construction of energy bases and the coordinated development of the economy and society.

3.1.2.3 Efficiency

By using advanced technology and innovative management, make energy resources recovery, processing, transformation, transport and distribution more efficient. Promote the integrated utilisation of energy resources to reduce the wastage incurred in the energy exploration and utilisation process.

3.1.2.4 Diversity

Promote the diversity in the types of energy, and increase the proportions of high quality and clean energy in the newly added energy supplies. Break away, as quickly as possible, from the situation of relying mainly on coal and oil to increase energy supplies, and gradually push forward the replacement of traditional fossil fuels by clean energy, and thereby improve and readjust the energy production structure. At the same time, energy sources must be diversified by making use of both domestic and international markets and resources, and actively exploiting and utilising overseas energy resources.

3.1.2.5 Cleanliness

Intensify the exploration and utilisation of clean energy and renewable energy, and increase the share of clean energy in the energy structure. Reinforce the clean utilisation of fossil fuels like coal, and reduce as much as possible the effects and damage that the exploitation and utilisation of fossil fuels have on the environment. In exploiting and using all types of energy, make sure that all environmental protection requirements are met.

3.1.3   Focus of Energy Exploration and Utilisation

The exploration of energy in China must take into full consideration factors such as energy demand, resource conditions, environmental protection and economic benefit. The transformation of the mode of energy exploration and utilisation from an extensive but basic mode to an intensive one should be accelerated. Based on scientific planning, coordinate and develop the exploration of both traditional and new energies, as well as fossil and nonfossil fuels. With electricity as the core, and focusing on large-scale bases of coal-fired, hydropower, nuclear and renewable energy power plants, China should use coal efficiently and cleanly, use hydropower proactively, use nuclear energy safely and develop new and renewable energies actively. The exploration and exploitation of oil and gas resources should be strengthened, and overseas energy resources should be fully utilised to ensure the sustainable supply of energy.

China's key areas of energy exploration and utilisation are as follows.

3.1.3.1 Coal

Strengthen resource integration and coordinate the vigour of exploration. Improve exploration by building large-scale bases of coal-fired power plants in resource-rich regions. Low-grade coal should be processed and transformed on-site to generate electricity and the scale of coal-fired power plants in the east should be controlled. Promote the research and application of the technology for the green and clean use of coal, and raise the level of the integrated utilisation of coal.

3.1.3.2 Oil

Have a full understanding of the basic position of China's oil supply; intensify domestic exploration, control production reasonably and maintain the stability of supply. Intensify offshore exploration of resources and the extraction of unconventional oil resources.

3.1.3.3 Natural gas

Intensify the exploration of natural gas, and carry out large-scale extraction of unconventional gases like coal seam gas and shale gas, to increase China's capacity for supplying natural gas. Develop natural gas peak-shaving power to a moderate degree; improve the consumption structure of natural gas and increase the proportion of residential consumption of natural gas.

3.1.3.4 Hydropower

Plan the exploitation of hydropower resources scientifically. On the basis of protecting the environment and improving the resettlement of migrants, large-scale hydropower bases should be constructed. Construct large- and medium-sized hydropower plants well; develop small hydropower plants orderly and step up the construction of pumped storage power plants.

3.1.3.5 Nuclear power

Safety must be accorded priority in the development of nuclear power. Plan the distribution and construction schedule of nuclear power bases along the east coast. Accelerate the adoption and research of advanced nuclear technology, and enhance nuclear power safety standards and the monitoring and management systems. Intensify uranium exploration and improve the technology for nuclear fuel processing, to ensure the supply of nuclear fuel.

3.1.3.6 New energy

In resource-rich areas, construct large-scale wind power and solar bases using a unified and orderly approach. Resolve the problems of integration and accommodation. Actively develop biomass in an orderly fashion, and carry out exploration and utilisation of distributed energy according to local conditions. Increase the research on the exploration and utilisation technology of new energies like flammable ice, ocean energy, nuclear fusion energy and hydrogen energy, for a continued development.

3.1.3.7 Overseas energy resources

Continue to implement the 'go global' strategy to open up international energy resources. Use various ways to promote international energy cooperation, diversify the channels of sourcing overseas energy and ensure China's energy security.

3.2 The Exploitation and Utilisation of Coal Resources

Coal is China's basic energy resource, and its optimal exploitation and utilisation is critically important for China's sustainable development. To exploit and use coal resource, planning efforts must be better coordinated with rational distribution. More large-scale coal bases must be constructed, especially in the west and north. The onsite conversion of coal must be enhanced, using advanced technology to bring about the clean and green use of coal. The integrated utilisation of coal and the modern coal chemical industry must be developed. The role of coal as the basic guarantee of China's energy supply must be given full play.

3.2.1 Coordinated Planning of the Exploitation and Utilisation of Coal Resources

3.2.1.1 Features of China's Coal Resources Endowment

China has very rich coal resources. Excluding the Taiwan region, China's total coal reserves in vertical depths less than 2000 m amount to 5.6 trillion tonnes. For coal reserves in vertical depths less than 1000 m, the total volume is 2.9 trillion tonnes.

The geographical distribution of China's coal reserves is inversely proportional to the levels of economic development. The country's total coal-bearing area is over 600 000 km^2. Apart from Shanghai, all provinces, autonomous regions and municipalities have coal reserves, but their distribution is not balanced. The ten economically developed provinces and municipalities in the east (Liaoning, Hebei, Beijing, Tianjin, Shandong, Jiangsu, Shanghai, Zhejiang, Fujian and Guangdong) have less than 8% of the nation's total coal reserves. In contrast, the coal reserves in the seven provinces and autonomous regions in the west and north (Xinjiang, Inner Mongolia, Shanxi, Shaanxi, Ningxia, Gansu and Guizhou) account for almost 76% of the national total.

China's overall coal occurrence and coal quality are relatively poor. Coal beds are buried deep underground with complex structures, and only around 6% of coal reserves are suitable for open-pit mining (compared with 60% in the USA and 76% in Australia). Low-grade products like high-ash coal, high-sulphur coal and high-moisture coal take up around 40% of total reserves.

3.2.1.2 The Distribution and Scale of China's Coal Exploitation and Utilisation

For a long period of time, the exploitation and use of coal resources in China has been based on the principle of handling what is near and easy before tackling the faraway and difficult. Limited by single modes of transport, insufficient transport capabilities and so on, China achieves regional balance of coal supply and demand by first adjusting them within large regions. Nation-wide balance of supply and demand is adjusted mainly through transporting coal out of the 'Three West' region (including eastern Ningxia).

With reserves in coal-producing regions in the eastern and central China drying up, or with limited prospects for growth in production due to over-mining, the country's centre for coal development is going to shift further westwards and northwards. The five provinces and autonomous regions of Shanxi, Shaanxi, Inner Mongolia, Xinjiang and Ningxia will become the key locations of China's coal exploitation and utilisation. Of the 14 large-scale coal bases planned by the state—Shendong, northern Shanxi, central Shanxi, eastern Shanxi, northern Shaanxi, central Hebei, Henan, western Shandong, Huainan and Huaibei, eastern Inner Mongolia, Yunnan and Guizhou, Huanglong, eastern Ningxia and Xinjiang—nine are located in the abovementioned five provinces and autonomous regions. The future increase in China's coal production will depend mostly on these nine bases.

The exploitation and utilisation of coal are affected and constrained by many factors. After considering factors such as coal resources, geological mining conditions, water resources, the environment and so on, China's comprehensive effective coal supply capacities in 2020 and 2030 are estimated at 4 billion tonnes and 4.2 billion tonnes respectively. Of these capacities, Shanxi, Shaanxi, Inner Mongolia and Ningxia collectively account for 50%, compared with 27% for Xinjiang.

3.2.1.3 Transformation of the Mode of Coal Exploitation and Utilisation

Reasonably Adjusting the Pace of Coal Exploitation in Different Regions

With a focus on the overall distribution of coal resources, China should step up the construction of large-scale coal bases in the nine key areas of Shendong, northern Shanxi, central Shanxi, eastern Shanxi, northern Shaanxi, eastern Inner Mongolia, Huanglong, eastern Ningxia and Xinjiang. Apart from meeting the demand within these areas, the coal produced should also be supplied to other provinces and regions in the country that need coal imports. At the same time, the pace of coal exploitation in East and Central China should be suitably adjusted. Under the coal reserves framework, resources in the eastern coal-producing regions should be set aside as strategic reserves. Implement limited extraction in the coalmines in Central China, putting a stop to further expansion of production. Ensure that the coal it produces can—at the same time as meeting its own needs—be supplied to neighbouring provinces and regions with shortages. In addition, continue the restriction of coal exports out of the country. Strict limits must be imposed on scarce coal resources in terms of their extraction, exploitation, quantities and use, so that different types of coal can be optimally utilised.

Increasing the Proportion of In Situ Coal Conversion

The improvement of the allocation of electricity sources should be integrated with the construction of coal bases. The main coal-producing areas should be increased, especially the rate of in-situ coal conversion in large-scale coal bases. The industrial chain should be extended and the added value of coal increased to drive the local socioeconomic growth and maximise the added value of coal resources. In particular, pithead power plants must be developed to increase the proportion of coal that is converted on-site into electricity. The coal chemical industry should see moderate development. The structural surplus and redundant construction that are characteristic of the traditional coal chemical industry should be done away with. China should establish its position as a world leader in new areas of coal chemical industry like coal-based olefin, hydrogen from coal, coal-to-liquids, methanol from coal, and coal to substitute natural gas (SNG).

Boosting the Integrated Conversion and Use of Coal Resources
China should promote the application of technology for the eco-friendly and efficient conversion and use of coal. It must introduce and nurture a series of new projects or new industries to, where possible, reduce resource wastage and increase resource utilisation efficiency in the entire process of the exploitation and utilisation of coal. This will enhance the industry convergence and preparation for the development of the circular economy. China should resolve the problem of overlapping rights of coal seam gas and coal mining, and step up the simultaneous development of coal seam gas and coal resources. The efficiency of coal-fired generation should also be increased by remodelling industrial boilers (furnaces) into highly efficient and clean facilities, developing gasification poly-generation, and developing regional CCHP. The by-products should be reused from coal production, processing and utilisation such as gangue, coal slime, coal ash, coal cinder, mine water and geothermal energy from mines.

3.2.2 Construction of Large Coal-fired Power Bases in the West and North

3.2.2.1 More New Coal-fired Power Plants Should be Located in the West and North

Given China's limited primary resources, coal-fired power plants will continue to be the country's main source of electricity in the decades to come. The development of coal-fired power plants is directly related to the security of China's energy and electricity supplies. The overall scale and layout of coal-fired power plants in the future should take into consideration various factors like local coal resources, water resources and environmental capacities, local socioeconomic development, and integrated energy transport capacities. With the aim of building up a modern and integrated energy supply system that is safe, reliable, economical and efficient, and adhering to the principles of resources conservation, cost reduction, environmental protection and the coordinated advancement of local economic development, the distribution of coal-fired power plants in the whole of China is to be improved. The focus is to build large coal-fired power bases in the key coal-producing regions in the west and north, and control the scale of new coal-fired power capacity built in the east. This will result in more new coal-fired power plants shifting to the west and north. Improving the distribution of China's coal-fired power plants has significant implications for the security of energy supplies, advancement of local economic development, and promotion of energy conservation and emission reduction.

Reducing Coal Transport Pressures and Ensuring Electricity and Energy Supplies in the East and Centre

For a long time, China's coal-fired turbines have mostly been located in the load centres in the eastern and central China. The regions, however, are comparatively deficient in coal resources, and most of its coal needs must be imported from other regions. This has resulted in the railroads being preoccupied with coal transportation for a long period of time. With rapid socioeconomic development in the future, the energy and electricity needs of the central and eastern regions will continue to rise. However, limited resources will mean that local coal production will gradually decrease, with corresponding increases in coal imports and transport volume. With the focus of coal exploitation gradually shifting to the west and north of China, the transportation distances between coal-producing regions and eastern and central China will increase even further. This will impose increasingly greater pressure on electricity supply in the east and centre. With the shift of new coal-fired power capacity towards the coal-rich west and north, advanced electricity transmission technology like UHV can transmit the electricity to the east and centre. At the same time as ensuring the security of electricity supply in the east and centre, it will effectively reduce the pressure on coal transportation.

Implementing the Strategy to Develop the West and Promote the Balanced Development of Regional Economies

The main coal-producing areas in China's western and northern regions are economically backward. For a long time, constrained by the traditional exporting of coal, the raw coal extracted in these regions is directly or partly washed before being transported to areas that need them. This is the primary processing of coal, which is a low level of exploitation and utilisation, and has little effect on promoting local socioeconomic development. By building large coal-fired power bases in the coal-rich areas of the west and north, the in situ coal conversion process can be expanded and the industrial chain extended. In accordance with the development of the circular economy, the various waste products produced in coal exploitation can be reduced, transformed into resources, and recycled to enhance the level of coal exploitation and utilisation. At the same time, many jobs will be created to absorb the labour force and increase tax revenues for the areas. The economic and social benefits are very obvious. Therefore, the westward and northward shifts of new coal-fired power plants help implement China's Western Development Program. It will transform the resource advantages of the west into economic advantages, bringing about the socioeconomic development of the west and north. This will reduce the nation's fiscal burden of transferring payments and promote the balanced socioeconomic developments of the east and west regions.

Reducing Environmental Pressures on the East and Centre and Bringing About Centralised Management of Pollution and Emissions

Currently, more than half of China's coal-fired installed capacity is distributed in the east. On average, there is one power plant every 30 km along the Yangtze River, and between Nanjing and Zhenjiang the average distance between power plants is only 10 km. According to the data provided by environmental departments, the east is the region with the most acid rain occurrences, and atmospheric pollution is the most serious in East China, the Henan-Hubei-Hunan-Jiangxi region, the Beijing-Tianjin-Hebei region and the coast of Guangdong. The annual sulphur dioxide emission in the Yangtze River Delta is 45 tonnes per km^2, 20 times the national average. Basically, the east has no remaining environmental capacity, and is not able to bear the environmental pressure of the construction of large-scale coal-fired power plants. Therefore, by shifting the construction of new coal-fired power plants to the west and north, which have higher environmental capacities, it will reduce the environmental pressures on the east and centre. The intensive and efficient development and the centralised management in the coal-fired power bases will reduce pollution and emissions, and bring down the total environmental cost for the whole country.

3.2.2.2 Accelerating the Construction of Large-scale Coal-fired Power Bases

Based on the distribution of coal resources in China, the key areas in the country's future coal-fired power development will be in the main coal-producing regions in the west and north, where large-scale coal-fired power bases will be constructed. Currently, there are 16 areas with the right conditions for these bases: Shanxi (including its northern, middle and southeast regions), northern Shaanxi, eastern Ningxia, Junggar, Ordos, Xilin Gol League, Hulunbeir League, Huolin River, Baoqing, Hami, Zhundong, Ili, Binchang, Longdong, Huainan and Guizhou (Figure 3.1). Apart from Baoqing in Heilongjiang and Huainan in

Figure 3.1 China's main coal-fired power bases.

Anhui, the rest of the bases are located in the key coal-producing areas in the west and north like Shanxi, Shaanxi, Inner Mongolia, Xinjiang, Ningxia and Gansu.

In terms of development conditions, China's west and north regions have rich coal resources and ample supplies. The scale of future coal-fired power bases will mostly be constrained by the supply of water resources. To meet the needs for large-scale development of the bases, the water supply problem can be resolved by 'conserving the flow' and 'opening the sources'. On the one hand, large-scale intensive development using air-cooled power generation technology will reduce water usage in the bases by a very large extent;[3] on the other hand, the construction of reservoirs and water diversion projects, the exchange of water rights, the full utilisation of urban regenerated water, the recycling of wastewater and other measures will raise the nation's efficiency in using water resources and increase the water supply.

Shanxi Coal-fired Power Base

Shanxi is the traditional coal storehouse of China. The three state-planned large-scale coal bases of northern, central and east Shanxi have 266.3 billion tonnes of proven reserves. After considering the amount of coal reserves and environmental issues, Shanxi's coal production can reach 900 million tonnes a year.

Shanxi's total water resources are 12.38 billion m^3/year. Most of the water is distributed in the rims of basins and near its provincial borders. The water needs of Shanxi's future coal-fired power bases will be met mostly by hydraulic engineering, urban regenerated water and water runoffs from mine pits. In principle, underground water will not be extracted and used. By saving water and making full use of secondary water resources, the use of water in power generation is expected to be 710 million m^3/year by 2020.

In terms of both coal and water resources, the installed capacity of the three bases of northern, central and southeast Shanxi will be around 100 GW. After meeting local electricity needs, the volume of electricity exported from the Shanxi coal-fired power bases will be around 26.2 GW in 2015 and around 41 GW in 2020.

Northern Shaanxi Coal-fired Power Base

The coal-producing region in northern Shaanxi has rich resources of high quality coal with proven reserves of 129.1 billion ton. The production scale of the four coal-producing areas of Shendong, Yushen, Yuhuang and Fugu can reach 455 million tonnes/year. Given more in-depth exploration, the production scale of individual areas can increase further.

Northern Shaanxi is located in the loess plateau of China's northwest. River run-offs in the region are small and there is a lack of water supply facilities. Through comprehensive planning of water conservancy projects, using urban regenerated water and mine drainage, the Yellow River water diversion project and water conservation measures, the future supply and demand of water resources in this base can be balanced. Water for the coal-fired power base will be locally sourced in the short term. In the long term, the problem

---

[3] The average water consumption of a large air-cooled unit is only one-sixth of a wet cooling unit with the same capacity. Based on current matured technology and level of water management, the water consumption index of a large air-cooled unit can be controlled at 0.12 m^3/(second.GW). Considering the wear and tear of water purification devices and water loss during transportation, the average water consumption index of a large air-cooled unit is around 0.15 m^3/(second.GW).

will be resolved with the Yellow River water diversion project. It is estimated that the water needed for electricity generation will be 148 million m^3/year.

In terms of both coal and water resources, the installed capacity of the northern Shaanxi coal-fired power base will be around 43.8 GW. After meeting local electricity needs, the volume of electricity exported from this coal-fired power base will be around 13.6 GW in 2015 and around 27.6 GW in 2020.

Eastern Ningxia Coal-fired Power Base

The coal-producing region in eastern Ningxia has rich resources of high quality coal with proven reserves of 30.9 billion tonnes. Mining technology in the region is good, and under current mining conditions the production scale of eastern Ningxia can reach 135 million tonnes/year.

The eastern Ningxia coal-producing area is located in Yinchuan city, to the east of the Yellow River, which means water is readily available. The eastern Ningxia water supply project can provide enterprises that use water with a reliable supply of water. The water target for the industrial projects in this coal-producing area is obtained mainly through the exchange of water rights. Under the Ningxia Hui Autonomous Region's planned exchange of water rights for the Yellow River, the water target of the area irrigated by the diverted waters of the Yellow River is mainly transferred to the eastern Ningxia base project. The amount of water allocated to electricity can reach 167 million m^3/year. This guarantees the water supply needed for the coal-fired power base.

In terms of both coal and water resources, the installed capacity of the eastern Ningxia coal-fired power base will be around 48.8 GW. After meeting local electricity needs, the volume of electricity exported from the eastern Ningxia coal-fired power base will be around 14 GW in 2015 and around 18.4 GW in 2020.

Junggar Coal-fired Power Base

The average thickness of the coal bed in the Junggar coal-producing region is 29 m. It has proven reserves of 25.6 billion tonnes, mostly lignite and long flame coal. Based on the production capability of the various coal-producing areas, the scale of coal production in the Junggar region can reach 140 million tonnes/year.

The total volume of water resources in the Junggar region is 360 million m^3/year. Water for the coal-fired power base will be sourced mainly from groundwater extraction, water diversion from the Yellow River and the use of urban regenerated water. Based on an analysis of the balance of water resources demand and supply, the water needed for power generation in Junggar will reach 178 million m^3/year in 2020.

In terms of both coal and water resources, the installed capacity of the Junggar coal-fired power base will be around 60 GW. After meeting local electricity needs, the volume of electricity exported from the Junggar coal-fired power base will be around 30 GW in 2015 and around 43.4 GW in 2020.

Ordos Coal-fired Power Base

The Ordos coal-producing region has proven reserves of 56 billion tonnes, and its total water resource is 2.58 billion m^3/year. The water needed for power generation in Ordos will reach 181 million m^3/year. In terms of both coal and water resources, the installed capacity of the Ordos coal-fired power base will be around 60 GW. After meeting local electricity needs, the volume of electricity exported from the Ordos coal-fired power base will be around 2.4 GW in 2015 and around 4.8 GW in 2020.

Xilin Gol League Coal-fired Power Base

The Xilin Gol League is located in central Inner Mongolia. It has proven reserves of 48.4 billion tonnes, mostly lignite. Most of the mines in Xilin Gol League have thick coal beds, stable structures and excellent conditions for extraction. Large-scale open-pit mining is possible and development costs are low. Based on its resources conditions, the scale of coal production in the region can reach 340 million tonnes/year.

Water resources in the Xilin Gol League coal-producing region total 2.61 billion m^3/year. Water supply can be expected to show a relatively significant increase through the development of water conservancy projects and increased utilisation of urban regenerated water and mining drainage, among other measures. Based on an analysis of the balance of water resources demand and supply, the water needed for power generation in the Xilin Gol League will reach 152 million m^3/year in 2020.

In terms of both coal and water resources, the installed capacity of the Xilin Gol League coal-fired power base will be around 50 GW. After meeting local electricity needs, the volume of electricity exported from the Xilin Gol League coal-fired power base will be around 16.92 GW in 2015 and around 30.12 GW in 2020.

Hulunbeier League Coal-fired Power Base

The Hulunbeier League coal-producing region has proven reserves of 33.8 billion tonnes, mostly lignite. Most of the resources can be extracted by open-pit mining, and the region has the conditions to be a large-scale coal-fired power base. According to existing resources conditions, the scale of coal production in the Hulunbeier League coal-producing region can reach 156 million tonnes/year.

The Hulunbeier League has rich water resources estimated at 12.74 billion m^3/year. The water resources available for electricity generation are abundant and are expected to reach 124 million m^3/year in 2020.

In terms of both coal and water resources, the installed capacity of the Hulunbeier League coal-fired power base will be around 37 GW. After meeting local electricity needs, the volume of electricity exported from the Hulunbeier League coal-fired power base will be around 11 GW in 2015 and around 19 GW in 2020.

Huolin River Coal-fired Power Base

The Huolin River coal-producing region has proven reserves of 11.8 billion tonnes, mostly lignite. The shallow depths of the deposits, the thickness of the coal seam and the simplicity of the structure make open-pit mining a suitable option. The scale of coal production can reach over 80 million tonnes/year.

The total volume of water resources in the region is around 240 million m^3/year. Through measures to protect water resources, develop water conservancy projects, conserve water and open up new sources, and consume water from mine dewatering, the amount of water available for electricity generation can reach 42 million m^3/year.

In terms of both coal and water resources, the installed capacity of the Huolin River coal-fired power base will be around 14.2 GW. After meeting local electricity needs, the volume of electricity exported from the Huolin River coal-fired power base will be around 3.6 GW in 2015.

Baoqing Coal-fired Power Base

The Baoqing coal-producing region is an important resources-producing area in Heilongjiang with proven reserves of 5.2 billion tonnes, all lignite. Based on the various coal resources conditions in the region and its infrastructure, the scale of coal production can reach 65 million tonnes/year.

The total volume of water resources in the region is 3.46 billion m^3/year, which can supply the Baoqing coal-fired power base with 150 million m^3 of water per year. Outside the region, the usable volume of water from the Songhuajiang River is 73 million m^3/year. There is ample water available for electricity generation to meet all the needs for the coal-fired power base.

In terms of both coal and water resources, the installed capacity of the Baoqing coal-fired power base will be around 12 GW. After meeting local electricity needs, the volume of electricity exported from the Baoqing coal-fired power base will be around 8 GW in 2015.

Hami Coal-fired Power Base

The Hami region in Xinjiang has rich coal resources with proven reserves of 37.3 billion tonnes. The coal seam is shallow and the technical conditions are good for mining. The future scale of coal production in Hami can reach 180 million tonnes/year, with the potential for further growth.

The total volume of water resources in the Hami region is 570 million m^3/year. According to local water resources planning, reservoirs like Wulatai will be built by 2020 to increase water supply. After balancing and distributing the water resources across all sectors, the water available for electricity generation will reach 62 million m^3/year in 2020.

In terms of both coal and water resources, the installed capacity of the Hami coal-fired power base will be around 25 GW. After meeting local electricity needs, the volume of electricity exported from the Hami coal-fired power base will be around 21 GW in 2015.

Zhundong Coal-fired Power Base

The Zhundong region, located in Xinjiang, has proven coal reserves amounting to 78.9 billion tonnes. The coal seam is shallow with low amounts of gas, and the technical conditions for extraction are good. Based on the region's plans for infrastructure development planning, the scale of coal production can reach 120 million tonnes/year in 2020.

The total volume of the Zhundong region's water resources is 1.39 billion m^3/year. With the project that supplies water from the Irtysh River to Urumqi and the eastern extension of the '500' reservoir to regulate the water between river basins, the water supply problem for the Zhundong coal-fired power base can be resolved. The amount of water available for power generation will be around 84 million m^3/year in 2020.

Based on the sustainable socioeconomic development of the Zhundong coal-producing region, and the rational use of its coal and water resources, the installed capacity of the

Zhundong coal-fired power base will be around 35 GW. After meeting local electricity needs, the volume of electricity exported from the Zhundong coal-fired power base will be around 10 GW in 2015 and around 30 GW in 2020.

Ili Coal-fired Power Base

The Ili coal-producing region in Xinjiang has proven reserves of 12.9 billion tonnes. The shallow coal seam makes extraction easy. Based on the resources conditions in the region, a coal mining area that produces up to 100 million tonnes of coal a year can be built up.

The region has rich water resources with a total volume of 17 billion m³/year. After considering water usage across all sectors, the water available for electricity generation can reach 300 million m³/year in 2020.

In terms of both coal and water resources, the installed capacity of the Ili coal-fired power base will be around 87 GW. After meeting local electricity needs, the Ili coal-fired power base will begin exporting electricity in 2015 and it will export around 10 GW in 2020.

Binchang Coal-fired Power Base

The Binchang coal-producing region is located to the northwest of Xianyang City in Shaanxi. It has proven coal reserves of 8.8 billion tonnes. According to its resources endowments, current exploitation and technical conditions, the scale of the region's coal production can reach 40 million tonnes/year.

The total volume of the Binchang region's water resources is 1.51 billion m³/year. Based on Shaanxi's plans for the exploitation and utilisation of the water resources in the province's river basins, many water resources projects will be built in the future, which will mainly go towards the living needs of residents and the industrial needs of the Binchang coal-producing region. If water from mining drainage is recycled, the water available for electricity generation in the region will reach 42 million m³/year in 2020.

In terms of both its coal and water resources, the installed capacity of the Binchang coal-fired power base will be around 14 GW. After meeting local electricity needs, the volume of electricity exported from the Binchang coal-fired power base will be around 8 GW in 2015.

Longdong Coal-fired Power Base

The Longdong region in Gansu is located on the southwest rim of the Ordos Basin. The region is rich in coal resources, with a concentrated distribution of high quality coal in favourable conditions. The region has proven reserves of 14.2 billion tonnes, and planned production capacity is over 100 million tonnes/year.

The total volume of the region's water resources is 1.25 billion m³/year. It is a region where water is relatively scarce. To resolve this problem, Gansu Province plans to embark on several water projects in tandem with the development of coal exploitation for the Longdong energy base. By making full use of reclaimed water from sewage treatment plants in the cities and using water from mine dewatering, water resources can be rationally and scientifically distributed to ensure that the water needs of the thermal power and chemical industries are met. The water available for electricity generation can reach 79 million m³/year in 2020.

In terms of both coal and water resources, the installed capacity of the Longdong coal-fired power base will be around 26.6 GW. After meeting local electricity needs, the volume of electricity exported from the Longdong coal-fired power base will be around 4 GW in 2015 and around 8 GW in 2020.

Huainan Coal-fired Power Base
The Huainan coal-producing region has proven coal reserves of 13.9 billion tonnes. The region's distribution of coal is concentrated and the coal seams are thick, with the average thickness of extractable coal seams measuring between 20 m and 30 m. The water systems in the mining areas are plentiful and the total volume of water resources is 5.8 billion m^3/year. The water for the coal-fired power base is sourced mainly from the Huaihe River and its tributaries. Water for power generation is ample.

Based on both its coal and water resources, the installed capacity of the Huainan coal-fired power base will be around 25 GW. After meeting local electricity needs, the volume of electricity exported from the Huainan coal-fired power base will be around 13.2 GW in 2015.

Guizhou Coal-fired Power Base
The Guizhou coal-producing region has proven coal reserves of 54.9 billion tonnes, and the total volume of its water resources is over 100 billion m^3/year. There is sufficient water for electricity generation. With Guizhou's fast-growing electricity needs, power generated by the Guizhou coal-fired power base will mainly be accommodated within the province.

The total developable capacity of the above coal-fired power bases amounts to over 600 GW (Table 3.2), which can meet the need to develop coal-fired electricity throughout the entire country. In the future, with the accelerated construction of large-scale coal-fired power bases in China's west and north, over two-thirds of the nation's newly added coal-fired installed capacity will be in the west and north. This will amply support China's socioeconomic development.

3.2.3   The Clean and Integrated Utilisation of Coal

The clean and integrated utilisation of coal is the necessary choice for the development of China's coal industry. They are also the key to bringing about the country's clean, low-carbon and efficient energy development. By increasing the scale of coal washing and processing, promoting the clean burning of coal, developing the integrated utilisation of coal resources, and controlling pollutants and greenhouse gases produced in the process, the efficiency of coal exploitation and utilisation can be increased and the negative effects to the environment can be reduced.

3.2.3.1   Raising the Rate of Coal Washing

Coal washing and processing technologies that remove gangue and reduce ash and desulphurise can control coal pollution from source, which is important for the clean utilisation of coal. In recent years, China has stepped up coal washing and the rate of coal washing has increased exponentially. However, it is still low compared to the high rates in other

Table 3.2 Scale of development and power export of China's large-scale coal-fired power bases for each target year.

Coal-fired power base	Proven reserves of coal (billion tonnes)	Volume of water resources (billion m³/year)	Installed capacity of electricity that can be developed (GW)	Scale of export (GW)	
				2015	2020
Shanxi	266.3	12.38	100.00	26.20	41.00
Northern Shanxi	129.1	4.84	43.80	13.60	27.60
Eastern Ningxia	30.9	0.32	48.80	14.00	18.40
Junggar	25.6	0.36	60.00	30.00	43.40
Ordos	56.0	2.58	60.00	2.40	4.80
Xilin Gol League	48.4	2.61	50.00	16.92	30.12
Hulunbeier League	33.8	12.74	37.00	11.00	19.00
Huolin River	11.8	0.24	14.20	3.60	3.60
Baoqing	5.2	3.46	12.00	8.00	8.00
Hami	37.3	0.57	25.00	21.00	21.00
Zhundong	78.9	1.39	35.00	10.00	30.00
Ili	12.9	17.00	87.00	0.00	10.00
Binchang	8.8	1.51	14.00	8.00	8.00
Longdong	14.2	1.25	26.60	4.00	8.00
Huainan	13.9	5.80	25.00	13.20	13.20
Total	773.1	67.05	638.40	181.92	286.12

countries. In countries like the USA, South Africa and Russia, the washing rates of raw coal are all over 55% and the washing rates for thermal coal are around 40%. In China, the rates were 43% and 20% respectively in 2009, lower by 12 and 20 percentage points respectively compared with advanced international standards. Vigorously developing coal washing and processing, and increasing the raw coal washing rate are necessary requirements for the development of China's coal industry. Apart from further increasing the scale and rate of raw coal washing, China should increase the localisation rate and the reliability of large coal preparation equipment to spur the industrial development of modular coal washing technology. Based on the regional characteristics of China's various coal-producing areas, suitable technologies like water-saving and dry coal cleaning should be developed.

3.2.3.2 Developing Technology for Clean Coal Burning

Promoting Ultra Supercritical Generating Units
Ultra supercritical generation technology is quite advanced, and these units are over 5% more efficient than conventional supercritical units. As they emit less pollution, ultra supercritical generating units are a practical and effective means of increasing the efficiency of coal-fired electricity generation. The current trend in the development of ultra supercritical generating units is to increase the steam parameters to further reduce coal

consumption in power generation. The steam temperature of the new generation of ultra supercritical generating units will be raised to 700°C, and the thermal efficiency will reach 50–55%. The volume of sulphur dioxide emission will be around 15% lower than the ultra supercritical generating units that are currently in operation. During the period of the 12th Five-Year Plan, China will develop the designs, production and operating technologies of its own 600 MW and 1000 MW level ultra supercritical units with independent intellectual property rights. It will also gradually research on the development of 1200–1500 MW level units.

Promoting Circulating Fluidised Bed Technology

Circulating fluidised bed technology has the advantages of high fuel adaptability, high burning efficiency, easy pollutant management and good adjustability. It can be used for burning low quality fuels like gangue, coal slime and coal middlings, and it is internationally recognised as a clean coal-burning technology that has been commercialised. The world's largest single power generating unit that uses circulating fluidised bed boilers has a capacity of 600 MW. China has independently developed its own 600 MW capacity ultra supercritical circulating fluidised bed boiler technology, and construction for the demonstration unit had begun in July 2011. The next key research objective is to develop circulating fluidised bed boiler technology with bigger capacities and supercritical steam parameters. The wide application of circulating fluidised bed technology will result in the more efficient use of resources like coal gangue, coal middlings, lignite and high-sulphur coal in China's northwest, northeast and southwest regions.

Developing IGCC Power Generation

By turning coal into gas and combining the fuel with a steam cycle, IGCC is a clean coal-fired electricity generation technology. IGCC brings about the cascade utilisation of energy, and increases thermal efficiency by a large degree. With current technology, thermal efficiency has reached 43-45%, with the potential to go higher. At the same time, IGCC units are very eco-friendly. The level of pollutants emitted is only one-tenth of that of regular coal-fired power stations. In the rest of the world, large-scale IGCC units have been in operation for over 10 years, and China has begun constructing its first IGCC demonstration power plant. The 400MW-level IGCC electricity generation facility is expected to begin its commercial demonstration operation around 2020. Following a review of the operating experience gained, IGCC will be developed on a large scale in China's northwest, northeast and southwest.

3.2.3.3   Actively Developing Integrated Utilisation of Coal Resources

Technologies that bring about the integrated utilisation of coal resources include combined cooling, heating and power (CCHP), waste heat and top gas pressure recovery turbine (TRT) power generation, coal waste recycling, etc.

Developing CCHP

At the same time as combined heat and power units are generating electricity, using the spent steam to provide users with heat can increase the overall thermal efficiency of the units. Where there is a cooling load, using the heating produced by heat and power cogeneration as energy in tandem with absorption or compression refrigeration technology can bring about the combined generation of heat, power as well as cooling. This will further increase the utilisation efficiency of fuels like coal and the economic efficiency of combined heat and power units. There is a great potential in the future for combined heat and power generation technology in urban areas where heating and cooling loads are more stable, such as central heating zones, industrial parks and large-scale residential areas. It can replace many small coal-fired boilers used solely for heating.

Promoting Waste Heat and TRT Power Generation

Waste heat and TRT generation transforms the excess heat or differential pressures produced in the production process into electrical energy. Industries like iron and steel and construction materials are big consumers of coal in China. In the production processes of these industries, a massive amount of low-grade excess heat is produced, e.g. the low-temperature exhaust emitted from cement kilns, or various differential pressures are generated, such as the top pressure generated by the steelmaking blast furnace. Making good use of the excess heat and pressure will help raise the efficiency of the integrated transformation and utilisation of coal, and result in high economic efficiency. This technology is suitable for application in industries that use large amounts of coal, e.g. steelworks, cement plants, etc.

Recycling Coal Waste

The wastes that are produced in the processes of coal mining, processing and utilisation include gangue, coal ash, and flue gas desulphurisation (FGD) gypsum. These wastes contain many useful substances, which should be recycled and reused in accordance with the requirements of developing a circular economy. Gangue that contains a higher coal content can be used directly in gangue-fired power plants for electricity generation, while those with a lower thermal value can be used as raw material for producing construction materials. Gangue containing specific mineral content can be used as carriers in the production of inorganic fertilisers and microbial fertilisers. Coal ash is a good raw material for cement production, and it can also be used to improve the soil structure. FGD gypsum can be used in the manufacture of construction materials, as well as in the improvement of soda saline-alkaline soil. Promoting the reuse of coal waste as renewable resources will reduce the long-term accumulated damage done to the environment and turn waste into good use. The long-term prospects are excellent.

3.2.3.4 Controlling the Emissions of Pollutants and Greenhouse Gases during Coal-burning

Strengthening Pollutant Emissions Controls

The pollutants produced during the burning of coal, such as particulates, sulphur/nitrogen compounds and heavy metals like mercury, have an enormous effect on the ecological environment. Their emissions should be controlled using various forms of advanced technology. The application of technologies to remove particulates like using bag filters and electronic filters should be accelerated. At the same time as developing technologies for gas desulphurisation, denitration and mercury removal technologies should also be improved on and promoted in thermal power plants.

Conducting Research on CCS Technology

Carbon capture and storage (CCS) technology is an important cutting-edge technology to reduce greenhouse gas emissions from coal-fired power plants. Currently, developed countries are conducting various research and developing pilot projects on CCS technology. The security and reliability of carbon dioxide sequestration, and its effects on the environment, are also being studied. China has also taken initial steps in conducting CCS technology research. Low-energy consumption and low-cost CCS technology is the focus of future research and development. CCS technology may have large-scale application after 2030. China is actively following the development of this technology and conducting research to better meet the requirements of a low-carbon economy.

3.2.4 Scientifically Developing the Coal Chemical Industry

Apart from being used in generating power and as a direct end-use fuel, coal is also an important raw material in the chemical industry. With the increasing shortage of oil resources, the development of the coal chemical industry has excellent market prospects. The coal chemical industry has three main technologies: coal gasification, coal liquefaction and coal coking. Based on technological development and market trends, the future of China's coal chemical industry should feature these three aspects: aggressive development of coal gasification and poly-generation, moderate development of coal liquefaction and control of coal coking capacity growth.

3.2.4.1 Aggressive Development of Coal Gasification and Poly-generation

Coal gasification is the key technology both in the production of various coal-based chemicals and also in the extension of the coal chemical industrial chain. By introducing and utilising advanced technologies from overseas, China's coal gasification technology is gradually maturing. It is the top producer of key products like synthetic ammonia and methanol. Looking forward, China should focus on coal gasification and poly-generation and aggressively promote the coal gasification industry. Coal gasification and poly-generation technology cleans and transforms the gases generated from coal gasification to produce high value-added chemical products and fuels. The exhaust will be transmitted to combined cycle generating unit to generate electricity. This brings about the coupling of coal gasification combined cycle generating technology with coal chemical technology. Coal gasification and poly-generation can produce many products like electricity, chemicals, heat and gas. In the production process it is easy to separate pollutants like sulphur and nitrogen from carbon dioxide. At the same time as raising the efficiency of the integrated utilisation of coal, it reduces environmental pollution. This is the direction in which the development of coal gasification technology is heading. Looking forward, China should rely on various backbone enterprises to invest more in the basic research of coal gasification and poly-generation, and launch key technological research and demonstration projects, in order to bring about industrial application as soon as possible.

3.2.4.2   Orderly and Moderate Development of Coal Liquefaction

China is the only country in the world to have mastered the key technology of direct liquefaction of 1 million tonnes of coal. It also has good technological foundations in indirect coal liquefaction. The development of coal liquefaction and partial replacement of oil is an effective measure to enhance China's energy independence and reduce its dependency on foreign oil. However, the technology currently available makes coal liquefaction expensive, and the production process consumes large amount of coal and water, and emits high levels of carbon dioxide. It is thus very damaging to the environment. Therefore, coal liquefaction should be moderately developed based on a comprehensive assessment of the economic efficiency and social impacts of the technology. Coal liquefaction projects can be moderately developed in areas with abundant coal and water resources and where conditions permit so as to provide technical reserves of coal-to-liquids (CTL) technology.

3.2.4.3   Strict Control of Coal Coking Capacity

China is the world's biggest producer and consumer of coal coke. In 2010 the total coke production in China reached 388 million tonnes. Its coking technology is also among the most advanced in the world. In recent years, China's coke production expanded enormously resulting in acute problems of excess capacity. With the upgrading of China's industrial structure, especially in the improvement of steelmaking technology and the higher proportion of electric arc furnace steelmaking, the problem of coke oversupply will intensify. Therefore, China's coal coking capacity should be strictly controlled. The focus of development should switch from simple expansion of scale to structural adjustment. Accelerate the pace of structural upgrade, eliminate outdated coke ovens and impose strict controls on new capacity. Increase industry concentration and competitiveness to promote the healthy development of the coal coking sector. At the same time, the various by-products of the coking process should be fully utilised. The simultaneously development of coal coking and coal chemicals will enhance economic and social benefits.

3.3   The Exploitation and Utilisation of Hydropower Resources

China has the biggest reserves of hydropower resources in the world. Since the implementation of its economic reform and opening up policy, the country has continued to intensify the construction of hydropower facilities, and hydropower development continues to scale new heights. At the end of 2011, China's hydropower installed capacity reached 230 GW to become the country with the world's biggest hydropower installed capacity. Compared to other clean energies, hydropower is low-cost and operates on mature technology with better output characteristics. It is the preferred form of energy to help China achieve the target of 15% nonfossil fuel consumption by 2020.

3.3.1   Construction of Large-scale Hydropower Bases

3.3.1.1   Constructing Large-scale Hydropower Bases Helps Accelerate the Exploitation and Utilisation of Hydropower Resources in China

Currently, the level of exploitation and utilisation of hydropower resources is quite low in China. According to the results of a nationwide hydropower resources review conducted in 2005, the installed capacity of China's economically feasible hydropower was 400 GW, while its technically feasible hydropower capacity was 540 GW. In developed countries, the exploitation rate of hydropower is usually above 60%. In China, it is only around 43% at the end of 2011. The potential for future development is huge.

Constructing large-scale hydropower bases is the main form of exploiting and utilising China's hydropower resources. China has rich hydropower resources found in large rivers such as the Yangtze River, Jinsha River, Nu River and the Yellow River, and the conditions for intensive development and large-scale export are excellent. China's hydropower technology continues to improve, and the design and construction of its hydropower plants are second to none in the world. The manufacture and installation of hydropower generating units are of international standards. The batch of 700 MW units that were independently designed and manufactured has been in operation for many years in the hydropower plants like the Three Gorges and Longtan. There are thus less technical obstacles in the construction of large-scale hydropower bases.

The total planned installed capacity of 13 large-scale hydropower bases is around 320 GW, or 60% of China's technically feasible hydropower installed capacity. The 13 bases are located in the main stream of the upper Yangtze River, Jinsha River, Dadu River, Yalong River, Wu River, the Red River tributary of Nanpan River, Lancang River, the upper Yellow River, the northern main stream of the Yellow River, the Northeast, west of Hunan, Fujian-Zhejiang-Jiangxi and Nu River (Figure 3.2). As of July 2011,

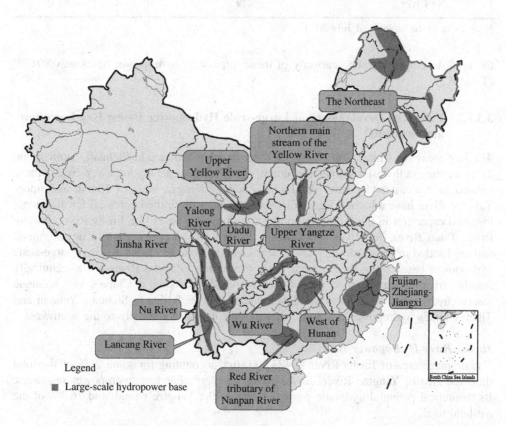

Figure 3.2 The distribution of large-scale hydropower bases in China.

Table 3.3 The development of large-scale hydropower bases in China.

No.	Hydropower base	Planned installed capacity (GW)	Developed installed capacity (GW)	Level of development
1	Main stream of Upper Yangtze River	34.41	21.17	61.5%
2	Jinsha River	76.52	1.20	1.6%
3	Dadu River	26.73	6.30	23.6%
4	Yalong River	29.06	3.30	11.4%
5	Wu River	11.27	3.58	31.8%
6	Red River tributary of Nanpan River	13.74	6.13	44.6%
7	Main Stream of Lancang River	31.98	8.97	28.0%
8	Upper Yellow River	20.83	7.97	38.3%
9	Northern main stream of the Yellow River	6.40	1.20	18.8%
10	The Northeast	18.69	6.17	33.0%
11	West of Hunan	6.10	6.10	100.0%
12	Fujian-Zhejiang-Jiangxi	10.93	6.95	63.6%
13	Nu River	36.39	0	0

Note: Data in the table as of July 2011.

the total developed installed capacity of these large-scale hydropower bases was 80 GW (Table 3.3).

3.3.1.2 Enhanced Development of Large-scale Hydropower Bases: Key Locations and Schedule

The key areas for China's future hydropower development are in Sichuan, Yunnan and Tibet in the southwest. Among the country's 13 large-scale hydropower bases, those located in the east and centre like Fujian-Zhejiang-Jiangxi, west of Hunan and upper Yangtze River have a higher degree of development, with limited potential for additional installed capacities in the future. In contrast, hydropower bases like Jinsha River, Yalong River, Dadu River, Nu River and Lancang River have a lower degree of development, and are located in the southwest. China needs to accelerate the construction of large-scale hydropower bases along these rivers. Also, the Yarlung Zangbo River has a technically feasible hydropower installed capacity of 89.66 GW, making it China's key strategic reserve hydropower base. With the development of river basins in Sichuan, Yunnan and Tibet, China's hydropower development will shift its focus gradually to the southwest.

Jinsha River Hydropower Base

The drainage area of Jinsha River is 473 200 km^2, accounting for some 26% of the total drainage area of Yangtze River. Jinsha River has very abundant hydropower resources. Its theoretical potential hydraulic power is 42% of the Yangtze's total, and 16.7% of the national total.

Twenty-five power plants are planned for the Jinsha River Basin, with a total installed capacity of 76.52 GW. Of the 25, 13 are on the upper reaches (planned installed capacity of 13.92 GW), 8 on the middle reaches (planned installed capacity of 20.90 GW), and 4 on the lower reaches (planned installed capacity of 41.7 GW). Based on the plans for the hydropower bases in Jinsha River, the installed capacity in commission in 2020 will reach 61.6 GW and 73.52 GW in 2030.

Yalong River Hydropower Base

Yalong River is located in the northeast of the Qinghai-Tibet Plateau. Its drainage area is around 136 000 km^2 and has a natural drop of 3830 m. It has abundant hydropower resources, and a technically feasible capacity of 34.61 GW. The Yalong River hydropower resources are characterised by abundant water, many large-scale power plants, small inundation losses and good overall adjustability, among other attributes. The prospects for development are good.

Twenty-two power plants are planned for the Yalong River Basin, with a total installed capacity of 29.06 GW. Of the 22, 11 are on the upper reaches (planned installed capacity of 2.8 GW), 6 on the middle reaches (planned installed capacity of 11.56 GW), and 5 on the lower reaches (planned installed capacity of 14.70 GW). Based on the plans for the hydropower bases in Yalong River, the installed capacity in commission in 2020 will reach 24.60 GW and 26.06 GW in 2030.

Dadu River Hydropower Base

Dadu River is the biggest tributary of the Min River System in the upper reaches of the Yangtze River. Its drainage area is around 77 000 km^2; its main stream is 1062 km long; and its natural drop is 4175 m. It has abundant hydropower resources.

Twenty-seven power plants are planned for the Dadu River Basin, with a total installed capacity of 26.73 GW. The installed capacity in commission in 2020 will reach 23.00 GW and 26.73 GW in 2030.

Nu River Hydropower Base

The source of Nu River is in the southern foot of the Tanggula Mountains in Tibet. The river passes through China's Tibet and Yunnan before entering Myanmar. Its drainage area within China is 138 000 km^2 and the natural drop of its main stream is 4848 m. Its water volume is abundant and stable. The topographical and geological conditions are good for hydropower development. The number of people to be relocated is small.

Twenty-five power plants are planned for the Nu River Basin, with a total installed capacity of 36.39 GW. Of the 25, 12 are on the upper reaches (planned installed capacity of 14.64 GW), 9 on the middle reaches (planned installed capacity of 18.43 GW), and 4 on the lower reaches (planned installed capacity of 3.32 GW). Its installed capacity in commission in 2020 will reach 4.68 GW and 26.39 GW in 2030.

Lancang River Hydropower Base

Lancang River originates in the northern foot of the Tanggula Mountains. It passes through China's Qinghai, Tibet and Yunnan before entering Laos. Its drainage area within China is 164 000 km^2 and its natural drop is 4695 m.

Twenty-two power plants are planned for the Lancang River Basin, with a total installed capacity of 31.98 GW. Of the 22, 13 are on the upper reaches (planned installed capacity of 15.52 GW), 5 on the middle reaches (planned installed capacity of 8.11 GW), and 4 on the lower reaches (planned installed capacity of 8.35 GW). Its installed capacity in commission in 2020 will reach 26.00 GW and 31.58 GW in 2030.

Yarlung Zangbo River Hydropower Base

Yarlung Zangbo River is the biggest river in Tibet, and the world's highest river. Its main stream measures 2057 km long and its drainage area is around 240 000 km^2. The technically feasible capacity of its main stream/hydropower resources is 89.66 GW, 95% of which are in its lower reaches. It is scheduled to enter the intensive development phase around 2030.

Taking together China's plans for the construction of large-scale hydropower bases and for the development of its major river basins, the installed capacity of the country's large- and medium-scale hydropower stations in 2020 will be around 270 GW, which is double the installed capacity of 2010. The hydropower bases in the southwest will move into the initial stage of development, while the mainstream hydropower developments in the Yalong and Dadu Rivers will basically be completed.

By 2030, the installed capacity of China's large- and medium-sized hydropower stations will increase 70 GW from 2020. All the planned hydropower bases in the southwest will be completed, with all main reservoirs in the rivers in operation. With the exception of the Nu and Yarlung Zangbo Rivers, the mainstream hydropower development in the rivers will basically be completed.

By 2050, the installed capacity of China's large- and medium-scale hydropower facilities will be 30 GW more than in 2030. The mainstream hydropower base in Nu River will basically be completed. The straightening of the lower reaches of Yarlung Zangbo River and the development of cascade hydropower stations will be completed.

3.3.1.3   The Transmission and Consumption of Electricity Generated in Large-scale Hydropower Bases

Most of the electricity generated by large-scale hydropower bases must undergo long-distance transmission for consumption by the eastern and central regions of China. Looking forward, China's new hydropower units will mainly be concentrated in the southwestern hydropower bases. The southwestern hydropower bases are located in areas where the economy is relatively backward, and where the power load is quite small. Water runoff should be kept to the absolute minimum. After large-scale hydropower bases have met local electricity needs, the power generated is to be transmitted to the eastern and central load centres, where market demands are high. Considering various factors like the balance of primary energy, distance of power transmission and efficiency of resource utilisation, the electricity generated in the lower Jinsha River will mainly be transmitted to Central China and East China. The electricity generated in the Yalong River will mainly be transmitted to East China. After meeting Sichuan's electricity needs for its development, the rest of the electricity generated in Dadu River will be transmitted to Chongqing, the Henan-Hubei-Hunan-Jiangxi region and East China. The electricity generated in the Nu and Lancang Rivers will mainly be transmitted to South China.

The hydropower bases in the southwest lie between 2000 and 3000 km from the load centres in eastern and central China. From the economical standpoint and in terms of the operational safety of the power grids, transmission must be done through UHV transmission lines. To ensure the smooth development of China's large-scale hydropower bases, the planning of power sources and power grids must be coordinated. At the same time as reserving market space in eastern and central China for hydropower consumption, the construction of UHV transmission networks must be stepped up in order to achieve optimal allocation of hydropower resources nationwide. Ample consideration must be given to the development schedule and scale of cascade hydropower plants. Electricity transmission capacity must be reserved to avoid repeated constructions and resource wastage.

3.3.2   Development of Small Hydropower

China has abundant resources for small hydropower systems. The orderly development of small hydropower systems can resolve the lack and shortage of electricity in rural areas. This will stimulate socioeconomic development in rural China, and ensure emergency power supplies.

3.3.2.1   Small Hydropower Resources and their Current Development

China's small hydropower resources are widely distributed across the country, with a potential capacity of 128 GW. This is the highest in the world. The potential capacity in the west is 80.22 GW, which accounts for 62.7% of the national total. The most resource-rich areas in terms of small hydropower are in Guangxi, Chongqing, Sichuan, Guizhou, Yunnan and Tibet in southwest China. The total potential capacity is 62.05 GW, or 48.5% of the national total. Inner Mongolia, Shaanxi, Gansu, Ningxia, Qinghai and Xinjiang in the northwest have relatively concentrated resources of small hydropower. The total potential capacity is 18.17 GW, or 14.2% of the national total. The small hydropower resources in the northeast and central China are mainly concentrated in Jilin, Heilongjiang, Hunan and the mountainous area of Hubei. The total potential capacity is 25.74 GW, or 20.1% of the national total. The small hydropower resources in the east are mainly concentrated in Zhejiang, Fujian and Guangdong. The total potential capacity is 22.09 GW, or 17.3% of the national total.

As of 2010, the number of small hydropower plants in China exceeded 45 000, with a total installed capacity of 56 GW and annual output of around 160 TWh, accounting for around 26% of China's total hydropower installed capacity. Among the small hydropower plants, the units in Fujian, Sichuan, Hunan, Zhejiang and Jiangxi are larger. From the current distribution of small hydropower systems, one can see that the development of small hydropower is still concentrated in areas with abundant small hydropower resources and dense populations.

3.3.2.2   Key Locations for Future Development

Based on the resources conditions for small hydropower and the nation's power needs for its socioeconomic development, China's small hydropower development will experience phenomenal growth followed by a plateau. Before 2030, many small hydropower plants will be put into operation. It will be the crucial period for small-scale hydropower development. After 2030, there will be fewer new hydropower plants. In terms of distribution, development of small-scale hydropower resources will be concentrated in the west.

By 2020, China's small-scale hydropower installed capacity will reach 75 GW. The exploitation rate will be 58.6%. Small-scale hydropower development will be concentrated in Sichuan, Yunnan, Fujian, Hunan and Guangdong, with Sichuan's installed capacity reaching almost 10 GW.

By 2030, China's small hydropower installed capacity will reach 93 GW. The exploitation rate will be 72.7%. The west will have 14 GW of additional installed capacity.

Before 2030, the small hydropower resources in China with better potential would have been developed, so small hydropower installed capacity between 2031 and 2050 will experience slower growth. The additional installed capacity will only be 7 GW, of which central and western China will account for 5 GW.

The development of small hydropower systems in China should be tailored to local conditions and planned scientifically. The disorderly construction of small hydropower plants must be avoided. Planning for small hydropower plants must be coordinated with planning for river development. Under the condition that the environment must be protected, the development and construction of small hydropower systems must be undertaken properly to ensure the sustainable use of small hydropower resources. Also, based on the principle of physical proximity for electricity consumption, the development of small hydropower systems must be closely linked to local power grids and, based on the results of grid upgrading and revamp in the rural areas, connection and consumption of small hydropower must be ensured.

3.3.3   Planning and Construction of Pumped Storage Power Plants

Pumped storage power plants are operationally flexible and responsive. They are capable of multiple functions like load shifting, frequency modulation, phase modulation, emergency reserves and black starts. The rationally planned construction of pumped storage power plants has important significance for ensuring the quality of electricity supplies, promoting the consumption of new energies, ensuring the security and stability of the power system and ensuring the effective operation of the economy. At the end of 2010, the installed capacity of China's pumped storage power plants is 16.93 GW, a mere 1.7% of the country's total installed capacity. This cannot meet the increasing peak-shaving needs of the system, and development must be accelerated.

3.3.3.1   Layout of Pumped Storage Power Plants

China has a large number of sites suitable for pumped storage power plants. Based on pre-site selection findings, the capacity of pumped storage power is over 130 GW nationwide. The economically developed load centres, as well as the west and the north, areas with large-scale development of new energies, are the key regions for the development of China's pumped storage power plants.

Planning for and Constructing Pumped Storage Power Plants in the Load Centres
Given that the eastern and central regions lack resources, they rely on the plentiful hydropower and coal-fired power generated in the west and north to meet their energy needs. To increase the utilisation rate of long-distance power transmission lines and to adapt to the need of the west and north to bundle and transmit wind power, the margin for adjustment in long-distance power transmission is relatively small. Also, the distribution of future nuclear power plants will be concentrated in eastern and central China. Given economic and safety considerations, nuclear power will not take part in load shifting. As there are many sites in the eastern and central regions suitable for pumped storage power plants, with excellent construction potential, there is a need to plan for and build a batch of such power plants in the eastern and central load centres to ensure the stable and economic operations of the power grids. They will provide services like load shifting, frequency modulation, phase modulation and emergency reserves to ensure the safe and economical operation of the system.

Constructing a Number of Pumped Storage Power Plants in the Western and Northern Regions
Power generation by renewable energy sources like wind and solar power is characterised by randomness and intermittence. The large-scale connection of renewables-generated electricity to grid is a big challenge to the flexible adjustability of the power system. China uses little oil and gas, power sources with good load shifting capabilities, and run-of-river hydropower accounts for a large part of hydropower. This means that the system has very limited means for load shifting. Pumped storage power plants have obvious load shifting benefits, and are good for improving load characteristics, and increasing the system's ability to receive renewable energies like wind and solar power. To adjust to the large-scale development of renewable energies, and increase the ability to utilise the power generated by wind and solar power on-site, the survey of sites for pumped storage power plants in the northeast and northwest must be stepped up. The development and construction of pumped storage power plants on excellent sites with relatively low unit investment should be accelerated.

Taking into consideration these factors—the economic viability of the entire system, the load shifting and frequency modulation needs of the system, the exploitation of renewable energies, the development of nuclear power—and analysing the current state of pumped storage power plants projects, the scale of China's pumped storage power plants will reach 60 GW by 2020, and 75 GW by 2030.

3.3.3.2 Bringing about the Sustainable Development of China's Pumped Storage Power Plants

The development of pumped storage power plants has important significance for the development of China's clean energy sources and in ensuring the safe and stable operation of the power system. To bring about the sustainable development of China's pumped storage power plants, we must start on areas like construction planning, operational system and technological innovation, and develop targeted measures to resolve the conflicts that arise during the development of pumped storage power plants.

Formulating a Scientific Development Plan for Pumped Storage Power Plants
In a power system, pumped storage power plants are marked by their scattered distribution. The reasonable scale and distribution of these generating units are closely connected with factors like power grid structure, power source structure, load characteristics, etc. To bring about the reasonable distribution of pumped storage power plants, we must step up the integrated planning for these plants. Preferred sites for the power stations should be chosen based on the needs of the power system and after taking into consideration various influencing factors. The reasonable proportion of electricity supplied by the pumped storage power plants and their construction schedules must be determined to give full play to the advantages of these plants.

Setting up a Reasonable Operation Model for Pumped Storage Power Plants
Currently, the operating costs of China's pumped storage power plants are included and calculated as part of the maintenance costs of power grids. Specific methods of cost verification and recovery have yet to be determined. Besides, the capacity price tends to be low and the return on investment is insufficient, which limit the self-development capacity of pumped storage power plants. In the long term, we must adhere to the principle of 'the beneficiary pays' and reasonably distribute the construction and operating costs of pumped storage power plants among power generating enterprises, power grid operators and power users. This will ensure reasonable profits for investors. At the same time, by applying on-grid-side time-of-use tariffs and two-part tariffs, price leveraging will result in the optimal utilisation of pumped storage power plants and increase their operational benefits.

Increasing the Degree of Technological Innovation of Pumped Storage Power Plants
Compared to developed countries, the design and manufacture of pumped storage power plant facilities in China still lags far behind. Without the benefit of the core controlling technology, facilities manufactured in China are less reliable and stable. Looking forward, more support should be given to the manufacturers of pumped storage facilities in financial terms as well as at the policy level. This will encourage technological innovation, improve the manufacturing standards of facilities made domestically, lower the construction costs of pumped storage power plants and meet development needs. As for the differences between pumped storage and regular power plants, construction and operating standards should be laid down to ensure the sustainable development of China's pumped storage power plants.

3.3.4   Environmental Protection and Migrant Relocation

Although China has the resource potential and technical advantages for large-scale development of hydropower, it faces many problems and challenges such as the submerging of lands, environmental protection and migrant resettlement. Looking forward, the large-scale development of China's hydropower must involve coordinated planning to properly manage the multiple relationships between migrant relocation, environmental protection and so on. This must be done to achieve harmonious development.

3.3.4.1 Strengthening Environmental Protection

The exploitation and utilisation of hydropower resources benefits the environment on the whole in that by replacing traditional fossil fuels, it reduces environmental pollution. Hydropower also has comprehensive benefits related to flood mitigation, shipping, etc. However, if environmental protection is not given due regard, the over-exploitation of hydropower can result in some ecological problems.

There is no conflict between hydropower development and environmental protection. As long as objective laws are respected and hydropower exploitation is undertaken in an orderly fashion with protecting the environment as a precondition, then there is every possibility of a coordinated development of hydropower exploitation and environmental protection. The principle of 'protection in development and development in protection' must be adhered to. Throughout the entire process of planning, exploration, design, construction and operation of hydropower plants, environmental protection measures must be reinforced and carried out. Hydropower development planning must be coordinated with river planning and environmental planning, and hydropower developments must be conducted based on the environmental assessments. During the construction of hydropower plants, development goals must be aligned with social development and ecological goals. Reasonable engineering measures and operation solutions must be adopted to reduce as much as possible the adverse effects on the ecological systems of the rivers. Hydropower plants should be allowed to play a more positive role in flood prevention, water replenishment, irrigation, reservoir area protection, etc.

3.3.4.2 Carrying out Proper Migrant Relocation

The development and construction of large-scale hydropower plants inevitably entails migrant relocation, which is a complex and systemic task that involves politics, economy, society, heritage, resources and the environment. To ensure the successful development of China's hydropower, population relocation must be carried out properly in tune with the times.

Strengthening the Overall Planning of Migrant Relocation
Ensure that plans for migrant relocation and project design are simultaneously introduced. The planning of migrant relocation must take into consideration the local governments, the hydropower operators, as well as the interests of both the resettled population and the host population. Hydropower development should be carried out in combination with promoting local economic growth and enriching and stabilising the relocated population. The relocation and settlement of residents should be carried out in combination with the construction of new socialist villages and towns. This is to ensure the livelihood and long-term development of the relocated populations while improving their production conditions and raising their living standard.

Strengthening the Management of Migrant Relocation
Set up a specialised organisation at the government level to conduct the task of migrant relocation, and unify the processes of compensation, production support, manpower

training and infrastructure construction at new locations. Set up a postmortem assessment system to sum up the migrant relocation experience and the lessons learnt, so as to improve management.

Based on the People-oriented Principle, Improving and Refining the System for Public Participation in Migrant Relocation
When planning for migrant relocation, take different opinions into consideration, especially those pertaining to the interests of the population to be relocated. Increase the transparency of the relocating process, and take special care to step up the publicity and guidance work to reduce conflicts and ensure the successful relocation of the population.

3.4   The Exploitation and Utilisation of Nuclear Power

China has a massive population with a huge demand for energy, but the country's per capita energy resources are low. The aggressive development of nuclear power is the natural choice to relieve China's energy shortage, increase its supply independence and help deal with climate change. Currently, nuclear energy is undergoing active development in China, with large nuclear units to be installed in the next 20 years. China must learn from a number of nuclear accidents in overseas countries to ensure the effective and safe development of its nuclear energy programme.

3.4.1   Construction of Large-scale Nuclear Power Base

Currently, the main global trend in the development of nuclear technology is the expansion of single unit capacity. This reduces per kW investment and increases the economic benefits of nuclear energy. Third-generation nuclear power generating units are usually on the scale of 1 GW and above. In addition, for the sake of operational safety, the site selection for nuclear power plants is very strict with due considerations given to factors and conditions like earthquakes, geology, hydrology, transport and climate. Therefore, the nuclear energy development in China needs to conform to the global trend in nuclear technology development and take full advantage of the country's many locations suitable for nuclear power plants for building economies of scale. Expand the scale and step up the standardisation of nuclear energy construction and operation to enhance the quality and benefits of nuclear power development.

China has quite a large number of sites available for developing nuclear power plants, and a large quantity of data and results have been obtained in preliminary explorations. Some of the locations have passed pre-feasibility assessments and possess the qualities to become sites for large-scale nuclear power plants. With more efforts in site selection, the future need for the construction of nuclear power plants with installed capacity of 300–400 GW can be met. Nuclear power plant sites are rare and valuable resources. By comparing, enhancing and assessing selected and alternate sites, and taking into consideration national economic development, energy needs and environmental protection, the orderly construction of nuclear power plants and rational distribution of nuclear power development can be achieved. In addition, given the long development cycle of nuclear

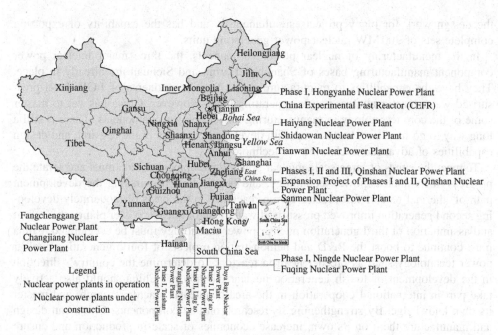

Figure 3.3 Distribution of nuclear power plants in operation or under construction in the Chinese mainland.

power, there is a need to strengthen the security of the power plant site beforehand and manage the relationship between nuclear energy development and the local economic development. Looking forward, China still needs to do more in studying and developing the plans for nuclear energy development, to take full advantage of 'cluster development, cluster management and cluster benefits'. Without prejudice to proper safety assessments, nuclear energy will be developed first along the eastern coastal areas, where power loads are growing rapidly and resources are scarce. At the same time, the development of nuclear power inland will be carefully studied. Figure 3.3 indicates the distribution of nuclear power plants in operation or under construction in the Chinese mainland as of July 2011.

3.4.2 Advancement of Nuclear Power Technology

In recent years, through the assimilation of imported technology and innovation, China has built a number of nuclear power plants on its own. The construction of these nuclear power plants lays a solid technological foundation, and allows the accumulation of substantial operational and management experience, for building large-scale nuclear power bases in the future in China.

In the area of nuclear power plant design, China has formed a professional and rationally organised design team with capabilities of independently designing 300 MW-, 600 MW- and 1 GW-level pressurised water reactor nuclear power plants. The team can take on the design work for many projects simultaneously and has the capability of exporting complete sets of 300 MW nuclear power generating units.

In the manufacturing of nuclear power components, the three major nuclear power component manufacturing bases of Shanghai, Harbin and Sichuan are already in place. They have the capabilities to manufacture most of the components of 1 GW-level pressurised water reactor nuclear power generating units. However, China has yet to master some of the core technology in manufacturing nuclear power components. There is still a long way to go for China to catch up with developed countries in the research and design capabilities of advanced nuclear power technology.

To meet the needs of large-scale nuclear power development, China must accelerate the progress of its nuclear power technology, and be steadfast in pursuing the development path of 'thermal reactor-fast reactor-fusion reactor'. On the basis of moderately developing second generation improved pressurised water reactor nuclear power plants, the import and assimilation of third generation nuclear power technology must be accelerated. China must continue to boost the R&D and demonstration projects of fourth generation nuclear power technology such as breeder neutron reactors, to determine the country's direction in the development of fourth generation nuclear technology. China should also actively take part in international cooperation in the area of nuclear reactor research, and master its own knowledge. By strengthening its research on key components, China can design and manufacture them on its own, increase economies of scale in production and ensure domestic availability of the main raw materials and key components. Also, China must intensify its research on nuclear waste disposal, and develop its own spent fuel reprocessing technology. This will bring about the maximum utilisation of nuclear resources and minimisation of nuclear wastes and nuclear proliferation. It will also reduce the negative effects nuclear power development has on the environment.

3.4.3   Building up a Nuclear Energy Safety System

Nuclear safety is the crux of nuclear energy development. Nuclear energy is a clean and economical form of energy that can replace fossil fuels to a large extent. However, the use of nuclear power carries huge risks. The Three Mile Island accident in 1979 sparked massive public panic in the USA, while the Chernobyl accident in the former USSR in 1986 caused massive human and economic losses, driving nuclear energy development worldwide to a low ebb. In March 2011, as a result of the ecological disaster due to the serious nuclear incident in Fukushima, Japan, global nuclear energy development, which had begun to show signs of recovery, suffered another major setback. These nuclear accidents demonstrate that safety is the most fundamental issue in nuclear energy development. The focus on nuclear safety must include all processes, from power plant design and component manufacture to operations and decommissioning.

China must accelerate its nuclear energy development, and intensify the building up of a system of nuclear safety standards. By developing step-by-step and gradually building up experience, and by prioritising safety and quality, nuclear energy development and nuclear safety management can steadily move forward. To ensure the operational safety of nuclear power plants, the coordination of nuclear energy development with other energy development and power grid development should be intensified. China should strengthen the interconnections of power grids and build interregional grids to provide nuclear energy development with sufficient support and assurance, to ensure the availability of mutual back-up during emergencies and to prevent nuclear fallouts from spreading or related incidents from occurring.

Compared to countries with mature nuclear power systems, China's nuclear safety still lags behind in management systems, legislation, standards specifications and information disclosure. There is an urgent need to build up a complete nuclear safety framework to provide assurance for the large-scale development of nuclear energy. In the area of management systems, there is a need to rationalise government institutional mechanisms on nuclear management to further define the authorisation limits of nuclear energy management departments. This will lead to a separation between government and corporate duties. In the area of legislation, there is a need to speed up the drafting and introduction of laws governing atomic energy to increase nuclear safety monitoring and regulatory powers. The country's emergency response to grade 3 nuclear radiation accidents must be perfected, and a sound assessment system for nuclear safety must be set up. In the area of standards specifications, there is a need to accumulate practical experience and gradually build up China's own standards in nuclear industry and technology. In terms of information disclosure, we need to increase the transparency of nuclear safety information and educate the public on the issue in order to gain a public understanding of and support for nuclear energy development.

3.4.4   Supply of Nuclear Fuel

China is a vast country with excellent geological conditions for uranium mineralisation. It is among the few countries with rich uranium resources. The short history and low degree of uranium exploration in China means that exploration potential and prospects are excellent. Currently, proven uranium deposits are rather scattered and they are mainly medium and small deposits (accounting for over 60% of total reserves). Uranium ores are mainly of medium and low grades, with 0.05–0.3% grade uranium ores forming the vast majority of total resources. Within the area to be investigated, there are still 3.6 million km^2 of land with extremely low or zero exploration work. Most of the proven uranium ore bodies are buried less than 500 m below ground; there has not been sufficient deep exploration work. In terms of range and depth of exploration, there is a considerable potential for growth in China's uranium reserves.

Based on the quantity of proven and economically recoverable uranium resources in China at the moment, and its current production capabilities, the country's reserves and production volume can meet the needs of its nuclear power development before 2020. However, they are still insufficient in meeting the forecasted medium- and long-term needs. To support major nuclear energy development in the future, China must ensure an ample and secure supply of nuclear fuel. It needs to work on the areas below.

3.4.4.1   Continuing to Intensify Uranium Exploration within China

Intensify the search for a super large deposit of between 10 000 and 100 000 tonnes as the main deposit to steadily raise the proven uranium reserves of China.

3.4.4.2   Strengthening Cooperation with Uranium-rich Countries

Actively participate in the international uranium trade. By making full use of both domestic and global markets and resources, China can increase the fuel supply for its nuclear energy development. China can enter into nuclear cooperation agreements with uranium-rich countries to strengthen the joint exploration and technical cooperation in uranium resources. Through various means such as global trade, uranium products in the international market can be purchased or uranium mining rights obtained so as to ensure the supply of natural uranium in China.

3.4.4.3   Accelerating Technological Innovation in Production Processes of Nuclear Fuel

Conduct experiments and research in the area of uranium mining and smelting as soon as possible to achieve breakthroughs in mining and smelting technologies. This will increase the utilisation rate of uranium resources and increase the quantity of economically recoverable uranium. In the area of uranium enrichment, China needs to step up the construction of larger centrifuge plants and domestic centrifuge uranium enrichment plants to enable enriched uranium production capabilities that are in tandem with the needs of major nuclear energy development.

3.4.4.4   Increasing the Utilisation Rate of Uranium Resources

Increase the pace of research and development of fast breeder neutron reactors; build up advanced nuclear fuel closed-loop systems; and reduce radioactive waste emissions. This will promote sustainability in the use of nuclear energy in China.

3.5   The Exploitation and Utilisation of New and Renewable Energies

With increasing shortages in the supply of fossil fuels, and the growing concern over global environmental problems like climate change, it has become a consensus among different countries in the world to develop new and renewable energies for the sustainable development of energy. As a major energy consumer, China's exploitation and utilisation of new and renewable energies will have important implications for ensuring the stable supply of energy, adjusting and improving the energy structure, protecting the environment, reducing the emission of greenhouse gases, and bringing about economic change and industry upgrading.

China has abundant new and renewable energy resources, and the potential for development is huge. The key means to bringing about the large-scale development of new and renewable energies is electricity generation. In the short and medium terms, China should vigorously develop renewable energies like wind and solar power. In areas with concentrated resources and favourable development and utilisation conditions, China should apply intensive development and build large-scale renewable energy power generation bases. Giving priority to local consumption, the power bases can be integrated into larger power grids to supply electricity nationwide. At the same time, small-scale power generators powered by wind, solar power and biomass should be built according to local conditions and connected to power grids in the vicinity. In the short term, China needs to intensify its research on the exploitation and utilisation of new energies like methane ice and nuclear fusion. This serves as a foundation to ensure the sustainable supply of future energy needs.

3.5.1 Building Large-scale Renewable Energy Power Bases

3.5.1.1 Building Large-scale Wind Power Bases in the 'Three North' Region and the Eastern Coast

The State of China's Wind Resources and the Development of its Wind Power
China's wind energy resources are mainly concentrated in the 'Three North' on land and along the eastern coast (Figure 3.4). According to the latest wind energy resources exploration and its assessment and analysis undertaken in 2009 by the China Meteorological Administration, China's wind energy development potential is over 2500 GW. Of this, grade 3 winds at 50 m high over land (wind power density greater than or equal to 300 W/m^2) have the development potential of around 2380 GW. For Grade 3 winds at 50 m high over water that is between 5 m and 25 m deep, the development potential is about 200 GW. The total usable amount of wind energy in China that is technically

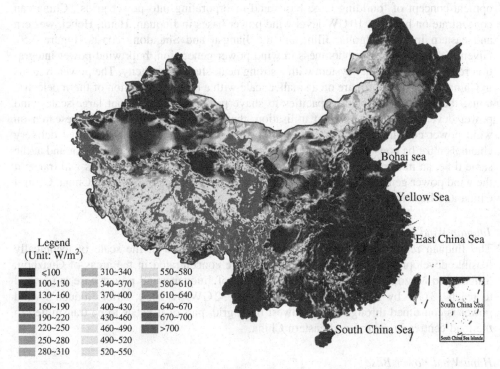

Figure 3.4 Distribution of wind power resources in China.
Source: Wind and Solar Energy Resources Centre, China Meteorological Administration.

feasible is 700–1200 GW, of which 600–1000 GW is the actual feasible amount from onshore sources. According to marine regional planning, for coastal waters with depths of less than 20 m, the development potential is about 150 GW based on the assumption that 20% of the surface is usable. The eastern and western regions of Inner Mongolia, Xinjiang's Hami, Gansu's Jiuquan, Hebei's Bashang, western Jilin, and the Jiangsu and Shandong coasts are the areas with the richest wind power resources in China. Grade 3 and above winds at 50 m high have the potential to generate 1850 GW. This accounts for 77.7% of the total onshore wind power resources in the country.

In recent years, China has seen sustained rapid growth in wind power. During the 11th Five-Year Plan, the annual average growth of the installed wind power capacity reached 100%. At the end of 2011, China's installed wind power capacity connected to grids amounted to 45.05 GW, 49 times higher than that in 2005. This accounted for 4.3% of China's total power generation capacity. The installed wind power capacity operated by the State Grid Corporation of China was around 42 GW. In 2011 China's wind power generating capacity was 73.2 TWh, or 1.6% of the country's total. A number of contiguous GW-level wind power bases have been developed, and the pattern for scale development is beginning to take shape.

Distribution of Large-scale Wind Power Bases in China

The distribution of wind power resources in China is concentrated in the west, north and coastal areas, very suitable for large-scale intensive development. Based on the development concept of 'building large bases and incorporating into power grids', China can concentrate on building 10 GW-level wind power bases in Jiuquan, Hami, Hebei, western and eastern Inner Mongolia, Jilin, and the Jiangsu and Shandong coasts (Figure 3.5). Given the volatility and randomness in wind power generation, bulk wind power integration requires an electricity system with a strong peak-shaving capacity. The power systems in China's west and north are on a smaller scale with a higher proportion of thermoelectric units; they have insufficient capacities to shave peaks. To bring about large-scale wind power development and efficient utilisation, there is a need to step up the research on wind power consumption. The construction of wind power bases and external delivery channels must be considered together. A consumer market must be provided for and at the same time, an interregional UHV transmission network must be set up. This will transmit the wind power generated in the west and north to load centres in Northern China, Central China and Eastern China.

Jiuquan Wind Power Base

The Jiuquan region has abundant wind power resources, with the scale of technically feasible development at around 40 GW. They are concentrated in the areas of Guazhou, Yumen and Mazongshan. The installed capacity of Jiuquan wind power base is planned to reach 13 GW by 2015, 20 GW by 2020 and 32 GW by 2030. After Jiuquan's wind power is consumed through the northwest main grid, part of it needs to be transmitted to the load centres in central and eastern China.

Hami Wind Power Base

The Hami wind power base is located at the wind zone of Santanghu-Naomaohu and the wind zone of southeastern Hami in Xinjiang. Its scale of technically feasible development

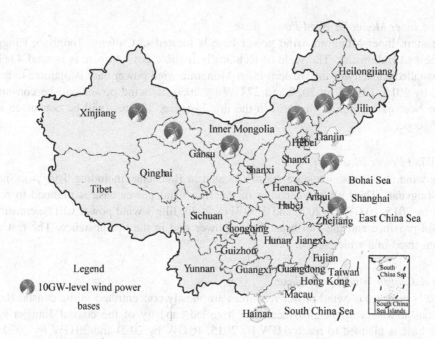

Figure 3.5 Distribution of 10GW-level wind power bases in China.

is around 65 GW. The installed capacity of the Hami wind power base is planned to reach 5 GW by 2015, 10 GW by 2020 and 20 GW by 2030. Apart from a small quantity of local consumption, most of the power needs to be transmitted to the load centres in central and eastern China.

Hebei Wind Power Base
Hebei's wind power resources are mainly distributed in Zhangjiakou, the Bashang area in Chengde and the coastal Qinhuangdao, Tangshan and Cangzhou. The installed capacity of the Hebei wind power base is planned to reach 11 GW by 2015, 16 GW by 2020 and 18 GW by 2030. Hebei's wind power will be consumed by the Beijing-Tianjin-Tangshan power grid and the southern Hebei power grid in the first instance. The rest will be consumed in a wider scope.

Western Inner Mongolia Wind Power Base
The western Inner Mongolia wind power base is mainly located in Ulanqab, Xilin Gol League, Bayannur, Baotou and Hohhot. The scale of technically feasible development is around 107 GW. The installed capacity of the western Inner Mongolia wind power base is planned to reach 13 GW by 2015, 27 GW by 2020 and 40 GW by 2030. Western Inner Mongolia's wind power will be consumed by the western Inner Mongolia power grid and northern China power grid in the first instance. The rest will be consumed in a wider scope.

Eastern Inner Mongolia Wind Power Base
The eastern Inner Mongolia wind power base is located in Chifeng, Tongliao, Hinggan League and Hulunbeier. The scale of technically feasible development is around 43 GW. The installed capacity of the western Inner Mongolia wind power base is planned to reach 7 GW by 2015, 12 GW by 2020 and 27 GW by 2030. Its wind power will be consumed by the Northeast China power grid in the first instance. The rest will be consumed in a wider scope.

Jilin Wind Power Base
Jilin's wind power resources are mainly located in Baicheng (including Tongyu), Siping and Songyuan. The installed capacity of the Jilin wind power base is planned to reach 6 GW by 2015, 10 GW by 2020 and 27 GW by 2030. Jilin's wind power will be consumed by Jilin province and the Northeast China power grid in the first instance. The rest will be consumed in a wider scope.

Coastal Jiangsu Wind Power Base
Jiangsu's reserves of wind power resources are mostly concentrated in the coastal shoals and coastal waters of the province. The installed capacity of the coastal Jiangsu wind power base is planned to reach 6 GW by 2015, 10 GW by 2020 and 20 GW by 2030. To support the peak shaving of the East China power grid, Jiangsu's wind power will be mostly consumed within the province. The rest will be consumed in a wider scope.

Coastal Shandong Wind Power Base
Shandong's wind power resources are mainly distributed in its eastern coast and some of its headlands, cape islands, islands, and wind-gaps in its mountain ridges. The installed capacity of the coastal Shandong wind power base is planned to reach 8 GW by 2015, 15 GW by 2020 and 25 GW by 2030. Shandong's wind power will be mostly consumed within the province. The rest will be consumed in a wider scope.

When the following factors are considered together—resource and development conditions at the wind power bases, interregional power grids construction plans, the research into transmission capacities—the scale of development of China's 10 GW-level wind power bases is planned to reach 120 GW in 2020 and 209 GW in 2030 (Table 3.4).

Resolving Problems Inherent in China's Large-scale Wind Power Development

Apart from hydropower, wind power is the renewable energy with the most mature technology and lowest development costs. Building large-scale wind power bases and bringing about the large-scale development of wind power in China bear a significance to improve China's energy structure and control its greenhouse gas emissions.

In recent years, China's wind power development has seen stellar achievements due to the hard work from all segments of society. The scale of wind power development has doubled annually over the last few years at a stretch. There has been rapid growth in the manufacturing capability of wind power equipment, and the various management systems are gradually being perfected. However, there are still great challenges. These problems must be assiduously dealt with to achieve China's wind power development goals.

Table 3.4 Scale of future development of China's main wind power bases. Unit: GW.

Wind power base	Installed capacity in 2020	Installed capacity in 2030
Jiuquan	20	32
Hami	10	20
Hebei	16	18
Western Inner Mongolia	27	40
Eastern Inner Mongolia	12	27
Jilin	10	27
Coastal Jiangsu	10	20
Coastal Shandong	15	25
Total	120	209

Unified Planning

There is a lack of unified planning in China's wind power development, with problems in development layout and construction schedules. The planning process for large-scale wind power bases tends to be disjointed and approvals are given in piecemeal fashion. This has caused difficulties in transmission and consumption of wind power. To achieve orderly development, China must strengthen the planning and control of wind power nationwide. The scale and schedule of development in each region and province must be determined through unified planning, so that the link-ups with national and regional planning can be effective. China also needs to set up a system to enable synchronised planning and production of wind power projects and their auxiliary grid construction, and the timely transmit the wind power. By incorporating a wind power consumption scheme, peak-shaving plans can be developed at the same time.

More Power Grid Construction

As an important infrastructure and medium of optimal resource allocation, power grids are irreplaceable in the process of wind power development in China.

Firstly, the characteristics of wind power resources determine that China has to build large bases and incorporate wind power into grids for its scale development. Most of China's wind power resources are located in the 'Three North' and coastal regions. The wind power generated by the eight major 10 GW-level wind power bases in construction cannot be consumed locally. Higher voltage power grids and large-scale transmission over long distances must be employed for the nationwide consumption of wind power. Also, a large power grid can make full use of the complementary natures of the wind power in different locations, and smooth over the volatility in wind power generation. This will reduce the impact on power grids and help maintain the safe and stable operations of the electricity system.

Secondly, raising the operational and management levels of power grids can significantly increase the scale of wind power development and consumption. Wind power has different generation characteristics from traditional power sources, and its large-scale integration into the grid will increase the operational difficulties of the power system. By raising the operational and management levels of power grids, there will be more accurate predictions of wind power generation, highly smart grid dispatching and more flexible power grid operations. Building a wind power friendly power grid can effectively enhance the ability to consume wind power. Currently, China's power grids still cannot adjust to the demand for large-scale integration and consumption of wind power. There is a need to intensify power grid construction to bring about the coordinated development of wind power and power grids.

Upgrading Wind Power Technology
Compared to countries with very developed wind power systems, China's wind power industry is still quite inadequate in terms of technology accumulation and R&D investment. The country still has much to catch up with in the core technology and the performance of key parts. The following must be done to solve the above-mentioned problems: improve the mechanism for long-term investment in wind power research and development, innovate in incentive mechanism and commercialising results, increase the support for independent R&D in wind power technology and equipment, intensify the key technology research in wind power equipment, and upgrade China's technology and self-sufficiency in wind power equipment.

Improving Relevant Standards and Policy Systems
Standard formulation in China's wind power industry obviously falls behind wind power development. This has resulted in extremely low thresholds for entry into the industry, low quality of development and slow improvements in technology. There is a need to raise the awareness of the importance to develop standards in the industry. In line with having a strong sense of responsibility towards the industry, standardisation work should be given top priority. Not only should the focus be on the manufacturing standards of wind turbines, but also on the development of wind farms and integration standards; not only on the design of individual wind turbines, but also on the overall design of wind farms and system operations. The standards should be developed in such a way as to facilitate the elimination of outmoded production capacity, the advancement of technology and the healthy development of the industry. There is a need to further improve the supporting policies for wind power development, reinforce the building of monitoring capacity, develop a time-of-use peak-valley tariff system on the grid-side and user-side, improve the wind power delivery pricing compensation system, and reform the tariff system for pumped storage power plants. These will promote the scientific development of China's wind power.

3.5.1.2 Building Large-scale Solar Power Bases

China's Solar Power Resources and Construction of Solar Power Bases
China has abundant solar power resources. It is estimated that the amount of solar radiation that falls on China's land surface every year is equivalent to 4.9 trillion tonnes of standard coal, which is roughly equivalent to the total electricity generated annually by 10 000

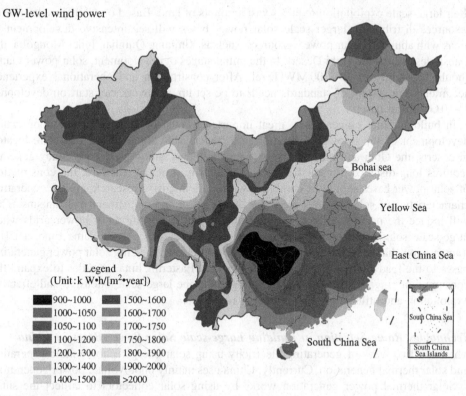

GW-level wind power

Bohai sea

Yellow Sea

East China Sea

Legend
(Unit : k W•h/[m²•year])

900~1000	1500~1600
1000~1050	1600~1700
1050~1100	1700~1750
1100~1200	1750~1800
1200~1300	1800~1900
1300~1400	1900~2000
1400~1500	>2000

South China Sea

South China Sea

South China
Sea Islands

Figure 3.6   Distribution of China's solar power resources.
Source: Wind and Solar Energy Resources Centre, China Meteorological Administration.

Three Gorges Dams. Around 70% of China's solar power resources are distributed mainly in the west and north like Tibet, Qinghai, central and southern Xinjiang, central and western Inner Mongolia, Gansu, Ningxia, western Sichuan, Shanxi and northern Shaanxi (Figure 3.6).

From the standpoint of development potential, China has 1 731 100 km² of desertified land, or 18.03% of the country's total land area, mainly distributed in the west and north, where there are abundant sunlight resources. Based on 5% of China's desertified land being used, the solar power installed capacity can reach 3460 GW, with an annual generating capacity of 4800 TWh. These areas have excellent land and solar power resource conditions for building large-scale solar power bases.

In recent years, China has quickened the intensive exploitation and utilisation of solar power. It has started construction on 10–100 MW-level solar power photovoltaic grid power generation bases in Gansu, Qinghai, and etc. Construction of a 10 MW-level solar thermal power generation pilot project has started. At the end of 2011, China's solar power grid installed capacity was 2.14 GW, or 8.2 times the level at the end of 2010. Solar power generation has progressed from infancy to large-scale development.

The construction of China's solar power bases should be carried out in orderly phases under the guidance of unified planning. Solar power resources have lower densities and their large-scale exploitation requires vast amounts of land. Based on China's solar power resources distribution, larger scale solar power bases will see intensive development in areas with abundant solar power resources, such as Xinjiang, Qinghai, Inner Mongolia, the Gansu deserts and the Gobi Desert. In the initial stages of development, solar power plants should be mainly of 10–100 MW level. After construction and operational experience accumulates, a system of standards needs to be set up, and work can start on developing 1–10 GW-level bases.

In building solar power bases, great importance must be attached to the coordinated development of solar power generation and power grids. Solar power bases are located in deserts, the Gobi and other sparsely populated areas, and the electricity they generate requires long-distance transmission in large capacity. Looking forward, the construction of solar power bases should be combined with more intensive research into the generating characteristics of solar power plants and their control and adjustment mechanisms. This will reduce the negative effects on the safe and stable operations of power grids when large-scale solar power plants are integrated with the grids. At the same time, a UHV transmission channel should be planned and constructed connecting solar power generating bases to the load centres in Northern, Central and Eastern China, in order to expand the consumption range of solar power and bring about the large-scale intensive and effective exploitation and utilisation of solar energy resources.

Technology Route Selection in Building Large-scale Solar Power Bases in China
There are two ways of generating electricity using solar energy: photovoltaic generation and solar thermal generation. Currently, China uses mainly photovoltaic power generation.

Solar thermal power generation works by using solar collectors to collect the sun's radiation. Hot steam or hot air is then produced to drive traditional steam generators or turbines to generate electricity. Solar thermal power generation offers more development potential for large-scale utilisation of solar power. Compared to photovoltaic generation, thermal generation has the advantage of better regulation during generation. Photovoltaic generation has the obvious disadvantages of being intermittent and uncertain. In the day-time, changes in the weather will cause large fluctuations in photovoltaic power generation. Besides, generation is impossible on rainy days or in the night time. Solar thermal power generation has more developed configurable technology and lower-cost high-volume thermal storage devices. The resulting stability and controllability of power generation means that its capabilities are close to those of conventional thermal power plants. Not only are additional allocations for peak shaving not required, it itself is a power source for peak shaving that provides ancillary services to wind power and photovoltaic power. In addition, solar thermal power possesses the inertia typical of generating units, which is positive for the stable operations of the power system.

Currently, as thermal solar power generation is experiencing the bottleneck in integrated optimal design and the manufacturing and maintenance of high-temperature components, there have only been several pilot projects. It has not entered the large-scale commercial development phase. From the economic standpoint, development plans and research results of organisations like the US Department of Energy and Solar Energy Industries Association show that solar thermal power generation has a bigger potential for cost reduction than photovoltaic power generation. Based on considerations of the characteristics of solar thermal power and operational features of the power system, China should attach more importance to the strategic position of solar thermal power generation as one of the technological choices in the development of China's large-scale solar power bases, increase the intensity of research, and accelerate the progress of industrialisation and commercialisation.

Analysing the features of both solar photovoltaic power generation and solar thermal power generation, the development of China's large-scale solar power bases can be focused on photovoltaic generation in the short term, but in the medium and long term, China should adopt both methods of power generation. In areas where conditions are suitable, large-scale solar thermal power plants should be developed. The next step is to increase investment in the research and development of solar thermal power generation technology and facilities, at the same time as making progress in developing solar photovoltaic power generation. Pilot power plants should be built and the solar thermal power generation market should be developed as early as possible. The progress made in technology and facilities will result in a positive effect of solar thermal power generation on the adjustment of the energy structure in China and in the country's energy conservation and emission reduction efforts.

3.5.2 Various Forms of Renewable Energy Development

3.5.2.1 Developing and Building Small and Medium-sized Wind Farms

On the basis of accelerating the building of 10 GW-level wind power bases, China should actively pursue the dispersion of its onshore wind power resources exploitation and use. In inland areas outside the 'Three North' region, the wind power resource evaluation and development work should be stepped up. The eastern and central area, which has relatively abundant wind power resources, should be encouraged to take advantage of its proximity to the load centres and good access to the power grids to build small and medium-sized wind farms of 10–50 MW-level that are suited to local conditions. This will support the distributed access of wind turbines to the distribution network, and on-site power consumption. The distributed exploitation and application of wind power should be actively encouraged to resolve the issue of power use in remote regions, mountainous areas and on islands. This will create a situation in which centralised and distributed developments are being carried out at the same time.

3.5.2.2 Promoting Extensive Application of Solar Energy

Apart from solar photovoltaic power generation and solar thermal power generation, heat generation is also an important utility of solar energy. The most widely applied technology in the use of solar heat is the solar-powered water heater. China is the biggest user of solar generated heat in the world. At the end of 2009, the total heat collecting surface area in operation in China's solar powered water heaters is 145 million m^2, with annual production reaching 42 million m^2. The country's utilisation and annual production tops the world, accounting for more than half of the world's total. The focus of the future development of solar generated heat is to expand its technological application and its market. By incorporating solar heat collecting systems into building design, solar energy can be used in multiple ways in buildings to supply hot water and to run heaters and air-conditioners.

An effective way of using solar energy in urban areas is to connect photovoltaic equipment directly to the end-users and using the light falling on building surfaces to generate electricity. It is estimated that the total area of rooftops in China is around 40 billion m^2. If 1% is installed with photovoltaic equipment, the installed PV power capacity will be 35.5–66.2 GW, and the annual power generation will be 28.7–54.3 TWh.

In addition, China can also take advantage of the complementary nature of wind and solar power in terms of geography and their respective characteristics, and build up a complementary wind and solar power generation system. In China, areas with abundant wind power resources also enjoy rich solar power resources. There is a certain level of complementariness between wind power and solar power generation. Combining the two has the advantage of smoothing over the output curve of wind power and solar power generation facilities. It will lessen the impact on the operations of power grids, reduce the need for peak-shaving power and increase the utilisation rate of land resources.

3.5.2.3 Exploiting Biomass

China has abundant biomass resources, giving it the material basis for large-scale exploitation and utilisation. It is estimated that with the continued socioeconomic development and technological progress, China's usable waste biomass resources will reach around 500 million tonnes of standard coal per year by 2030. Biomass can be used to generate electricity, or it can be changed into liquid, gaseous or solid fuel to be reused.

Biomass power generation has good development prospects in the country. China is a major agricultural nation, with an abundance of biomass resources like straw and forestry waste. For a long time, villagers would burn large amounts of leftover straw that cannot be otherwise processed. Not only is this a waste of resources, it causes severe smogs and fire hazards. Building biomass power plants in villages can make good use of the unprocessed straw and thus protect the environment; it can also vastly improve the way villagers use energy and improve living conditions in rural areas. At the same time, it also increases employment opportunities in rural areas and increase the income of villagers. It has important significance in serving Agriculture, Countryside and Farmers, as well as the development of new villages. In the villages in the north, the construction of a 25 MW biomass power plant can consume all the usable straw within a radius of 50 km, and supply some 400 000 rural households with a year's supply of electricity. Apart from utilising straw, tree branches and forestry waste in the rural areas, centralised waste processing in urban areas together with the construction of wastepower plants is also an important form of biomass power generation. The building of waste power plants can reduce the amount of land used as landfills and promote the use of urban waste as resources. It has great potential for development. It is estimated that by 2020, China's biomass power generation installed capacity will reach 15 GW and 25 GW by 2030.

Biomass fuel as an oil substitute is widely used in Brazil and the USA. Many areas in China are also promoting the use of ethanol gasoline. There is some potential in developing biomass fuel but China's large population and limited land means the potential will not be huge. Producing biogas from forestry waste recycles resources, and helps protect the environment and improve villagers' lives. It has much development potential in the extensive rural areas in China. Solidifying biomass into solid fuel facilitates storage and transport, and increases its burning efficiency. Replacing coal in the cities to provide heat and fuel power plants is one of the ways to utilise biomass.

All in all, as an important renewable energy, the active exploitation and utilisation of biomass can upgrade China's energy structure, increase energy security and safety, enhance coordinated development of rural and urban areas, and better protect the environment. At the same time, we must bear in mind that China has a huge population with a small amount of arable land per capita. The exploitation and utilisation of biomass should be undertaken according to the principle of 'not competing with humans for food and not competing with food for land'. Ensuring food security must be the prerequisite and bottom-line of biomass development. In addition, the development process must be accompanied by sound environmental efforts to prevent new problems like soil erosion and ecological deterioration.

3.5.2.4 Intensifying Development of Geothermal Energy

Rich thermal resources are found beneath the earth's surface. According to estimates, the total amount of thermal energy in the earth's core is 170 million times that of the world's total coal deposits. The amount of geothermal energy lost through the earth's crust every year is equivalent to the heat generated by burning 100 billion barrels of oil.

Mankind's use of geothermal energy has had a long history. Current utilisation, however, is concentrated on the geothermal energy close to the earth's surface, whose development focus is on combined cooling, heat and power (CCHP) and providing heat for cultivation. Huge amount of deep geothermal energy is found 1000 m below the earth's surface and beyond. The main method of their exploitation and utilisation is to make use of enhanced geothermal systems to convert them into electricity.

China has abundant geothermal resources and the potential for their development is huge. In terms of shallow-level geothermal energy, the area of warming (cooling) at the end of 2010 was 140 million m^2, with 35 million m^2 of warming by geothermal heat. The total installed power capacity for high-temperature geothermal energy was 24 MW. The total amount of various types of geothermal energy was equivalent to 5 million tonnes of standard coal. In terms of deep-level geothermal energy, China has good conditions for resource development, especially in the southwest and the southeast coast, areas where the earth's crust temperature is abnormally high. Specialised research on deep-level geothermal energy exploitation should be organised and conducted in the future to focus on resolving technological issues of resource assessment and site identification. This will lay the foundation for deep-level geothermal energy exploitation and utilisation.

3.5.3 Distributed Energy Development

3.5.3.1 Concept and Categories of Distributed Energy

Distributed energy has yet to have a definition that is agreed upon internationally. Organisations like the US Energy Department and the World Alliance for Decentralised Energy (WADE) have conducted research on distributed energy. Based on the actual circumstances both within and outside China, distributed energy refers to renewables-based integrated generating facilities or energy cascade multigeneration systems with power output that

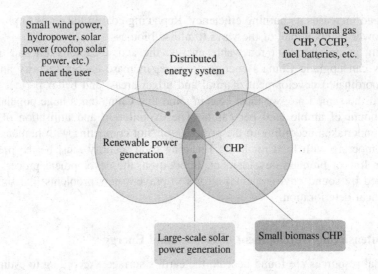

Figure 3.7 Relationship between distributed energy system, renewable power generation and CHP.

are independent of or connected to LV distribution networks for priority supply of energy to user locations.

Based on the type of fuel utilised, the distributed energy systems can be grouped into the following categories: distributed energy system that uses traditional primary energy such as natural gas or diesel as fuel; distributed energy system that uses renewables such as wind power and solar power as primary energy; distributed energy system that uses secondary energy such as hydrogen as fuel; and energy storage systems including batteries and so on.

Based on the products provided, the distributed energy systems can be grouped into the following categories: decentralised power generating system that supplies only electricity; combined heat and power (CHP) cogenerating system; and combined cooling, heating and power (CCHP) tri-generating system.

There are differences between distributed energy systems and concepts like renewables-based power generation and CHP. Large-scale renewable energy projects (e.g. wind power plants and photovoltaic power bases) and large-scale on-grid CHP units are not distributed energy systems (Figure 3.7).

3.5.3.2 Development Orientation of Distributed Energy

The distribution of China's energy resources and productivity determines that energy resources must be better distributed in the whole country. A centralised energy supply system is the main method of supplying energy in China. The development of distributed energy cannot replace the traditional method of energy supply, but is positioned as a positive complement to centralised energy supply. China should build up an energy supply system with centralised energy supply as the core, supported by distributed energy supply. The main function of distributed energy in China's energy supply framework is to enhance energy utilisation efficiency for energy conservation and emission reduction, supply electricity and energy to remote villages and offshore islands, and provide emergency reserves and support in the event of major accidents and breakdowns occurring in the major power grids.

Based on energy resource conditions and the current situation of energy utilisation, the development focus of energy in China is natural gas-based CCHP tri-generating systems and distributed renewable energy systems, which have the benefits of energy conservation and emission reduction. Natural gas CCHP tri-generation and decentralised renewable energy systems (like photovoltaic power in buildings) are developed mainly in large and medium-sized cities with high load concentrations. This will increase the utilisation efficiency of energy and the reliability of energy supply. Other decentralised renewable energy systems like small hydropower, decentralised wind power and photovoltaic solar power generation, and biomass power generation should be developed in rural and remote areas where the loads are low. This will improve the energy supply in rural areas and solve the problem of access to energy in remote areas.

3.5.3.3 Towards Scientific Development of Decentralised Energy in China

Driven by policies pertaining to small hydropower, combined utilisation of resources and renewable energy utilisation, the overall development momentum of distributed energy in China is positive. As at 2009, the installed capacity of China's distributed energy power generation was over 45 GW, the highest in the world. This accounted for nearly 5% of the country's total installed capacity. Currently, China's distributed energy faces acute problems in the areas of independent technology and equipment manufacture, operational efficiency, power generation and grid integration. A targeted development strategy must be formulated to guide the scientific development of China's distributed energy.

Overall Planning and Development is Necessary

Distributed energy is situated directly on the user-side, and is closely linked to urban development planning, power source and grid planning, heat energy planning and natural gas planning. The development and planning of distributed energy need to coordinate and integrate effectively with the abovementioned special plans. This is especially the case for natural gas distributed energy, where ample considerations must be made regarding factors like natural gas supply capacity and network allocations before making plans for construction. Care must be taken to coordinate the development of distributed energy with the development of power distribution networks and electricity sources. The effects of decentralised energy on electricity demand, reserve margins, and the planning and operation of power distribution networks must be carefully considered to avoid duplication. The decentralised development of renewable energy and natural gas should be promoted through pilot projects to gain experience and explore solutions. On the basis of successful pilot projects, application should be promoted with reference to the maturity levels of various decentralised energy technologies.

Standardised and Orderly Integration with the Power Grid is Necessary

To enhance operational reliability and efficiency, decentralised energy should be integrated into the public power grid where possible and when conditions allow. The integration of distributed energy with the power grid will have a profound effect on the planning and design, quality of power, and dispatch operations of the electricity system. Therefore, distributed energy systems must undergo a process of standardisation as required, regarding its technical standards and management regulations. Integration with the power grid must be undertaken through a strict procedure to prevent any negative impact on the safety and stability of the public power grid.

More R&D is Necessary

China lags far behind the developed countries in terms of distributed energy core technology. Key facilities such as gas turbines are basically imported. There is insufficient independence in manufacturing capabilities, and not enough experience in system design and improvement. These have resulted in high investments and operational costs of distributed energy to adversely affect economic benefits. There is an urgent need to accelerate technological R&D and quicken the adaptation so that China can manufacture more of these facilities on its own. At the same time, technical standards suited to China's conditions need to be laid down to give impetus to the technological progress and industry upgrading of distributed energy in the country, so that an industry system with independent manufacturing and innovation capabilities can take shape as quickly as possible.

3.5.4   Exploitation and Utilisation of New Energy

Types of energy with good potential for future exploitation and utilisation include primary energies like methane ice, ocean energy, nuclear fusion energy and space solar power, as well as secondary energies like hydrogen energy. China should intensify its follow-up study on exploiting and utilising new energies to cope with energy shortage, and to acquire technical and resource reserves for energy upgrading and replacement.

3.5.4.1   Methane Ice

Methane ice, also known as natural gas hydrate, has the advantages of abundant reserves, high energy density, and low pollution and emission levels when burnt and utilised. It is usually found on the slopes of marine continental shelves, deep seas, deep lakes and permafrost areas. The main component of methane ice is methane (over 90%) and $1\,m^3$ of methane ice can produce $164\,m^3$ of methane gas. Its energy density is ten times that of coal. According to estimates, the total amount of methane in the world's methane ice reserves is around 2.0×10^4 trillion m^3. The carbon content is twice that of the proven reserves of fossil fuels.

China has abundant methane ice resources and the prospects for exploitation and utilisation are good. Since 1991, China has surveyed and studied methane ice resources, and they were discovered in the South China Sea, East China Sea and the Qinghai-Tibet Plateau permafrost. The volume of methane ice reserves in the northern part of the South China Sea alone is equivalent to half of China's onshore oil reserves. Prospective reserves for onshore methane ice are equivalent to over 35 billion tonnes of oil. However, current technology for methane ice exploitation is still immature and there is still a long way to go before it can be used commercially.

Looking forward, China should intensify methane ice exploration and step up the research on the technology for its exploitation. It should include methane ice exploitation and utilisation in the country's energy development strategy, and increase the capital and manpower input to quickly master the technology for mining methane ice. Given the greater technical difficulty and complexity of seabed methane ice mining compared to onshore exploitation, the exploration and use should begin on land. After the gradual accumulation of experience, methane ice mining can be extended to offshore and deep sea mining.

3.5.4.2 Ocean Energy

Ocean energy refers to the energy produced by the various physical and chemical processes in the ocean. Its main sources are solar radiation and the gravitational changes between astronomical bodies like planets. Ocean energy can be divided into tidal energy, wave energy, current energy and temperature difference energy. It has the advantages of renewability, abundance of resources and low negative impact on the environment. However, its disadvantages include instability, low energy density, hostile operational environment, and low economic benefits of exploitation and utilisation. Ocean energy is mainly used in electricity generation. Countries with rich ocean resources have built test facilities for power generation using wave energy, tidal energy and temperature difference energy. The installed capacity is usually less than 500 kW. With the increasing importance that various countries are placing on ocean energy resources, the exploitation technology will continue to mature. Ocean energy has the potential to become an important component of mankind's energy supply system in the future.

China's long coastline contains abundant ocean energy resources, mainly distributed in the eastern and southern coasts. The amount of energy that can be exploited and used is around 1000 GW. Currently, China has several tidal energy power stations built in Zhejiang, Shandong, Fujian and Guangxi. The country has also made some progress in the study of wave energy and current energy technologies, and pilot projects have been initiated. The technology for temperature difference energy is still at the laboratory and theoretical stage.

Looking forward, China should step up its R&D and import of various technologies to exploit ocean energy. It should increase investment in innovation and technical upgrades. China should set up an effective management system to exploit ocean energy, initiate the surveying and protection of resources, and formulate medium- and long-term strategic plans for development, which should be coordinated with the construction plans of ports and terminals, and the development of offshore islands. More should be done in the development of pilot projects in ocean energy electricity generation so as to increase competitiveness in this area.

3.5.4.3 Nuclear Fusion Energy

Nuclear fusion energy is another form of nuclear energy apart from nuclear fission. In fusion reactions, two light atomic nuclei (deuterium and tritium) fuse together to form a heavier atomic nucleus (helium), releasing energy in the process. Nuclear fusion energy is a clean and safe form of nuclear energy. Compared to current nuclear fission reactions, it produces almost no radioactive substances and nuclear waste, which makes it much safer. The raw material needed for nuclear fusion can also be found in huge quantities. The amount of deuterium in 1 litre of seawater produces, by nuclear fusion, an amount of energy equivalent to the heat energy contained in 300 litres of oil. If all the 40 trillion tonnes of deuterium in seawater were to be used in nuclear fusion, it would be enough to meet mankind's energy needs for billions of years to come.

Mankind has already achieved uncontrolled nuclear fusion (the hydrogen bomb), but to make use of fusion energy effectively it is necessary to control the speed and scale of nuclear fusion reactions, so that the energy outflow can be continuous and stable. Studies have been conducted on controlled nuclear fusion since the 1950s. The USA, Europe, Japan and China have built experimental devices to achieve controlled nuclear fusion through the magnetic confinement method. Given the great technical difficulties and capital investments in controlled nuclear fusion research, various countries are moving towards joint studies. In May 2006, a joint agreement was signed by representatives from the USA, Japan, the EU, China, Russia, South Korea and India to jointly develop the International Thermonuclear Experimental Reactor (ITER) for testing and verifying the feasibility of electricity generation by nuclear fusion. Within international nuclear fusion circles, it is agreed that the commercialisation of nuclear fusion power generation is still 50 years away.

Nuclear fusion energy is a strategic energy resource. China should make an early start on the long-term strategic arrangements for the development and application of nuclear fusion. It should participate fully in the ITER project and bear the research responsibilities, and accumulate experience in design and manufacture, to keep up with advanced international developments. At the same time, the country should make full use of its experimental facilities to strengthen basic technology research and build up a team of experts, so as to master the key technologies in the design, manufacture and construction of nuclear fusion reactors.

3.5.4.4 Space Solar Energy

Solar energy is the primeval source of most of the energies exploited and utilised by mankind. Radiating across the universe in the form of electromagnetic waves, the amount of solar energy that reaches the earth is only 1/2.2 billion of total solar radiation. If the solar energy in space can be fully utilised, then mankind will be in possession of an inexhaustible source of energy.

Compared to the utilisation of solar energy on earth, space solar energy has many advantages. It can generate electricity for 24 hours nonstop, and there is more solar energy to be utilised. Power generation facilities can be much bigger in a zero-gravity environment, and they can be cleaner and require less maintenance. Energy can be transmitted to different terrestrial receivers to meet different regions' peak energy demands better. However, the utilisation of space solar energy faces many challenges. Building a space solar energy power station is a gargantuan undertaking. Tens of thousands of tonnes of equipment must be launched into space, which means space delivery technology of a larger scale must be developed. Solar panels must be made from special materials. There will be higher demands on the reliability of space solar power station equipment and in-orbit maintenance technology. Measures must also be taken to prevent damage from other flying objects and space junk.

Despite the many challenges, the continuous acceleration in the pace of mankind's exploration of outer space means that these problems are not insurmountable. Space solar energy power stations will open up a new path in mankind's fuller use of solar energy. China's aeronautical technology is among the most advanced in the world, and so it should make an early start in the utilisation of space solar energy. By proposing a development roadmap, conducting research on key technologies, and using the development and building of space solar energy power station as a starting point, China can achieve important technological innovations and breakthroughs in areas like new energy and new materials.

3.5.4.5 Hydrogen Energy

Hydrogen is a clean energy medium, which can be transformed into usable electricity through fuel batteries or be used as fuel for transport and everyday life. Its advantages are abundance, high calorific value and zero-pollution. Hydrogen can be produced by fossil fuels, but this consumes energy and emits greenhouse gases, making it unsuitable for large-scale application. Hydrogen can also be produced by water decomposition. The energy needed can come from clean energies like wind power, solar power and nuclear power, causing almost no environmental pollution. This is the future direction of hydrogen production. In addition, biomass production will also be an important method of hydrogen production in the future.

Developed countries in North America and Europe attach great importance to the exploitation and utilisation of hydrogen energy. The USA is an advocate of the hydrogen energy economy, planning to achieve the total transformation into a hydrogen energy society by 2040. Europe has intensified its research and development into fuel batteries and hydrogen-fuelled vehicles, and plans to make the transition into a hydrogen energy economy by 2050. The supply, transport and utilisation of hydrogen energy have become important components of Japan's 'New Sunshine Project'. China has also begun its hydrogen energy research, launching various studies on exploitation and utilisation technology, as well as making test models of hydrogen-fuelled vehicles.

Faced with problems like the immaturity of hydrogen production and storage technologies, high costs and commercial nonviability of fuel batteries, China should develop a technology roadmap for hydrogen energy development as soon as possible, to guide the direction of hydrogen energy research, development and utilisation. Looking ahead, the research into hydrogen production technology through nonfossil fuels should be stepped up; existing hydrogen production technology should be improved and perfected; and new hydrogen production technology should be sought. China should take advantage of its resource advantages and develop high performance rare earth hydrogen storage materials. On the basis of test models, the overall planning and development of infrastructure such as hydrogen-fuelled vehicles, hydrogen refuelling stations and hydrogen pipeline networks should be gradually promoted.

3.6 The Exploitation and Utilisation of Oil and Gas

The total volume of China's conventional oil resources is quite high, but their richness and quality are comparatively low. Looking ahead, China should step up its exploration efforts and increase its reserves. The country also has some nonconventional oil resources, which are of low grades and at the initial stage of exploration. It is estimated that by 2030, or even by 2050, the amount can only be supplementary to conventional oil supplies. Most of China's key oil fields that have been developed are already entering a period of declining production, so there is little possibility of high growth in reserves. From the perspective of long-term oil security, the peak production of China's crude oil in the future should be controlled in a reasonable way.

China's natural gas exploration is in its early stage, and there is much potential for growth in reserves. However, the exploration rate of gas fields in the nine major basins is still quite low at the moment. Exploration and development should be intensified. Nonconventional natural gas reserves are in abundance, and they are a good supplement to conventional natural gas. However, the technology to exploit nonconventional natural gas is still immature. Increased state investment and innovation is required to raise the levels of exploitation and utilisation. Meeting the domestic needs of the people will be the first priority. After that, areas with secure gas sources can develop gas-fired peak-shaving power to a suitable degree.

3.6.1   Exploration and Development of Oil Resources

3.6.1.1   Intensifying the Exploration of Conventional Oil Resources

A dynamic evaluation of nationwide oil and gas resources reveals that as at 2009, China had 88.1 billion tonnes in geological oil resources, with cumulative proven reserves of 29.95 billion tonnes. The percentage shares of oil reserves in the northeast, the eastern coast, the central-western region and offshore waters are 32.6%, 27.0%, 30.4% and 10.0% respectively (Table 3.5).

China's oil resources do not have a high discovery degree. According to the figures issued by the Ministry of Land and Resources, the discovery degree of China's oil resources is only 34%. Exploration is at the middle stage and there is much potential for resource discovery.

Oil resources to be discovered in China are mainly found in the western region and offshore waters. Looking ahead, there is a need to step up the efforts and investments in oil exploration and development to find high quality oil resources, both on the land and in

Table 3.5   Regional distribution of China's proven oil reserves. Unit: billion tonnes.

Region	Cumulative proven reserves			Remaining technically recoverable reserves	Remaining economically recoverable reserves
	Total	Extracted	Not extracted		
Northeast	9.77	7.54	2.23	0.88	0.61
Eastern Coast	8.09	6.17	1.92	0.64	0.46
Central-West	9.09	5.57	3.52	0.97	0.69
Offshore	3.00	1.59	1.41	0.39	0.35
Nationwide	29.95	20.87	9.08	2.88	2.11

Source: The data in the table are based on figures from the Ministry of Land and Resources.

the sea, to ensure a stable supply of oil for China. On land, oil exploration will focus on the four major basins: the Songliao, Ordos, Tarim and Junggar Basins. Offshore, it will be undertaken in Bohai Bay and the South China Sea. Deep sea oil exploration technology will be developed to enable China to manufacture deep sea exploration equipment on its own. The construction of deep sea oil fields will be accelerated to assert China's dominance in oil exploration in the South China Sea.

3.6.1.2 Intensifying the Exploration and Development of Nonconventional Oil Resources

Nonconventional oil resources refer to oil resources that, with current economic and technological capabilities, cannot be extracted, developed or processed for utilisation using conventional methods and technologies. These mainly refer to shale oil and oil sands.

According to the latest nationwide evaluation of oil resources, China has 53.61 billion tonnes of recoverable nonconventional oil resources, of which 47.64 billion tonnes are geological resources of shale oil and 5.97 billion tonnes are geological resources of oil sands. Based on total volume, China has considerable potential in non conventional oil resources. Compared to global nonconventional oil resources that are currently economically recoverable, however, China's shale and oil sands contain lower amounts of oil. Besides, the layers are shallow, individual sites are small and they are found buried deep beneath the earth, making extraction very difficult.

Based on an analysis of China's resources situation and the future trends in technological development, it is estimated that the production of shale oil and oil from oil sands can reach 8.5 million tonnes by 2030, and 25 million tonnes by 2050. Overall, the supply capacity of China's nonconventional oil resources may be limited, merely able to supplement conventional oil supplies before 2050, but they have important implications for providing some relief to China's tight oil supply and reducing the country's dependence on imported oil.

To give more impetus to the exploitation and utilisation of its nonconventional oil resources, China should first intensify the exploration and evaluation of shale and oil sands resources. The real situation of China's nonconventional oil resources must be ascertained in order to set up as quickly as possible standards for the evaluation and calculation of the country's shale and oil sands reserves. Secondly, technological research and extraction experiments should be stepped up. There is a need to focus on developing shale mining, as well as intensifying research on new transport technology, oil sands oil separation and oil synthesis technology, and oil sands retorting and modification technology. Finally, support policies must be implemented with regards to the exploitation and utilisation of nonconventional oil resources. Extraction of shale oil and oil sands oil in China is difficult and production costs are high. The state therefore needs to promote exploitation and utilisation of nonconventional oil resources by initiating policy incentives such as tax reduction and subsidies to make mining slightly profitable.

3.6.1.3 Rational Control of Oil Production to Maintain Supply Stability

China's oil resources are distributed in varied ways with reserves maintaining steady growth over a long cycle period. Currently, China's oil resource mining is at the middle stage and oil reserves are at a stage of steady growth that looks set to continue for some time, possibly over 30 years. However, most of China's current key oil fields have entered high water cut and high recovery levels, with ever-diminishing production. If there is no input of considerable reserves, the stability and growth of the country's oil production will face serious challenges.

Based on an analysis of future trends in China's oil reserves, the growth of the country's proven recoverable oil reserves can maintain stable growth on a high base in the long term. Additional proven recoverable oil reserves will be kept at around 200 million tonnes before 2035, followed by subsequent slowdown in growth. By around 2050, China's additional proven recoverable oil reserves can be kept at around 150 million tonnes.

An analysis of the characteristics of China's oil and gas resource distribution, the growth trend of crude oil reserves, and the production situation of mature oil fields from a sustainability point of view, shows that it is in the interest of China to maintain the stability of crude oil production in the long term. From an overall perspective, China will do well to keep its crude oil production stable at the current level of 180–200 million tonnes or slightly higher.

3.6.2   Exploitation and Utilisation of Natural Gas Resources

3.6.2.1   Speeding up the Exploration and Development of Conventional Natural Gas Resources

China has abundant natural gas resources, but current exploration levels are quite low and more exploration and development work ought to be done in the future. Dynamic evaluation of nationwide oil and gas resources reveals that as at 2009, China had 52 trillion m^3 in geological natural gas resources, with cumulative proven reserves of 7.04 trillion m^3. The percentage shares of natural gas reserves in the northeast, the eastern coast, the central-western region and offshore waters are 6.5%, 2.3%, 83.8% and 7.4% respectively (Table 3.6). The discovery degree is at 18%, obviously lower than that of countries with highly developed natural gas industries. It is roughly equivalent to the early stage of rapid growth in natural gas reserves in the USA. China's natural gas reserves have tremendous growth potential.

Since 2000, China's natural gas reserves have entered a stage of rapid growth, with new proven recoverable reserves maintained at over 260 billion m^3 for 10 years in a row, reaching an average of 300.6 billion m^3. Looking ahead, China's natural gas production is likely to continue to grow rapidly, entering a production peak at around 2030. By then, annual gas production is likely to reach 250 billion m^3, which will be sustained until after 2050.

3.6.2.2   Speeding up the Exploration and Development of Nonconventional Natural Gas Resources

Nonconventional natural gas resources refer to gas resources that, with current economic and technological capabilities, cannot be extracted, developed and utilised using conventional methods and technologies. These mainly include coal bed methane, shale gas, tight gas, methane ice, etc. China has a huge abundance of nonconventional natural gas

Table 3.6 Regional distribution of China's proven natural gas reserves. Unit: trillion m³.

Region	Cumulative proven reserves			Remaining technically recoverable reserves	Remaining economically recoverable reserves
	Total	Extracted	Not extracted		
Northeast	0.46	0.09	0.37	0.19	0.12
Eastern Coast	0.16	0.10	0.06	0.05	0.01
Central-West	5.90	2.06	3.84	3.00	2.40
Offshore	0.52	0.18	0.35	0.28	0.23
Nationwide	7.04	2.43	4.62	3.52	2.76

Source: The data in the table are compiled based on oil and gas reserves figures from the Ministry of Land and Resources.

resources like coal bed gas and shale gas, and prospects for exploration and development are excellent. The recoverable volume of shallow coal bed gas buried less than 2000 m in the earth is 36.81 trillion m³, while the potential recoverable volume of onshore shale gas is 25.08 trillion m³. If the development is rationally directed, nonconventional natural gas can become an important component of China's natural gas supply, which can go about alleviating the country's relative insufficiency in high quality energy resources. For this reason, China should actively embark on the exploration of nonconventional natural gas resources. Determine as quickly as possible the real potential and distribution of nonconventional natural gas resources, and designate coal bed methane, shale gas, etc. as mineral resources so that they can be placed under 1-grade national management. Given that nonconventional natural gas resources require large investments for development, and given their low production levels per gas field, as well as their lack of immediate results and high profits, suitable incentives and support policies can be given to nonconventional natural gas producers.

Coal Bed Gas
Coal bed gas is the most practical gas source among nonconventional natural gases in China in recent times. It has huge potential for development. After more than a decade of actual exploration and theoretical study, coal bed gas has entered the stage of large-scale development. However, several problems remain: the overlapping of coal bed gas mining rights and coal mining rights, lack of delivery facilities and utilisation equipment, and the lack of management standards for the industry. All these problems require the state to issue relevant industry policy for a solution. If the policies to promote the development of coal bed gas are put in place and work well, it is estimated that by 2020 the production volume of coal bed gas will reach 50 billion m³, becoming an important component of China's natural gas supply.

Shale Gas
Globally, the technology for shale gas extraction is already quite mature. The USA has already entered a stage of rapid growth, and its shale gas production is already close to one-fifth of the country's total natural gas production. Through intensive mining of

shale gas, the USA has achieved sustained growth in natural gas production. In 2009, it became the world's biggest producer of natural gas, and its demand for imported gas was significantly reduced. Compared to countries like the USA, China's shale gas mining has yet to reach a breakthrough in key technologies. Although it has already achieved strong technical and production capabilities in manufacturing equipment like drills, it still has a long way to go in terms of technology for system sets and individual units. Most of China's shale gas is buried deep underground and extracting them poses great difficulty. To promote China's shale gas development, it should refer to the models of countries such as the USA. A mature mining area is selected and through technological cooperation with foreign companies, a pilot test is developed. By building up a development framework for shale gas technology, shale gas exploration matching technologies that suit China's shale gas mines can be mastered as quickly as possible. At the same time, support should be provided in the areas of resource surveying, mining rights management, and policies such as financing and taxation, to quicken the pace of China's shale gas development. The aim is for shale gas production to reach 60–100 billion m^3 by 2020.

3.6.2.3 Moderate Development of Gas-fired Peak-shaving Power

Natural gas is mainly used in urban gas supplies and electricity generation, and as a raw material in the chemical industry and industrial fuel. With greater urbanisation and the continuous growth of gas pipeline networks, future demands for urban gas supplies, such as gas for domestic use, will continue to grow rapidly. Compared to coal, natural gas has greater environmental benefits, but gas resources are comparatively scarce with weak supply security and uncertain gas prices. Therefore, China's natural gas development should meet the people's needs in the first instance, and natural gas should be mainly used for everyday use, industrial fuel and as a raw material in the chemical industry. With this as the basis, natural gas for power generation should be developed to an appropriate extent.

In recent years, China has seen rapid growth in the demand for peak-shaving power. With the quickening pace of the structural adjustment of China's economy and the continued increase in everyday electricity use by urban and rural residents (manifested in the increased use of air-conditioners in summer), the average daily minimum loads of power grids in load centres like northern China and eastern China are showing a declining trend, and the peak-valley difference of power grids is widening every year. The peak-shaving pressure of the systems is also rising annually, resulting in an urgent need to increase peak-shaving power. The large-scale exploitation and utilisation of intermittent energy sources like wind power and solar power has further increased the need for peak-shaving power. Given its flexible regulatory and excellent peak-shaving capabilities, gas-fired power generation can adapt to the peak-shaving needs of the power system. It should be developed at a moderate pace in the future.

Considering factors such as resource conditions and power generation costs, China's gas-fired power generation should be focused on peak-shaving power. It is estimated that the installed capacity of China's natural gas-fired power generation will reach around 70 GW by 2020, 100 GW by 2030, and 170 GW by 2050.

3.7   The Exploitation and Utilisation of Overseas Energy Resources

The current state of the world is both advantageous and disadvantageous to China in its efforts to expand the exploitation of overseas energy. China benefits from the slowdown in global economic growth brought on by the international financial crisis. Industries in some countries are facing acute problems with a sharp drop in their market values, and to ease their cash flow they have no choice but to sell part of their quality assets in the market. Some of these include energy assets. In order to attract foreign investment to bring about an economic recovery, some countries have lowered their investment thresholds and opened up more investment areas previously closed to foreigners. Certain energy infrastructure industries, where foreign capital could not even get a foot in previously, are now opened to foreign enterprises. This has provided an exceptional opportunity for Chinese energy industries to acquire overseas assets and participate in international energy cooperation. At the same time, the Chinese economy continues to maintain its stable growth, which, together with the country's ample foreign reserves, will provide a strong backing for Chinese enterprises that venture out to acquire quality assets overseas. There are, however, factors that are not to China's advantage. China is a latecomer in the international energy market, where most of the quality energy assets are already controlled by other countries. There is a higher degree of difficulty and risk for China to acquire overseas energy resources. The ever increasing competitive pressures from emerging nations like India and traditional energy importing countries like Japan and South Korea have inadvertently pushed up the price China has to pay to acquire foreign energy resources. Overall speaking, however, the advantages to China override the disadvantages. The immediate future will be a period of important strategic opportunity for China to expand the scale of its exploitation of foreign energy resources. China's energy enterprises should fully grasp this great opportunity to actively utilise overseas energy resources and increase the level of energy security.

3.7.1   Development and Import of Overseas Oil and Gas Resources

Since the 1990s, China has actively taken part in international energy cooperation and seen some marked improvements in the area of oil and gas resources cooperation. Through its state-owned oil enterprises, China has launched oil and gas joint projects in many countries around the world, including countries in Africa, the Middle East, Central Asia, the Americas and the Asia-Pacific, as well as Russia. A global framework in international oil and gas cooperation is taking shape.

Following the two oil crises, the major oil consuming countries became very aware of the risks involved in over relying on the resources from one particular region, and actively implemented diversified oil acquisition strategies. For now and a long period of time afterwards, China will see growing oil consumption. Given limited domestic resources and an ever widening gap between supply and demand, a diversified foreign investment and acquisition strategy for oil and gas resources to expand the sources of import is a strategic choice for raising China's oil and gas supply security.

3.7.1.1 Diversifying Sources of Oil and Gas Imports

The world's oil and gas resources are concentrated in a circular region stretching from North Africa to the Persian Gulf, the Caspian Sea and Russia. The countries and regions in this vast land have become the supply centre of the world's oil and gas resources. According to a 2010 survey on the global source of crude oil exports, the Middle East accounted for 44.2% of the world's total oil exports, with Africa accounting for 18.7%, and Russia and Central Asia 17.0%.

China's oil imports are mainly sourced from the Middle East and Africa. In 2010 China's total crude oil imports amounted to around 239.31 million tonnes, and the top ten sources of import accounted for 79% of total imports. Among the top ten import sources, five were countries in the Middle East and two were African nations (Table 3.7). In recent years, the proportion of crude oil imports from the Middle East and Africa has remained little changed at around 70%. The overly focused import structure and the complex political situations in these import sources have brought considerable risk to China's energy security.

To reduce the risks in oil imports, China needs to actively expand its import sources to create a diverse import structure. Considering the world's oil and gas resources distribution, the global competition for these resources, and the international cooperation and agreements that China has entered into, China's target regions for future oil and gas imports should be the Middle East, Africa, Central Asia, Russia and the Americas. Its development strategy should be: stabilising the Middle East, bolstering Africa, expanding Central Asia and Russia, opening up the Americas. The main areas of cooperation will be in resource development and trade, bringing about mutual cooperation and joint development.

Table 3.7 The top ten sources of China's oil imports. Unit: 1000 tonnes.

Rank	2009		2010	
	Country	Import volume	Country	Import volume
1	Saudi Arabia	41 953	Saudi Arabia	44 642
2	Angola	32 172	Angola	39 381
3	Iran	23 147	Iran	21 319
4	Russia	15 304	Oman	15 867
5	Sudan	12 191	Russia	15 240
6	Oman	11 638	Sudan	12 599
7	Iraq	7163	Iraq	11 238
8	Kuwait	7076	Kazakhstan	10 054
9	Libya	6344	Kuwait	9830
10	Kazakhstan	6006	Brazil	8047
Import volume for the whole year	203 790		239 310	

Source: The data in the table are compiled based on figures from the General Administration of Customs of China.

Stabilising Oil and Gas Imports from the Middle East

The Middle East, with its extremely rich oil and natural gas resources, is China's main source of imported oil. To reduce their export risks and ensure more stable export earnings, and to offset the effects of reduced demand in the West, oil and gas exporters in the Middle East are actively seeking new export markets. China should take advantage of this to step up its energy cooperation with major oil exporters like Saudi Arabia and Iran. It should participate more in the upstream exploration of oil and gas resources in the region, and set up oil and gas production and supply bases to stabilise the volume of oil and gas imports from the Middle East.

Bolstering Oil and Gas Imports from Africa

The share of Africa in China's oil and gas imports is growing, and the continent has the biggest potential for China in this area. China should leverage on the excellent political relationships it enjoys with most countries in Africa to expand the scope for cooperation. It should invest more in the infrastructure of major oil-producing countries like Angola, Sudan and Nigeria and increase its manpower training to bolster the country's position in the development and production of African oil and gas. This will meet the challenges posed by Western countries that are accelerating their exploration and development of African oil and gas.

Expanding Oil and Gas Imports from Central Asia and Russia

The geographical proximity of Central Asia and Russia, and in consequence their oil and gas resources, is of great advantage to China. Whether it is leveraging on the advantage of their resources or quickening the pace of their economic recovery, the countries in this region have shown more willingness for external energy cooperation. At the moment, China enjoys good political relationships with these countries, establishing an excellent foundation for oil and gas development cooperation. Looking forward, the focus will be on strengthening the oil and gas development and cooperation with countries like Kazakhstan and Turkmenistan, to make Central Asia an important strategic base to ensure the stable supply of oil and gas to China.

Opening up Oil and Gas Imports from the Americas

The Americas have rich oil and gas reserves, and the region is a world leader in exports. Looking forward, China should gradually expand its political and economic influence in the region to create the right conditions for deeper oil and gas cooperation. The countries to be focused on are Brazil, Venezuela, Mexico and Canada. The Americas will be an important supplement to China's oil and gas imports.

3.7.1.2 Diversifying Transport Channels of Oil and Gas Imports

The transport channels of China's oil and gas imports are exposed to high risks . Currently, apart from importing via a small number of pipelines and rail links from neighbouring regions and countries like Central Asia and Russia, over 90% of China's oil imports are shipped by sea, the majority of which take the Strait of Hormuz-Indian Ocean-Strait of Malacca-South China Sea route. Whether it is oil from the Middle East or Africa, the shipping lanes they take come under the military control or influence of countries like the USA and India. In the event of unexpected incidents, China will find it hard to secure its transport channels of crude oil imports. At the same time, China has insufficient marine transportation capacity for oil imports, especially due to a severe shortage of very-large crude carriers (VLCCs). This means that most of China's oil and gas imports are carried by foreign tankers, which makes China very vulnerable to external manipulation when there are changes in market conditions.

To ensure the security of oil and gas transport, China must further diversify the transport channels of its overseas energy resources. On the one hand, it must strengthen its protective and control capabilities over its marine transportation channels, proactively develop marine security cooperation with the relevant countries, and set up a security framework for transport. At the same time, it must build up its own VLCC fleet, and utilise more Chinese vessels to import oil and gas. On the other hand, it must proactively exploit its geopolitical advantages to open up overland channels, open up more oil and gas pipelines with oil- and gas-rich neighbours, and reduce its dependence on marine transportation channels.

3.7.1.3   Diversifying the Trading Methods in Oil and Gas Imports

To avoid oil import risks, China should focus on the diversification of trading methods in oil and gas imports. It should also adopt a reasonable composition ratio, increase the weight of long-term supply contracts, and proactively make use of futures trading. International oil trade is conducted through several methods: long-term supply contracts, spot trading, quasi-spot trading, futures trading, etc. To a certain extent, long-term supply contracts remove the contingencies and uncertainties of spot trading, especially for a country like China, a large oil consumer with a limited voice in international oil prices. They can ensure the stability and reliability of oil supplies. Oil futures have become the biggest commodities futures in the global futures market. The robust development of oil futures trading will help avoid the risk brought about by fluctuating prices, and increase China's influence on world oil prices.

In the settlement method for oil transactions, China should gradually work towards the greater use of the RMB as the currency for settlement instead of the US dollar. As the US dollar continues to depreciate, there is obviously more diversity in the settlement currencies for international oil trading. There have been increased calls from oil exporting countries to use currencies other than the US dollar for the settlement of oil transactions. Countries like Iran have already begun using the euro and Japanese yen to settle oil exports. Russia, using its clout as a major oil exporter, has been pushing for the internationalisation of the Russian rouble. With China's growing status in the world oil trade, conditions are becoming conducive to using the RMB as the settlement currency for international oil transactions. Using the RMB as settlement currency can help China avoid losses due to currency fluctuations, and it also gives China the initiative in international oil price settlement.

3.7.2 Import of Overseas Coal and Electricity

3.7.2.1 Increasing the Volume of Coal Imports

Coal is the fossil fuel with the most extensive geographical distribution and biggest deposits in the world, with a reserve-production ratio far higher than oil and natural gas. Seventy-five percent of the world's coal deposits are found in five countries: the USA, Russia, China, Australia and India.

Global production of coal has continued to increase over the last few years. In 2010 the world's total coal production was 5.33 billion tonnes of standard coal. China accounted for 48.3% of this volume, while the combined production of the six countries of the USA, Australia, India, Indonesia, Russia and South Africa accounted for 39.8%. Table 3.8 shows the coal production and consumption in 2010 of the world's major coal producing nations (excluding China). Countries like Australia, Indonesia, Russia and South Africa have rich coal deposits but low domestic demand. They are the world's major coal exporters.

Constrained by factors such as resources conditions and environmental capacity, future growth in China's coal production will be limited. To meet socioeconomic development needs, China must proactively tap the international market and increase its coal imports from overseas.

Currently, China imports most of its coal from Australia, Indonesia, Vietnam and Russia via overland and sea routes. In 2010 China imported 165 million tonnes of coal, over 80% of which originated from these four countries. Considering various factors like the geographical location and resources situations of coal producing countries, trends in coal production and development, and export policies, China should choose Australia, Mongolia, Russia and Indonesia, which have shorter transport distances, rich resources and high potentials for export growth, as its main sources of coal imports in the future.

3.7.2.2 Actively Promoting the Import of Electricity

With their rich power generating resources like coal and hydropower, China's neighbouring countries, after having met their own domestic energy needs, have the potential to

Table 3.8 Coal production and consumption in 2010 of the world's major coal producing countries (excluding China). Unit: million tonnes of standard coal.

Rank	Country	Production	Consumption
1	US	789	749
2	Australia	336	62
3	India	309	397
4	Indonesia	269	56
5	Russia	213	134
6	South Africa	204	127
7	Kazakhstan	80	52
8	Poland	79	77
9	Columbia	69	5
10	Germany	62	109

Source: The data in the table are from BP, *Statistical Review of World Energy 2011*.

build power generating bases to export electricity to China. Based on reasonable pricing, economic feasibility and supply reliability, China can gradually and systematically develop international electricity cooperation for the import of clean power from overseas. The benefits are manifold: it will effectively improve China's energy supply structure, reduce domestic pressures in energy resources development and the environment, encourage energy conservation and emissions reduction, and better protect China's energy security.

The results of related studies have shown that countries to the north of China like Russia, Mongolia and Kazakhstan have abundant power generating resources, with great potential for exporting electricity. Based on an analysis of factors like energy resources, water resources, physical distance to China and the economic feasibility of development, it is estimated that by 2020, the Russian Far East and the region of Siberia have the potential to export around 31 GW of coal-fired electricity and around 3 GW of hydropower to China. Mongolia and Kazakhstan have the potential to export around 40 GW and 30 GW of coal-fired electricity to China respectively.

In short, the prospects for international electricity cooperation between China and the three countries of Russia, Mongolia and Kazakhstan are excellent. It is estimated that by around 2020, the electricity bases of these three countries will have the potential to export over 100 GW of electricity to China. From a geographical standpoint, the Russian Far East and Siberia are better positioned to transmit electricity to China's Northeast and North China, while Mongolia and Kazakhstan can transmit electricity to North and Central China respectively.

4

Energy Transport and Allocation

The imbalanced distribution of energy resources and energy consumption means that China's high-capacity and long-distance energy transport and optimisation of energy allocations over large areas is inevitable. In the future, we must change the ways of energy allocation to fully leverage the advantages of various modes of transport, and build up a comprehensive modern energy transport system that is coordinated, supplementary, coherent and integrated so as to remove the bottleneck in China's energy transport, raise the efficiency of energy allocations, optimise the pattern of energy allocation and comprehensively maximise the economic, social and environmental values. We will transport coal and transmit power concurrently, speed up the development of power transmission and optimise the modes of coal transport to change the former excessive reliance on rail for coal transport. We will press ahead with greater efforts in the development of Strong and Smart Grids (SSGs) with UHV grids as the backbone so that an energy resource allocation platform characterised by an optimal grid structure and strong energy allocation capacity will be established. We will also strengthen the construction of oil and gas pipeline networks and optimise the network topology and structure, with a view to realising network-based energy transmission and allocation and harmonious development between the upstream and downstream sectors.

4.1 Modern Comprehensive Energy Transport System

The energy transport in China mainly covers coal, oil, natural gas and electric power. Hydropower, nuclear power, wind power and large-scale solar power are transmitted and utilised after conversion into electric power. Coal is mainly transported by railways, highways, waterways and a combination of different modes of transport. Oil and natural gas are mainly transported by pipelines, railways, highways, waterways and a combination of different modes of transport. Electric power (thermal power, hydropower, nuclear power, wind power, solar power, etc.) is transmitted through power grids (see Figure 4.1).

At present, more than half of the coal that China consumes is used to generate electricity and China's energy development is mainly hindered by the problems arising from the transport of thermal coal. Viewed from a technical perspective, there are two ways to move thermal coal from the coal production bases of the 'Three West' Region

Electric Power and Energy in China, First Edition. Zhenya Liu.

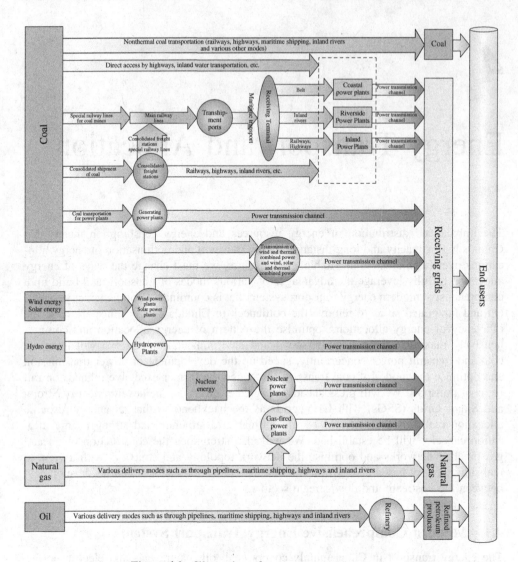

Figure 4.1   Illustration of energy transport systems.

(including eastern Ningxia) to the end users in eastern China. We can (i) ship coal by rail or a combination of rail and waterway transport to the coal-fired power plants in the east to generate and then transmit electricity to the end users; (ii) build mine-mouth power plants at the coal production bases to generate electricity on the spot for transmission to the coastal regions in the east through long-distance power transmission grids. In this sense, the power grid forms another important mode of coal transport.

In recent years, the problem of tight energy supply has kept cropping up in China, caused mainly by the mismatch between energy distribution and modes of transport, the imbalance between coal transport and power transmission, as well as insufficient energy transport capacity. All of these problems are closely related to the construction of energy transport systems. Driven by China's growing economy, the demands for energy in developed regions will increase dramatically and the imbalanced distribution of energy supply and consumption areas is bound to become more notable. That makes large-scale, cross-regional or long-distance allocation of energy resources inevitable. In the future, China will generally transport energy from the west to the east and from the north to the south, and the scale of energy flow will expand considerably. On the one hand, coal will continue to be transported from the west to the east and from the north to the south; on the other hand, the scale of power transmission from the west to the east and from the north to the south is expected to be substantially expanded. Oil will continue to be transported from the west to the east and from the north to the south; the scale of petroleum import will continue to grow. Natural gas will be transported from the west to the east and from the north to the south while offshore gas will be transported onshore. In conclusion, the scale of energy transport will become bigger and bigger, which will impose higher demands on energy transport. Building a highly modernised comprehensive energy transport system has become a major strategic task.

A modern comprehensive energy transport system refers to an integrated energy transport system consisting of various optimally structured and properly balanced modes of transport such as railways, highways, waterways, pipelines, power grids and their transmission routes and stations. It is a system where various coherent and integrated modes of energy transport are coordinated and complement each other by adopting modern and advanced technologies so as to realise secure and efficient energy transmission in line with the economic and geographical conditions as well as the distribution of energy endowments and productive resources in China.

4.1.1 The Significance of Establishing a Modern Comprehensive Energy Transport System

4.1.1.1 Ensuring Secured and Reliable Supply of Energy

Coal reserves in China are concentrated in the northern and northwestern regions, with northern China taking up 37.7%, northwestern China 33.1% and other regions 29.2%.[1] Hydro resources are mainly distributed in Yunnan, Guizhou, Sichuan, Chongqing and Tibet in the west, where 66.7% of the hydro resources nationwide are situated. 90% of the oil resources are distributed in northwestern, northern, northeastern China and the continental shelves. Natural gas resources are mainly distributed in Sichuan, the northwestern regions and the continental shelves. In terms of economic development, the 10 provinces (and municipalities) in the eastern coastal areas, including Liaoning, Hebei, Beijing, Tianjin, Shandong, Jiangsu, Shanghai, Zhejiang, Fujian and Guangdong, are economically relatively well-developed, and the GDP, population and power consumption in these regions account for 57%, 41% and 54% of the national total respectively. However, these regions are comparatively poor in energy resources, with water resources accounting for less than 8% and coal resources about 8% of the national total. Because of

---

[1] It refers to the basic reserves of coal which are calculated according to the 2009 National Mineral Reserve Report. Northern China includes 5 provinces (municipalities): Hebei, Beijing, Tianjin, Shanxi and Shandong while northwestern China includes another 5 provinces (autonomous regions): Xinjiang, Inner Mongolia, Ningxia, Gansu and Qinghai.

this nationwide condition, energy transport in China is characterised by being concentrated along a corridor (from the west to the east and from the north to the south) and spreading long distance (In 2010 the average distance of coal transport by rail was 622 km, that of petroleum products was 917 km and the average distance of coal transport by waterways was 1255 km), imposing high demands on the construction of an energy transport system. In the future, we must strengthen the construction of our energy transport system in a coordinated way when it comes to exploiting and utilising the coal resources in the western and northern regions, the water and power resources in the southwestern region, the wind power resources in the 'Three North' Region and the inland and offshore oil and gas resources, or when it comes to importing various energy resources such as coal, oil, gas and electric power. Otherwise, energy supply in China would be badly affected. In recent years, we have been alarmed by tight supplies of coal and electric power resulting from insufficient transport capacity for thermal coal. The exploitation and utilisation of natural gas, an important and high-quality resource with huge development potential, is mainly hindered by transport corridors and pipeline network construction. The large-scale exploitation and utilisation of such new energy resources as wind and solar energy also need to be supported by modern power grids with a stable structure, operational flexibility and a strong capacity for resource allocations.

4.1.1.2   Raising Energy Transport Efficiency

Apart from such primary energy resources as hydro, nuclear and wind power, a majority of energy products in China, such as coal, oil and gas, are carried from resource-developing enterprises to end users by a combination of two or more modes of transport. Different combinations of transport modes may result in different levels of efficiency. Even if delivery can be completed by only one mode of transport, we still have to optimise the choices of different modes of transport or try to enhance transport efficiency by leveraging more advanced technology and scientific management. For a long time, integrated planning for the development of energy transport has been lacking in China, while the exploitation and transport of energy resources has been uncoordinated, and so have been the different modes of transport. Also, the role that the power grid plays in energy transport has not been understood fully. Those are the reasons why the energy transport system fails to meet the requirements of sustainable energy development, and we still have to perfect this system and improve the efficiency of energy transport. The potential of energy transport may be maximised and its efficiency enhanced effectively by establishing a modern comprehensive energy transport system on the basis of scientific planning to fully leverage the comparative advantages of various modes of transport, optimise the combinations of different modes of transport, and implement scientific allocation and scheduling. The modern comprehensive energy transport system is an energy transport system where different advanced technologies are availed of and energy production, transport and deployment are seamlessly connected. The development in technology and management will definitely help improve the efficiency of energy transport. Taking long-distance power transmission as an example, the transport capacity of a technologically advanced 1000 kV UHV AC transmission line is 4 to 5 times that of a 500 kV AC transmission line, while its per unit power transmission losses are only 1/4 to 1/3 of that of a 500 kV AC transmission line.

4.1.1.3 Maximising the Overall Value

China's land resources are increasingly strained. By establishing a modern comprehensive energy transport system, we are able to make full use of the valuable resources of energy transport channels so as to reduce the footprint on land resources. Moreover, by strengthening the construction of an energy transport system and guiding a rational allocation of energy, we can raise the conversion ratios of primary energy resources processed on-the-spot in the resource-rich western regions, extend the local energy industry chains, boost local economic and social development, reduce pollution caused by the transport of primary energy resources, and ease the severe environmental pressures resulting from conversion between primary and secondary energy resources in the load centres in eastern China. By establishing a modern comprehensive energy transport system, we can also remove power grid bottlenecks of large scale development and utilisation of such clean energy as hydro, wind and solar power, which will have far-reaching implications for combating climate change. Over the past few years, 50% of the costs of thermal coal transported from the 'Three West' Region to eastern China were attributed to the intermediate links. Establishing a modern comprehensive energy transport system can reduce the intermediate links in energy transport, especially thermal coal transport, and thus can effectively counter the upward pressures on thermal coal prices caused by poor integration of the thermal coal production, supply and marketing processes.

4.1.2 The Guiding Principles for Developing a Modern Comprehensive Transport System for Energy

The development of a comprehensive transport system for energy is a huge systems engineering project and shall be pursued in an orderly manner under the guidance of unified planning.

4.1.2.1 Fully Leveraging the Comparative Advantages

In order to build a modern comprehensive transport system for energy, we need to have precise market positioning, scientific planning and appropriate allocation based on the distribution of and demands for resources so as to utilise the advantages of various modes of transport and ensure full utilisation of transport resources. Railway, featuring the large transport volume, all-weather operation and 'door-to-door' delivery advantages, plays a key role in the medium- to long-distance transport of coal and is an important complement for the medium- to long-distance transport of petroleum products. Pipelines have obvious advantages in the large-scale and medium- to long-distance transport of oil and gas. Town gas is mainly transported through gas pipeline networks. Road transport is flexible and convenient, but it is suitable mainly for short-distance movement or for linking up with railway and waterway transport. Waterborne transport plays an important role in energy transport in the coastal regions and areas along the rivers, and in particular sea transport is an important transport mode for energy imports and exports. Grid transmission of electric power is characterised by a minimum of intermediate links, high efficiency and wide coverage. Moreover, total transmission capacity is becoming larger while economically-viable transmission distance is lengthened with the rapid development of UHV transmission technology. So grid transmission will take on added significance to the development of modern comprehensive transport systems for energy. Firstly, it will optimise the transport structure for thermal coal and reduce the over-reliance on rail for coal transport. Secondly, it can deliver clean and renewable energy in an efficient way and promote the transformation of the energy structure into a green and low-carbon structure. Thirdly, we can benefit from extensive interconnections and promote the optimal allocation of energy resources.

4.1.2.2 Committing to System Optimisation

In order to develop a modern integrated transport system for energy, we must work to ensure proper integration of various modes of energy transport in terms of distribution, capacity and construction timing, while coordinating the planning and development of each mode of transport in all aspects. We must also coordinate and promote the development of relevant projects such as support infrastructures and control systems to ensure seamless integration of the entire energy transport system and achieve optimal performance and efficient operation as a whole. At the same time, we should incorporate the transport system for energy into the overall energy system, and give full consideration to the distribution, physical characteristics and demands of diversified energy resources and optimise the allocation of energy resources as well as enhance the overall efficiency of energy development and utilisation by building up a modern integrated transport system for energy. For example, as wind power is intermittent and subject to fluctuations, we need to fully consider the development and transmission of coal-fired power and hydropower while developing wind power in northwestern China on a large scale and transmitting the generated power to other regions. Since hydropower accounts for a large proportion of power development in central China and coal-fired power has commanded a dominant position in northern and eastern China, the development of synchronous UHV grids in northern, eastern and central China can achieve the mutual reinforcement of hydro and thermal power while sharing emergency standby systems and peak-shaving capacity to lower the total installed capacity and form a scientific grid structure.

4.1.2.3 Committing to Safety and Environmental Protection

Development of a modern integrated transport system for energy must play a positive role in the prevention and resolution of the various risks arising from energy transport. Judging from the current situation, the imbalance between coal transport and electricity transmission in the transport of thermal coal and the reliance on the Strait of Malacca for oil imports is a major systematic risk facing China's energy transport, and also a problem to be faced and solved when building a modern integrated transport system for energy. In addition, natural disasters, external damage and the risk inherent in a transport system also need to be dealt with and mitigated by means of technical progress. Meanwhile, a modern integrated transport system for energy must be environment-friendly, and we must fulfil the requirements of energy conservation and emission reduction in the construction process, reduce the footprint and the pollution caused by spillover and leakage and minimise the environmental impact. At the same time, it is necessary to fully consider the environmental conditions and ecological restoration capabilities of the areas at the two ends of and along the transport route as the base for adopting a rational structure and various technical measures to minimise the environmental costs arising from energy development, utilisation and transport.

4.1.2.4 Committing to Simultaneously Exploring Potential and Developing New Infrastructure

In order to build a modern integrated transport system for energy, it is necessary on one hand to further increase investment in infrastructure, with new or expansion modern energy transport projects to optimise the transport structure and strengthen transport capacity. On the other hand it is necessary to actively use modern advanced technology to upgrade the existing energy transport system and resolve the bottlenecks as well as rationalise the relationship between various links so as to tap deeply the potential of the existing transport systems. During the '11th Five-Year Plan' period, the transport capacity of railways has been improved more than 50% through six increases of train speed whereas State Grid Corporation of China has raised the transmission capacity of the existing grids by a total of 154 GW through technical innovation.

4.1.2.5 Committing to Appropriate Level of Development Ahead of Time

The development of an energy transport system is an important element of China's infrastructure development and also the intermediate link between energy resource development and energy users. Lagging behind in this area of construction will affect not only the development and utilisation of energy resources and reduce the efficiency of investment in energy development, but also socioeconomic growth as a whole. Due to the serious consequences and the numerous factors involved in the project and also the long construction period, we shall adhere to the principle of pursuing a level of development appropriately ahead of time in the construction of a modern integrated transport system for energy. Traditionally, China's investment in power grid construction has remained low; not only has power grid development failed to move ahead of electric power development, it has even lagged far behind the needs of socioeconomic development, particularly lacking in transregional, long-distance power transmission capacity to the detriment of power supply security and stability. It has become a very urgent and difficult task to strengthen the development of UHV power grids at all levels, and to fully leverage the role of power grids in the modern integrated transport system for energy.

4.2 Optimisation of the Modes of Coal Transport

Coal is the major product for China's energy transport. The current mode of power generation development focused on achieving local balance has exacerbated the imbalanced distribution of China's coal production and consumption areas. Cross-regional transport of coal has increased year after year, repeatedly giving rise to periods of tension in coal, electricity and transport capacities. With the continuous westward and northward migration of China's coal development centres, the transport demand for coal has further increased. In order to resolve the bottleneck in coal transport in China, we should accelerate the development of power transmission to achieve a balance between coal transport and power transmission in the future. Compared with the modes of coal transport, the modes of power transmission have a strong complementary nature in terms of economy, ecological impacts, land use and promoting coordinated regional economic development. We will vigorously develop power transmission systems and fully leverage the role of power grids in the optimal allocation of resources, which is conducive to optimising the overall structure of China's coal transport and to fundamentally resolving the problems of coal, electricity and transport in China.

4.2.1   The Present Situation of Coal Transport

4.2.1.1   Overall Pattern of China's Coal Production and Consumption

The distribution of coal resources in China is extremely uneven, and most of the coal resources are distributed in the western and northern regions. Currently, Shanxi, Shaanxi, Inner Mongolia and Ningxia have become major energy production bases and output areas for coal, where coal production grows strongly and accounts for a rapidly rising share of the country's total output. The country produced 3.24 billion tonnes of raw coal in 2010. More specifically, Shanxi, Shaanxi, Inner Mongolia and Ningxia produced 1.81 billion tonnes in total, or 55.9% of the country's total output. Local coal production in these regions increased by 1.192 billion tonnes over 2000, adding 14.1 percentage points to their share of the country's total output (see Table 4.1).

Table 4.1   Raw coal output in different regions and their respective shares in recent years.

Region	2000		2005		2010		YoY growth 2000–2010 (%)
	Output (1000 tonnes)	Share (%)	Output (1000 tonnes)	Share (%)	Output (1000 tonnes)	Share (%)	
Shanxi, Shaanxi, Inner Mongolia and Ningxia	618 000	41.8	989 000	44.9	1 810 000	55.9	11.3
Beijing, Tianjin and Hebei	68 000	4.6	95 000	4.3	94 000	2.9	3.2
Northeast	141 000	9.5	185 000	8.4	196 000	6.0	3.3
East China	217 000	14.7	297 000	13.5	350 000	10.8	4.9
Central South	194 000	13.1	267 000	12.1	277 000	8.5	3.7
Southwest	194 000	13.1	291 000	13.2	365 000	11.3	6.5
Xinjiang, Gansu and Qinghai	47 000	3.2	81 000	3.7	148 000	4.6	12.0
Nationwide	1 479 000	100	2 205 000	100	3 240 000	100	8.2

Source: Data in this table are compiled based on the *China Energy Statistical Yearbook* published by the China Statistics Press over the years and data provided by the Coal Industry Research Centre.
Note: Northeast refers to the three provinces of Liaoning, Jilin and Heilongjiang; East China refers to the seven provinces (municipalities) of Shanghai, Jiangsu, Zhejiang, Anhui, Fujian, Jiangxi and Shandong; Central South refers to the six provinces (autonomous regions) of Henan, Hubei, Hunan, Guangdong, Guangxi and Hainan; Southwest refers to the four provinces (municipalities) of Chongqing, Sichuan, Guizhou and Yunnan.

Table 4.2 Coal consumption of different regions and respective percentage shares in recent years.

Area	2000		2005		2010		YoY growth 2000–2010 (%)
	Output (1000 tonnes)	Share (%)	Output (1000 tonnes)	Share (%)	Output (1000 tonnes)	Share (%)	
Shanxi– Shaanxi–Inner Mongolia– Ningxia	230 000	16.3	369 000	15.9	617 000	19.1	10.4
Beijing– Tianjin–Hebei	166 000	11.8	234 000	10.1	296 000	9.2	6.0
Northeast	188 000	13.3	250 000	10.8	352 000	10.9	6.5
East China	364 000	25.8	689 000	29.7	937 000	29.0	9.9
Central South	254 000	18.0	429 000	18.5	600 000	18.6	9.0
Southwest	154 000	10.9	255 000	11.0	312 000	9.7	7.3
Xinjiang, Gansu and Qinghai	55 000	3.9	93 000	4.0	112 000	3.5	7.4
Nationwide	1 411 000	100	2 319 000	100	3 226 000	100	8.6

Source: Data in this table are compiled based on the *China Energy Statistical Yearbook* published by the China Statistics Press over the years and data provided by the Coal Industry Research Centre.
Note: Northeast refers to the three provinces of Liaoning, Jilin and Heilongjiang; East China refers to the seven provinces (municipalities) of Shanghai, Jiangsu, Zhejiang, Anhui, Fujian, Jiangxi and Shandong; Central South refers to the six provinces (autonomous regions) of Henan, Hubei, Hunan, Guangdong, Guangxi and Hainan; Southwest refers to the four provinces (municipalities) of Chongqing, Sichuan, Guizhou and Yunnan.

China's regional economic development is imbalanced, with coal consumption concentrated mainly in the eastern and central regions like Beijing–Tianjin–Hebei, East and Central South China. In 2010 the country's total coal consumption reached 3.226 billion tonnes. In particular, Beijing–Tianjin–Hebei, East and Central South China together consumed 56.8% of the country's total, an increase of 1.2 percentage points over 2000 (see Table 4.2).

In recent years, China's installed capacity of coal-fired generation and its electricity output have experienced rapid growth, leading to a sharp increase in the consumer demand for coal. In 2010, thermal coal (including heating supplies) in the region of 1.779 billion tonnes was consumed, representing 55.1% of the country's total coal consumption, up 9.3 percentage points from 2000 (see Figure 4.2).

4.2.1.2 Overview of China's Coal Transport

The imbalance in the distribution of coal production and consumption in China dictates the overall pattern of coal transport from the west to the east and from the north to the south, as well as the fan-shaped transport network extending eastward and southward, with coal bases in the 'Three West' Region (including eastern Ningxia) as the core. The overall pattern of China's coal transport is shown in Figure 4.3.

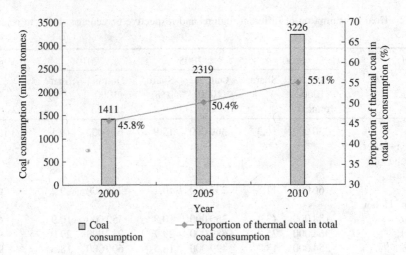

Figure 4.2   Changes in coal consumption in China.

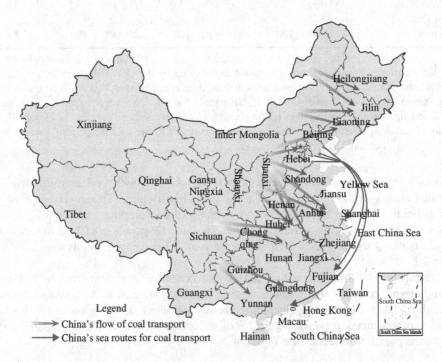

Figure 4.3   Overall pattern of China's coal transport.

With the worsening imbalance between coal production and consumption in China, the overall scale of China's cross-regional coal transport is constantly increasing. In 2000, the net outward shipment of coal was 428 million tonnes compared with net inward shipment of 360 million tonnes between China's major geographical areas. By 2010, the net outward shipment of coal between China's major geographical areas was 1.282 billion tonnes, or 3.0 times that of 2000, and the net inward shipment was 1.268 billion tonnes, or 3.5 times[2] that of 2000. Regarding the general situation of outward movement of coal, the regions of Shanxi, Shaanxi, Inner Mongolia and Ningxia are the most important regions in China for coal production and export, where the net outward shipment of coal was 1.193 billion tonnes in 2010, being 3.1 times that of 2000, or 93% of the total national outward shipment. East China is the most important area in China for coal consumption and inward shipment, where the net inward shipment of coal was 587 million tonnes in 2010, being 4.0 times that of 2000, or 46% of the total national inward shipment of coal.

China's coal is mainly transported by rail, supplemented by waterway and highway. Intermodal transport is the major mode of coal transport in practice. According to preliminary statistics, a total of 6.6 billion tonnes of coal was transported by highway, railway and waterway in 2010 in China, with shipment turnover of 3.09 trillion tonnes · km. China's Coal Transport-Consumption Coefficient[3] is 1.64, in which the Highway Transport-Consumption Coefficient is 0.87, and that of railway and waterway is 0.58/0.19 respectively. This means for every ton of coal consumed in China, an average of 1.64 tonnes of freight volume would need to be moved through a combination of different transport modes involving highway, railway and waterway over a distance of 467 km.

Railway

Railway is the major mode of transport for coal, and coal has been the most important cargo for railway transport. According to statistics, nationwide transport of coal by rail grew at an average annual rate of 10.6% from 2000 to 2010, with a total volume of approximately 2 billion tonnes shipped in 2010. The proportion of coal in total rail-based freight volume rose from 43.6% in 2000 to 55% in 2010. The 'Three West' Region forms the country's largest and most concentrated outward railway transport channel for coal, consisting of 12 rail lines (such as the Da-Qin line) divided into northern, central and southern corridors based on the geographical locations of the routes (see Figure 4.4). In particular, the northern corridor includes the 5 rail lines of Ji-Tong, Da-Qin, Feng-Sha-Da, Jing-Yuan and Shuo-Huang; the central corridor covers the 2 rail lines of Shi-Tai and Han-Chang; and the southern corridor consists of the 5 rail lines of Hou-Yue, Long-Hai, Tai-Jiao, Ning-Xi and Xi-kang. It is statistically shown that 920 million tonnes of coal were transported through the railway transport channels in the 'Three West' Region in 2010, or 46% of the nationwide coal transport by rail. Specifically, around 400 million tonnes of coal, or 43% of the total volume of coal in the 'Three West' Region, were transported through the Da-Qin line, which is China's most important outward transport corridor for coal.

Waterway

Coastal waterway transport is an important means of coal transport in the southeastern coastal areas in China. Inland river transport is an important mode of coal transport for inland river basins across the Yangtze River, the Beijing-Hangzhou Grand Canal, the Huaihe River and the Pearl River.

---

[2] Due to the impact of import/export and coal inventory, the data for total outward and inward shipments of coal may vary slightly.

[3] Coal Transport-Consumption Coefficient is the ratio of coal transport volume to total consumption volume.

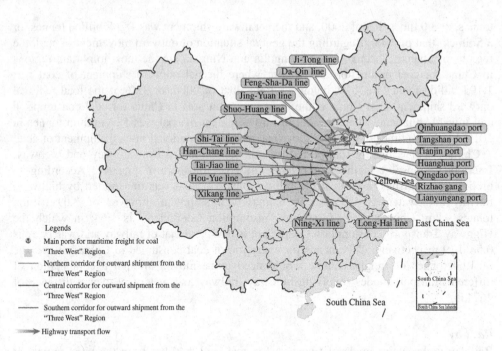

Figure 4.4 Diagram of railway transport corridors in the 'Three West' region.

Highway

Highway transport plays an important role in complementing interprovincial coal shipment by providing freight consolidation and distribution services for rail and water-based coal transport. Highway transport of coal in China is mainly outward bound from the 'Three West' Region and eastern Inner Mongolia to East and Central South China. In recent years, the highway has been compelled to assume a major role in meeting much of the interprovincial demand for coal transport amid China's tight transport capacity. According to preliminary surveys, about 400 million tonnes of coal shipments in the 'Three West' Region were transported by highways to other provinces in 2010, double that of 2005.

4.2.1.3   Major Current Problems in Coal Transport

Lack of Railway Transport Capacity for Coal and Insufficient Shipment Capacity in Major Ports

Since the '10th Five-Year Plan', China has devoted more efforts to develop its transport infrastructure, especially the channels of coal transport. However, amid the fast growing demand for coal, the capacity of the main railway transport lines is saturated and the development of rail-based consolidated freight systems is lagging, resulting in a heavy burden on road transport. The capacity of major loading ports is also strained and there are not enough ships in the coastal fleet as the ships are aging. Moreover, with the exception of the Da-Qin rail line, the Shuo-Huang rail line, Qinhuangdao Port, Huanghua Port and Caofeidian Port that are dedicated to coal transport, China's other coal transport channels are also burdened with providing passenger and cargo services. Currently, the transport capacity for coal is seriously lacking in flexibility and is susceptible to interference from other transport services.

The Scale of Road-based Coal Transport is Increasing, Resulting in Severe Congestion

In recent years, spurred by external factors such as robust demand for coal, insufficient rail capacity, complicated application formalities and higher surcharges as well as internal factors such as larger coal transport vehicles and steadily improving road conditions, the amount of coal transported by interprovincial highways has continued to rise, especially in the 'Three West' Region where the volume of coal transported to outside regions has continued to grow strongly. Coal transport by road has greatly increased the consumption of refined petroleum products, meaning that more advanced forms of energy are consumed in place of low-grade energy. Given the shortage of oil resources, the heavy reliance on roads for coal transport is uneconomical. Moreover, traffic congestions are often experienced in the main transport routes for coal due to the excessive load of coal being moved. For example, the Beijing-Tibet Expressway, Xuanda-Jingzhang Expressway, National Highway 109, National Highway 108, and the Shijiazhuang-Taiyuan section of National Highway 55 often suffer severe traffic congestion due to the long queues of vehicles stretching dozens of kilometres. Nearly all vehicles for coal transport are heavy trucks, with full load capacities of more than 60 tonnes and some even up to more than 120 tonnes with trailers. The service life of roads is significantly shortened due to the overloaded vehicles, which have also caused serious damage to other facilities such as bridges and tunnels. Meanwhile, due to the road capacity it has taken up, coal transport has seriously affected the transport of other goods.

Imbalanced Proportion between Coal Transport and Power Transmission, While the Role of the Grids has yet to be Fully Fulfilled

For a long time, China's coal-fired power has been dominated by the development mode of achieving local balance within the different provinces and prefectures. As a result, China's coal transport heavily relies on coal shipment handling, transregional transmission capacity is insufficient, and the power transmission ratio is quite low. In 2009, in the 'Three West' Region (including eastern Ningxia), the ratio of coal handling to power transmission is about 20:1, showing that the proportion of power transmission in coal energy delivery is less than 5%. Due to a lack of transport capacity, coal transport has become a major bottleneck in China's energy supply, especially during bad weather and peak passenger traffic in holidays when it often causes stretched coal and electricity supply in the eastern and central regions. This situation is particularly severe during the annual Spring Festival. Moreover, coal transport has such disadvantages as bulky volume, long distance movement and multiple intermediate steps, which are an important reason for the constantly rising thermal coal prices in the load centres of the eastern and central regions, resulting in sharp increases in power generation costs.

4.2.2   The Future Coal Transport Patterns

4.2.2.1   Outlook for Consumer Demand for Coal

Based on the distribution characteristics of China's coal resources and constraints on coal development nationwide, China's future coal development is guided by strategic thinking focused on controlling the development scale in the eastern region, stabilising the development scale in the central region, and expanding the development scale in the western region. The focus of coal development will continue to shift westward and northward, while emphasis will be placed on building up the large coal (coal-fired power) bases in, among other locations, Shanxi, Ordos, eastern Inner Mongolia (including Xilin Gol League) and Xinjiang.

According to the future development thinking and allocation plans for coal while taking into account factors such as China's economic prospects, energy restructuring and measures against greenhouse gas emissions, China's coal consumption is expected to reach 4.2 billion tonnes by 2020, or 58.4% of the nation's total primary energy consumption, representing a decrease of 9.6 percentage points from 2010. It is expected that China's coal consumption by 2030 will hit 4.3 billion tonnes, or 52.3% of the country's total primary energy consumption, representing a decrease of 6.1 percentage points from 2020 (see Table 4.3).

4.2.2.2   Outlook for Coal Transport Demand

According to an analysis of the consumer demand for coal and production/supply pattern, China's coal transport demand in the future will further increase, with the coal handling and power transmission to significantly expand. Also taking into account the development of China's future energy transport systems with reference to the production capacity and consumption of coal bases and the use of coal-fired power in different power-receiving regions, our analysis of future cross-regional demand for coal transport is as follows:

By 2020, coal production and local consumer demand for coal in the 'Three West' Region (including eastern Ningxia) is expected to reach 2.07 billion tonnes and 650 million tonnes respectively, with the demand for outward shipment of coal maintaining at 1.42 billion tonnes. The comparative figures for Xinjiang will be 350 million tonnes and 120 million tonnes respectively, with the demand for outward shipment of coal reaching 230 million tonnes, compared with the same demand for lignite in eastern Inner Mongolia of 60 million tonnes.

By 2030, the domestic coal production is projected to show limited growth from 2020. With the increasing consumption of coal in coal-producing areas, the outward shipment of coal across regions will go down in the future. By then, the demand for outward shipment

Table 4.3   China's future demand for coal consumption.

Category	2010	2020	2030
Coal consumption (raw coal, billion tonnes)	3.2	4.2	4.3
Proportion of coal in the primary energy consumption (%)	68.0	58.4	52.3

of coal in China's 'Three West' Region (including eastern Ningxia) is expected to be about 1.28 billion tonnes, compared with about 190 million tonnes of coal for Xinjiang and 50 million tonnes of lignite for eastern Inner Mongolia.

4.2.3   Equal Emphasis on Coal Transport and Power Transmission

There are capacity bottlenecks in China's coal transport caused mainly by the irrational distribution of power sources and the under-utilisation of the optimal resource allocation capacity of the power grids. If this problem cannot be solved as soon as possible, the sustainable development of China's economy and society will be severely constrained. Our strategic choice of optimising and adjusting China's energy structure and distribution is to pay equal attention to coal transport and electricity transmission by accelerating the development of power transmission, constructing UHV power grids and improving the large-scale, long-distance transmission capacities while continuing to strengthen the construction of railways and other coal handling facilities, which is in particular of great significance to optimise China's coal transport, improve the efficiency of coal production, conversion, transport and utilisation, reduce the adverse impact on the ecological environment, curb the use of precious land resources, promote the coordinated development of the regional economy and enhance the security, economy and reliability of energy supply.

4.2.3.1   Accelerated Development of Power Transmission is Conducive to Changing the Over-reliance of China's Energy Transport on Coal Handling

As China's coal-fired power plants are concentrated in the eastern and central regions, energy transport relies heavily on the transport of coal, which translates into ever-increasing pressures on railway systems. Despite the expansion and modification efforts made by railway authorities to increase the transport capacity for coal, the need for transporting coal required for power development in the eastern and central regions still cannot be sufficiently met, and the tight coal transport capacity continues to exist or even deteriorate. At the same time, the imbalance between coal transport and a much lower level of electricity transmission has led to rapidly rising coal prices in the eastern and central regions, and the supply of thermal coal cannot be assured during periods of peak passenger traffic.

In accordance with the development direction of separating the passenger and cargo flows in the rail transport sector, China will build and expand the major transport corridors for coal to directly transport the washed coal or high calorific coal from the coal-producing bases in the 'Three West' Region (including eastern Ningxia) to the eastern or central regions or to the ports, for onward shipment by rail and waterway to the eastern or central regions. Most of the coal is used to supply the coal-fired power plants in the regions for power generation in continuation or expansion of the existing coal transport modes. If we continue to build coal-fired power plants in the eastern and central regions to achieve a local balance of electric power, the stable supply of coal in the eastern and central regions must rely on the continuous expansion of rail transport capacity, and this pattern of development has been proven to have serious drawbacks which must be effectively rectified.

After several years of study and practice, China has fully and independently mastered the core technology of UHV electricity transmission and conditions are now in place for carrying out large-scale applications. The development of UHV electricity transmission technology provides a new, safe, stable, economical and clean way of large-scale transport of coal over long distances. Electricity can be transmitted to the eastern and central regions by converting the coal of the 'Three West' Region (including eastern Ningxia) into electricity in-situ through UHV electricity transmission technology. The in-situ balance of China's electric power can be fundamentally changed and the allocation of coal resources on a nationwide scale optimised.

Amid the continued phenomenal growth of the economy and society, there will be more room for the development of coal handling and power transmission in the future. There are great functional differences between the railway network and UHV power grids in terms of optimising energy structure and distribution, and the relationship between them is not one of simple substitution, but of a rational division of labour and mutual complementation for the modes of thermal coal transport. Different from electricity transmission that mainly transmits clean and quality power, coal transport involves moving coal that is mixed to a large extent with dust, soil, gravel and other impurities. Based on the volume of coal transported in China, hundreds of millions of tonnes of dust, soil, gravel and other impurities are transported over thousands of kilometres every year, causing a massive waste of resources. Therefore, washed coal with a high calorific value is more suitable for railway transport, while low calorific coal such as raw coal, lignite, washing coal and gangue should be used for in-situ power generation, with the generated power sent out through electricity transmission. By constructing UHV electricity transmission projects and a scientific and rational energy transport system, we can change the over-reliance of energy transport on coal handling and completely solve the existing shortage of coal, electricity and transport capacities. At the same time, the accelerated development of electricity transmission has improved the energy transport system's level of diversification, and reduced the number of intermediate steps. It can also significantly ease the impact of natural disasters on the energy transport system and effectively enhance the ability of the energy transport system to withstand the impact of natural disasters, with significant implications for improving the safety and security of electricity supply in the eastern and central regions.

According to the research findings on China's ratio of coal handling to electricity transmission in the future, given equal emphasis on coal handling and electricity transmission and the accelerated development of electricity transmission, the volume of coal directly exported from the 'Three West' Region (including eastern Ningxia) in 2020 is expected to be about 1.11 billion tonnes, and the output transmitted to other regions equivalent to about 300 million tonnes[4] of coal, and the ratio of coal handling to electricity transmission 3.7:1. The coal directly exported from Xinjiang is expected to be 100 million tonnes, with the electricity transmitted to other regions equivalent to some 130 million tonnes[5]

---

[4] When power transmission is translated into coal handling, the utilisation of coal-fired units is set at 5000 hours, the consumption of coal for power generation at 300 g of standard coal/(kWH). The calorific value of the raw coal from the 'Three West' Region (including eastern Ningxia) is set at 5000 kcal/kg. On this basis, 1 kW of power transmission capacity in the 'Three West' Region (including eastern Ningxia) involves about 2.1 tonnes of coal shipment.

[5] The power is mainly transmitted from Xinjiang by DC mode. The utilisation of coal-fired unit is set at 5500 hours, the consumption of coal for power generation at 300 g of standard coal/(kWH), and the calorific value of the raw coal from Xinjiang at 5500 kcal/kg since the coalfield resources in Xinjiang are generally better. On this basis, 1 kW of power transmission capacity in Xinjiang involves about 2.1 tonnes of coal shipment.

of coal, with a ratio of coal handling to electricity transmission at 0.8:1. It is necessary for northeastern China to regulate the proportions of certain coal types with the interior regions, west of the Shanhai Pass to achieve a balance between coal supply and demand. The net delivery of coal in northern China, eastern China and central China will reach 1.17 billion tonnes. By then, China will basically form an energy transport pattern of equal emphasis on coal handling and electricity transmission.

With the further development of transregional electricity transmission and the increase of local coal consumption, the coal directly shipped from the 'Three West' Region (including eastern Ningxia) is expected to drop to 880 million tonnes by 2030, and the electricity transmitted to other regions will approximately be equivalent to 400 million tonnes with the ratio of coal handling to electricity transmission at 2.2:1. The coal directly shipped from Xinjiang will decline to 66 million tonnes, and the electricity transmitted outside will be equivalent to about 130 million tonnes of coal, with the ratio of coal handling to electricity transmission at about 1:2. The demand for delivering coal in northeastern China will further increase and reach 130 million tonnes; the net delivery of coal to the receiving areas in northern, eastern and central China will reach 1.31 billion tonnes.

4.2.3.2 Accelerated Development of Electricity Transmission is Conducive to Coordinated Utilisation of the Country's Environmental Resources

China's eastern and central regions lack essential environmental capacity for the further development of coal-fired power plants. Over the years, a large number of coal-fired power plants have been built in the eastern and central regions, and the resulting massive coal combustion has led to serious pollution such as acid rain at the expense of the environment. According to the 2010 Report on the State of the Environment of China, the distribution of acid rain in China is mainly concentrated along and south of the Yangtze River, and to the east of the Tibetan Plateau. It is basically located in the load centre areas covered by the East China Grid, the Central China Grid and the China Southern Power Grid (see Figure 4.5). The emission of sulphur dioxide per unit of land area in the eastern and central regions is 5.2 times that in the western region, while the annual emission of sulphur dioxide in the Yangtze River Delta region is reaching 45 tonnes/km^2. Therefore, from an environmental perspective, there is basically no further room for coal-fired power development (see Figure 4.6).

Through developing pit-mouth power plants in the western coal-producing region and vigorously promoting the recyclable utilisation of coal resources, we can alleviate the environmental damage caused by coal mining. From the perspective of water utilisation, the joint development and integrated operation of coal mines and power plants can fully utilise the water resources in the mines, eliminate water pollution caused by the indiscriminate discharge of waste water from the mines, and maximise the precious water resources. From the perspective of coal utilisation, we can use the low calorific value coal and coal gangue for pit-mouth power generation to reduce the pollution caused by fuel stockpiling. From the perspective of ash utilisation, we can transport the waste residue produced by power generation to the mines for backfilling the mined-out areas, thereby reducing environmental pollution arising from the solid wastes of the power plants and preventing the mines from collapsing. From the perspective of land use, we should avoid constructing large-scale coal storage bases for the sake of effective land conservation.

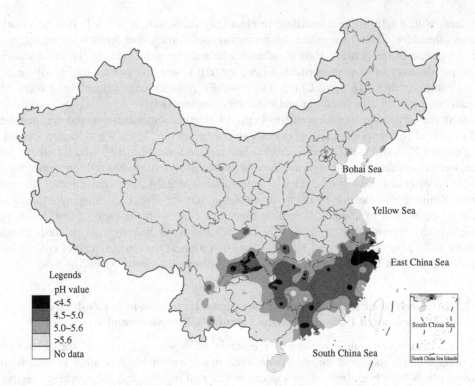

Figure 4.5   Acid rain map of China in 2010.
Source: 2010 Report on the State of the Environment of China, Ministry of Environmental Protection.

In regard to economic losses caused by environmental damage, the economically developed eastern and central regions have suffered much more serious economic losses from environmental pollution than the western and northern regions. Environmental losses caused by air pollution include lost health, reduced agricultural production and damaged materials, which are positively correlated to an area's population density and per capita GDP. Take sulphur dioxide emissions as an example, economic losses caused by sulphur dioxide emissions per unit in the eastern and central regions are 4.5 times the level of loss sustained by the western and northern regions.

A comparison of the eco-environmental impacts shows that power transmission can promote the optimal utilisation of China's environmental space and protection better than coal transport. In the future, more coal-fired power plants should be built in the western region where there is a larger environmental space.

4.2.3.3   Accelerated Development of Power Transmission is Conducive to Promoting Coordinated Development of China's Regional Economy

Compared with coal transport, power transmission plays a more effective role in boosting the economic and social development in the western region. Coal transport represents the outflow of primary resources and plays a limited role in driving the local socioeconomic

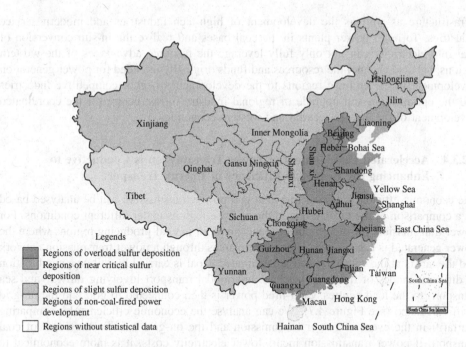

Figure 4.6 Space distribution of China's coal-fired power development.

development. Power transmission to other regions in coal-producing areas can extend the industry chain of coal exploitation and utilisation, which is more beneficial to the promotion of local socioeconomic development than mere coal transport. It is estimated that in terms of contribution to Shanxi Province's GDP, the combined coal transport/power transmission modes carry a ratio of about 1:6, compared with a ratio of about 1:2 in terms of contribution to job creation.

Increasing the scale of cross-regional power transmission is more helpful in controlling the power supply costs in the end-use regions and promoting the optimal structural adjustment to energy consumption. The introduction of lower-cost power from other regions helps strengthen the competitiveness of local industries and promote local economic development. Meanwhile, it can also boost power consumption, optimise the structure of energy consumption and improve the efficiency of the energy economy. In addition, enhanced cross-regional power transmission capacity and expanded interconnection capacity of energy transmission can also bring about such incremental benefits as contingency reserves, lower installed capacity, cross-river basin compensation and mutual reinforcement of hydro and thermal power.

Cross-regional power transmission is more conducive to a rational division of labour. From the perspective of promoting the economic development of the supply regions and the end-use regions, power transmission is a win–win strategic option. China's coal-producing areas enjoy the comparative advantage of abundant natural resources. The end-use regions are in the intermediate or advanced stages of industrialisation, where the focus of development lies in the adjustment, optimisation and upgrade of industry infrastructure as well as the development of high-tech industries and modern service industries. To build power plants in the coal bases and realise the in-situ conversion of coal into electricity can not only fully leverage the resource advantages of the western regions, but also channel the resources and funds originally intended for power generation development in the end-use regions to the development of other competitive industries and the optimisation and upgrade of regional industry infrastructure for the coordinated development of the eastern, central and western economies.

4.2.3.4 Accelerated Development of Power Transmission is Conductive to Enhancing the Economic Efficiency of Energy Transport

The economic efficiency of coal transport and power transmission can be analysed based on a comparison of the tariff levels in the end-use regions under different conditions. For power transmission, pit-mouth power plants are built in coal-producing regions where the power generated is transmitted to load centre areas through a power transmission network and the tariffs at the receiving end are estimated. Coal is carried by rail or a combination of different modes of transport (such as intermodal transport involving railway and sea transport) to the load centres. Coal-fired power is then connected to grid and the on-grid tariff calculated (see Figure 4.7). We can analyse the economic efficiency by comparing the tariff in the event of power transmission and the on-grid tariff in the event of coal transport. If power transmission incurs lower electricity costs, it is more economical to transmit power; otherwise, it would be more economical to transport coal. Based on the coal prices in June 2011, to transmit power from the coal-fired power base in the 'Three West' Region to the load centres in the eastern and central regions, the electricity cost incurred by power transmission is RMB0.06 to 0.13/(kWh) lower than the on-grid tariff based on coal transport. It is obvious that the economic efficiency of power transmission is superior to that of coal transport.

The economic distance of UHV transmission is able to cover most areas from the major coal bases to the load centres in China. There is a positive correlation between the critical economic distance of UHV transmission and coal price differences between the supply and end-use regions (see Figure 4.8). When the coal price difference between the supply and end-use regions is RMB200/ton of standard coal and the transport distance is within 800 km, UHV transmission is more economical than coal transport. When the coal

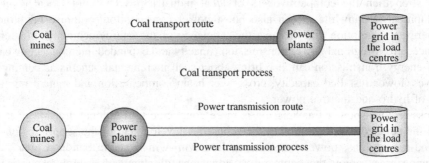

Figure 4.7 Schematic diagram of comparison between the economic efficiency of coal transport and power transmission based on coal price differences.

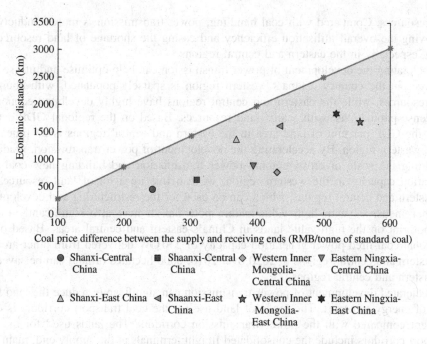

Figure 4.8 Critical economic distance of UHV transmission.

price difference is more than RMB450/ton of standard coal, the economic distance of UHV transmission may increase to more than 2000 km. When the coal price difference is RMB600/ton of standard coal, the economic distance of UHV transmission may reach around 3000 km.

Power transmission is more conducive to smoothing away the price fluctuations of energy transport and it is a 'nonstop' mode of energy transmission, with few intermediate links. Capital costs represent a fairly large proportion of transmission project costs, while operational and maintenance expenses account for a relatively small portion. After the completion of a power transmission project, very few factors can cause changes in power transmission costs and prices during its lifespan. So the process can be said to entail relatively stable transport costs. In contrast, the process chain for coal transport is longer, with many intermediate links in between. As variable operational and maintenance costs represent a comparatively large cost component of this mode of transport, the transport costs involved are more likely to fluctuate. Therefore, compared with coal transport, power transmission can more effectively ease the fluctuating costs of power supply in the end-use regions.

4.2.3.5 Accelerated Development of Power Transmission is Helpful in Improving China's Overall Utilisation Efficiency of Land Resources

China lacks land, especially arable land resources. Currently, the per-capita land area is less than 1/3 of the world's average. With a growing population, more land is used for mining, transport and urban development, resulting in an increasingly tight supply of land resources. Compared with coal handling, power transmission is more conducive to improving the overall utilisation efficiency and easing the shortage of land resources in China, especially in the eastern and central regions.

Accelerating the development of power transmission can help optimise and utilise land resources in the country. China's western region is sparsely populated, with abundant land resources, while the eastern and central regions have highly developed economies and dense populations, with scarce land resources. Based on the regional GDP data in 2010, the GDP per unit of land area in the eastern and central regions is 23 times that of the western region. By accelerating the development of power transmission, gradually expanding the scale of cross-regional power transmission and building new coal-fired generation capacity in the western region, we can free up a lot of land resources for the eastern and central regions, which can be used for the restructuring and development of other industries with high value-adding potential. It may also significantly reduce the footprint on the high-value lands in China's eastern and central areas. Based on an additional coal-fired power generation capacity of 230 GW delivered from other areas to the eastern and central regions in 2020 over 2010, 6000 hectares of land can be saved for the eastern and central regions.

Accelerated development of power transmission can significantly reduce the land footprint of energy transport. The mode of land use for the coal transport corridors is quite different compared with the power transmission corridors. The lands used for the coal transport corridors include the consolidated freight terminals at the supply end, main railway lines for coal handling, transhipment ports, destination ports and dedicated railway lines for the power plants at the destinations. Specifically, railways are land corridors for coal handling that directly occupy lands to the complete exclusion of other land uses. The power transmission corridors are air corridors, covering the tower bases and substations, switch stations (converter stations), line corridors and the land under the tower bases that can be used. Take transmission from the 'Three West' Region to eastern China as an example, we analyse the land used for delivering the same power capacity to the end-use regions by coal transport and by power transmission through a comparison of land footprint per unit of power capacity. After taking into account the land footprint for the two transport modes, we have learnt that the land space used for coal transport is 2–4 times that for the UHV AC transmission mode from the 'Three West' Region to eastern China. In addition, we can reduce the land used for coal yards and coal gangue by the integrated development of coal and power capacity and intensive development of coal-fired power bases.

Coal transport and power transmission are two important modes of energy transport and the comparison of the respective integrated benefits is subject to many boundary conditions. In the process of electric power development, the saying of 'coal transport for a long distance and power transmission for a short distance' began to appear in the 1960s as a reflection of the situation when coal prices and transport costs were relatively low, while grids were of a low grade and small in size, with insufficient transmission capacity. In recent years, coal transport costs have soared. At the same time, with the rapid development of power transmission technology, especially the gradual maturing of UHV transmission technology, grid structure and transmission capacity are greatly enhanced. As a result, the economic transmission distance has been extended greatly while transmission efficiency and economic benefits have improved remarkably. Based on considerations of security, the environment, the economy, technology and land resources utilisation, building large-scale coal-fired power bases in coal-rich areas with the generated power transmitted to the load centres by means of UHV transmission is a scientific and rational option. This option will be conducive to promoting the optimal allocation and efficient utilisation of China's coal and other energy resources.

4.3 Strong and Smart Grid Development

The power grid serves as an important foundation of and a guarantee for the economic development and social progress in modern times. Achieving the scientific development of power grids carries great significance in establishing the modern comprehensive energy transport system, optimising the allocation of energy resources on a wider scale, safeguarding energy security, stimulating technological innovations, combating climate change, and promoting sustainable socioeconomic development. With the current extensive application of UHV power transmission and intelligence technologies, stepping up the construction of Strong and Smart Grids (SSGs) and establishing a modern power grid system that is safe, environment-friendly, highly efficient and interactive have become a critical task for us in implementing the strategy of 'One Ultra Four Large' (1U4L) and promoting the transformation of energy development models and the strategic transition of energy in China.

4.3.1 Overview of Power Grid Development

4.3.1.1 Present Situation of Power Grid Development

Since the implementation of its economic reform and opening up policy, China has been speeding up the construction of power grids. As continuous breakthroughs have been made in power grid technology, China's energy allocation capacity has been strengthened significantly with constant improvements in security, reliability and economy.

Continuous and Rapid Expansion of Power Grids
The rapid socioeconomic development in China is driving up the demand for electricity. In 2010, the total electricity consumption nationwide reached 4,200 TWh, 14.0 times as much as that in 1980. Scaling a succession of new heights (see Figure 4.9), the total installed capacity exceeded 100 GW in 1987, rising to over 300 GW in 2000, 500 GW in 2005 and further to 1000 GW in 2011.

Given the need to commission large power capacity and the rising load demand, the power grids in China have continued to grow. By the end of 2010, the loop length of 35 kV or higher voltage transmission lines amounted to 1.337 million km and total transformation capacity reached 3.62 billion kVA, or 4.8 times and 23.2 times the level in 1980 respectively. The loop length of 220 kV or higher voltage transmission lines amounted to 445 600 km and total transformation capacity reached 1.99 billion kVA, or 14.9 times and 56.9 times the level in 1980 respectively (see Figure 4.10).

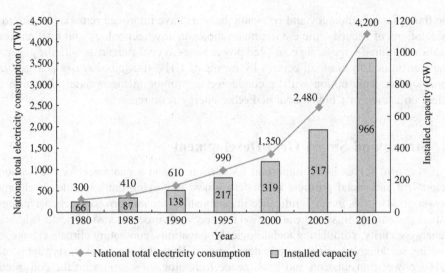

Figure 4.9 Growth of installed generation capacity and national total electricity consumption in China.

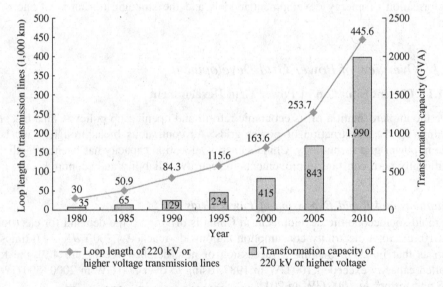

Figure 4.10 Growth of 220 kV or higher voltage power grids.

Constant Upgrade of Transmission Voltage

In order to facilitate the optimisation of resource allocation, the transmission voltage classes in China have been continually upgraded (see Figure 4.11). The highest transmission voltage has grown from 220 kV during the early years of the founding of new China to the current 1000 kV. As the construction of power grids continues, the functions of power grids at different levels are more clearly positioned and the voltage classes have

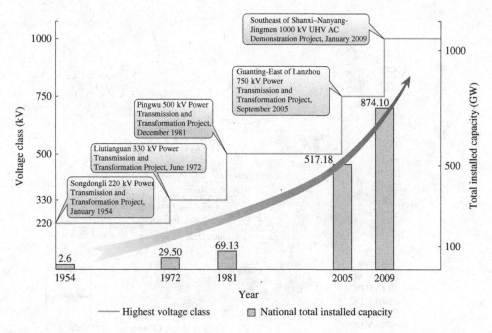

Figure 4.11 Development of transmission voltage classes in China since the founding of new China.

become more uniform and rational. The 110 (66), 220, 500 (330) 750 and 1000 kV AC and the ±500, ±660 and ±800 kV DC standard transmission voltage series have now been established in China, with the ±1100 kV DC transmission voltage series under research.

Formation of National Interconnection Network
With the fast-growing power loads and the continuous expansion of power systems, the active promotion of network interconnection and the optimisation of power resource allocations on a national level have become indispensable for the development of the electricity industry. By the end of 2011, apart from Taiwan, all provincial power grids in China had achieved AC/DC interconnection, forming a nationwide interconnection network (see Figure 4.12), with five synchronous power grids covering northern/central China, East China, northeastern China, northwestern China and South China. The north-eastern grid and northern/central grid achieved asynchronous interconnection through the Gaoling Back-to-Back Project, while the asynchronous interconnection between the north-ern/central grid and Eastern grid was realised through the three ±500 kV DC circuits of Gezhouba-Nanqiao, Longquan-Zhengping and Yidu-Huaxin and the Xiangjiaba-Shanghai ±800 kV DC circuit. The asynchronous grid interconnection between northwestern grid and northern/central grid was accomplished through the Lingbao Back-to-Back Project, the Deyang-Baoji ±500 kV DC circuit, and the Ningdong-Shandong ±660 kV DC circuit. The asynchronous grid interconnection between northern/central grid and southern grid was made possible through the Three Gorges-Guangdong ±500 kV DC transmission.

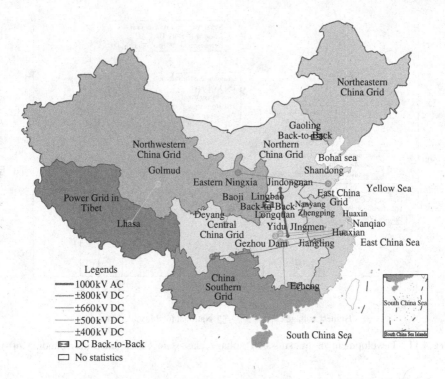

Figure 4.12   Layout of nationwide network of power grids.

Significant Improvement in Power Supply Reliability

Through more efforts in building and revamping the power grids in urban and rural areas, power supply capacity and reliability have improved remarkably in China. In 2010, power supply reliability in urban areas averaged 99.923%, an improvement of 0.034 percentage point from 2000, with the average annual outage time per customer at 6.72 hours, an improvement of 3 hours from 2000. The average power supply reliability for rural users reached 99.714%, or 0.321 percentage point higher compared with 2005, with the average annual outage time at 25.06 hours, a reduction of 28.14 hours compared with 2005.

Breakthroughs in Power Grid Technology

As China's power grid development becomes more technology-oriented, a relatively well-established technology research system with a team of technology talents is now in place. Thanks to independent innovation, we have managed to launch 1,000 kV UHV AC Demonstration Projects and associated expansion projects as well as the ±800 kV UHV DC Demonstration Projects. We have acquired the core technologies of UHV transmission and transformation and equipment manufacture. What's more, we have developed a portfolio of world-leading products using proprietary intellectual property, enabling us to secure a commanding position in the world's power grid technology.

Continuous Enhancement of Cross-regional Power Transmission Capacity

In order to fully leverage the optimal allocations of power grid resources, China has strengthened and improved the major grid structure in different provinces and regions in recent years, while speeding up the building of cross-regional power transmission channels to enhance transmission capacity. By the end of 2010, the capacity of cross-regional power transmission had reached 40.20 GW, up 2.3 times from 2005. The total volume of cross-regional power transactions for the whole year totalled 149.2 TWh, representing an increase of 85.9% compared with 2005.

4.3.1.2   Problems and Challenges Facing Grid Development

Grid Development is Generally Lagging Behind

For a long time, the development of electric power in China has been marked by a long-standing approach to putting power generation ahead of power supply, resulting in the development of power grids falling behind that of power generating facilities. From an investment perspective, the nation's cumulative investment in the electricity industry since the founding of the new China has totalled RMB6.9 trillion. Of this amount, RMB2.68 trillion or 38.8% has gone into grid development, falling far short of the international standard of 50–60% for power grid development. Over the past few years, China has significantly expanded its investment in power grids. During the '11th Five-Year Plan' period, cumulative investment in power grid development amounted to RMB1.46 trillion, while RMB1.69 trillion went into power generating facilities. With the investment in power grids taking up 46.4% of the electricity industry's total investment for the relevant period, the imbalance between power grid development and power generation has been eased to a certain extent. However, the situation of lagging power grid development has not changed fundamentally.

Insufficient power grid investment has contributed further to the weakness of both UHV and distribution networks in China. On the one hand, the construction of the large-scale, long-distance and high-efficiency power transmission corridors that extend from the resource bases to the load centres is lagging far behind and needs to be sped up. On the other hand, such problems as the irrational structure of the power distribution networks, inadequate power supply capacity and some other 'bottleneck problems' still exist, and power supply reliability has yet to be improved.

Growing Power Demand and Power Generation Development Dictate More Stringent Requirements on the Grids

In the coming 20 years, the electricity demands in China will continue to grow rapidly and the eastern and central regions will be the load centres in China. Studies show that China's peak loads in 2015, 2020 and 2030 will respectively reach 1010 GW, 1410 GW and 1940 GW, or respectively 1.5 times, 2.1 times and 3.0 times the level in 2010. Total power consumption nationwide will respectively top 6300, 8600 and 11 800 TWh, or respectively 1.5 times, 2.0 times and 2.8 times the level in 2010. Power consumption in northern, eastern and central China will respectively account for 66.3%, 65.1% and 64.1% of the national total.

In the future, China's power generation infrastructure will mainly be distributed in the energy resource-rich western and northern regions. In line with resource distribution, development potential and transport conditions, China needs to continue to strengthen development of the coal-fired power bases in western and northern China, the hydro power bases in southwestern China, the nuclear power bases in the coastal regions as well as major wind power bases and solar power bases. It is expected that during the '12th Five-Year Plan' period, 65% of the coal-fired generation will come from the coal-fired power bases in western and northern China, 80% of the additional installed capacity of hydropower will come from the hydropower bases in southwestern China, and 80% of the additional installed capacity of wind power will come from the wind power bases in the 'Three North' Region.

The growing electricity demands and the distribution of the newly installed power plants show that China will further increase the cross-provincial and regional power flows and extend the distance of power transmission in the future. To ensure the security of power supply and optimise resource allocation across the country, we must enhance power grid transmission capacity and achieve high efficiency in energy transport.

New Challenges from New Energy Development and New Modes of Consumer Power Services

Generation of new energy, such as wind and solar power, is characterised by randomness and intermittence and less controllable and predictable than conventional power generation using fossil fuels. The development and utilisation of new energy on a large scale has posed great challenges to grid coordination and control, so we must make use of advanced automation, coordinated control and energy storage technologies to accurately predict and control the various energy resources including new energy. We should also improve the power output features of new-energy power generation and ensure that the power grids can meet the massive generation and network interconnection requirements for new energy. By doing so, we may better promote the optimisation and adjustment of energy structure and reduce the reliance on conventional power generation using fossil fuels.

In the meantime, the progress of society and the development of new sources of power consumption such as electric automobiles and smart household appliances have placed higher demands on power quality and the content of power services. As a result, electricity suppliers have to provide power supply solutions that are safe and reliable, economical and high quality, flexible and interactive, user-friendly and open, while expanding the content and scope of power services to offer more variety, convenience and flexibility. This will bring about two-way interaction between users and grids, providing customers with access to more choice and control.

4.3.2   The Future Landscape of Power Flows

4.3.2.1   Size of End-user Market

The areas around Beijing, Tianjin, Hebei and Shandong, the eastern four provinces in central China as well as eastern China and Guangdong province are relatively well-developed, with a rather substantial demand for electricity. As future electricity demands will continue to grow rapidly, these resource-scarce areas will become important end-user markets of electricity in China.

Based on the power demands and the requirements for newly-added capacity in these end-user regions, and taking into consideration of all the following factors as the coal-fired power generation projects being under construction or having been approved or planned, the imported hydropower and coal-fired generation capacities from other areas, and the decommissioned units as well, the size available for the end-user markets in these regions can be worked out on an electricity balance calculation basis. In the years of 2010–2015, the total market demand in the end-user regions is estimated at 200 GW, before rising to 400 GW by 2020.

4.3.2.2 Development and Output of Large Energy Bases

According to the distribution characteristics of China's energy resources and from the standpoint of the sustainable development of energy and electricity, China will focus on the massive development of coal-fired power bases, hydropower bases, nuclear power bases and renewable energy power bases in regions richly endowed with resources in the future in order to meet the electricity demand for socioeconomic development. In terms of geographical distribution, China's large nuclear power bases will mainly be built in coastal areas and inland energy-scarce provinces close to the load centres, and the power generated will be consumed locally. However, the large coal-fired power bases and renewable energy power bases, such as wind power bases in the western and northern regions, and the large hydropower bases in the southwest are far away from the load centres, the power produced in these bases needs to be transmitted over long distances to the load centres for consumption.

Power generated in the large coal-fired power bases will be transmitted in the following directions: from western Inner Mongolia and Shanxi to central and northern China; from Xilin Gol League to northern and eastern China; from Hulunbeir League and Baoqing to northeastern and northern China; from Shaanxi to northern and central China; from Ningxia to northern and eastern China, and from Xinjiang to central and eastern China. The volume of power exported from the major coal-fired power bases in Xinjiang, Gansu, the 'Three West' Region, eastern Inner Mongolia and Heilongjiang Province is estimated at 189 GW by 2015, 286 GW by 2020 and 330 GW by 2030.

Power generated in the large hydropower bases will be transmitted from Sichuan to central and eastern China and from Yunnan to Guangdong. The volume of hydropower exported from the southwestern region will reach 54.5 GW, 76 GW and 120 GW by 2015, 2020 and 2030 respectively.

Power generated in the large-scale wind power bases will be transmitted in the following directors: from Northwestern Jiuquan and Hami mainly to central China; from western Inner Mongolia (including Xilin Gol League) and Hebei to northern, central and eastern China; from eastern Inner Mongolia mainly to northeastern and northern China. Power generated in Jilin will mainly be consumed in northeastern China, and the wind power of Jiangsu and Shandong will be mainly consumed locally. It is expected that the scale of transprovincial and transregional export of wind power of China will reach 34 GW by 2015, 74 GW by 2020 and 110 GW by 2030.

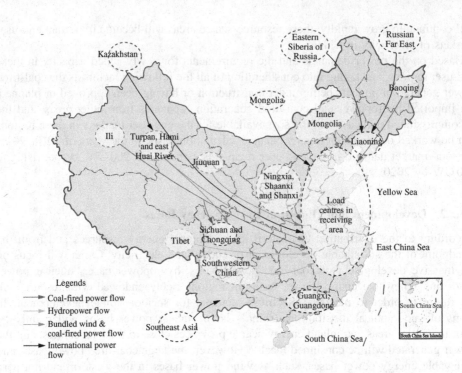

Figure 4.13 Future pattern of China's power flows.

4.3.2.3   The Overall Pattern of Power Flow

Given the load distribution and power sources allocation in China, a flow pattern of electricity being transmitted from 'the west to the east' and from 'the north to the south' on a massive scale will take shape (see Figure 4.13). The hydropower in southwestern China as well as coal-fired electricity and wind power in western and northern China will be fed into the power grids in the load centres of northern, central, eastern and southern China through the transregional power grids. At the same time, those countries and regions bordering with China and with rich energy resources for power generation, such as Russia, Mongolia, Central Asia and Southeast Asia, also have the potential to transmit power through the nearest routes to the load centres in China.

In the future, China's electricity flow pattern will dictate higher requirements for power grid transmission capacity, and there is an urgent task to enhance the transregional power transmission capacity to achieve network interconnection and optimal allocations of resources over larger areas. China's cross-border and transregional power transmission capacity is expected to reach 450 GW in 2020, or about 30% of the national total, among which the electricity inflow into the eastern and central regions will reach 350 GW. It is expected that the flow of coal-fired power and wind power will be further increased by 2030. Hydropower will be developed in Tibet and transmitted to other regions on a large scale. Efforts will be made to bring Kazakhstan's electricity to China. As a result, the volume of cross-border power transmission will grow and the transmission capacity of UHV and transregional and cross-border power grids will increase to about 500 GW.

4.3.3 The Thinking Behind SSG Development

4.3.3.1 The General Thinking

Given the functional role of power grid, the current situation of power grid development and the need for optimising the allocation of energy resources, the future development of China's power grids should be closely linked to national energy strategy deployment and be well-attuned to the requirements in terms of power development, user demand, energy conservation and emission reduction in vigorous promotion of the 1U4L strategy. By capitalising on advanced UHV power transmission and smart grid technologies, the construction of SSG should be accelerated with UHV power grids as the backbone and the coordinated development of power grids at all levels. The SSG development in China is designed to comprehensively enhance the resource allocation capability, security, stability and economic operational efficiency of power grids so as to build up a safe, environment-friendly, efficient and interactive modern grid system as a green platform for optimising energy resource allocation, a service platform to meet the varied demands of users, and an infrastructure platform to protect national energy security.

4.3.3.2 Basic Guiding Principles

China's SSG development should adhere to the following basic principles.

Equal Emphasis on the 'Strong' and 'Smart' Aspects

The high level of integration of the strong grid with smart technology reflects the inherent requirement and direction for the development of China's power grids. The fast-growing demand for power and the imbalanced distribution of power generation sources and power loads dictate the need for a strong grid structure, with large-scale power transmission and safe and reliable supply to meet the necessary optimal allocation of resources over large areas. Meanwhile, in order to meet the user demand for power quality and increasingly diversified power services, grid operation and control should be made more flexible and efficient, with a high level of automation and adaptive capacity. Therefore, the development of China's grids must reflect an equal emphasis on the 'strong' and 'smart' aspects to provide adequate power supply and more smart power services for consumers. 'Being strong' is the foundation of smart grids, and 'being smart' is essential to maximising the role of strong grids. The two requirements are supplement and complement to each other and form a dynamic unity.

Coordinated Development

Grid development should be coordinated with socioeconomic development. It should be incorporated into the overall plans for socioeconomic growth, with a level of construction appropriately ahead of time to support sustainable expansion of the economy. Grid development should be coordinated with power generation development, with centralised planning for power grids and power generation as the priority where the structure and distribution of power sources is optimised to collaboratively promote the development of UHV power grids and large-scale coal-fired, hydro, nuclear and renewable energy power generation bases. Power grids at all levels should feature coordinated development and effective convergence to shape a grid infrastructure with a clear structure, well-defined functions and reasonable compatibility, so that coordinated development can be achieved between transmission and distribution, between active and reactive power and between primary and secondary energy sources.

Efficient and Intensive Development

Technological innovation, key technology research and equipment development should be strengthened while the application of new technologies, new processes, new equipment and new materials should be promoted. The upgrading and transformation of existing power grids should be enhanced and grid losses reduced. We should promote the adoption of typical designs and carry out standardised and serialised development in line with the principles governing advanced technology, optimised design, unified standards and land conservation. The focus should be placed on the input-output analysis of the grid, the balance between business benefits and social benefits and the unity of advanced technology and economic efficiency so as to avoid duplicate investment and maximise the overall efficiency of grid infrastructure.

Safe and Environment-friendly Development

Efforts should be made to follow the objective laws of grid development, optimise the structure and distribution of the grids, and meet all security and operational stability standards specified in the Guide on Security and Stability. The development of automatic safety devices should be strengthened and a reliable 'Triple Bottom Line' approach[6] established. Advanced research on the response measures for failures should be strengthened, differential planning and development carried out, the rapid response capability of the grids and the ability to withstand severe natural disasters improved, and the safe and stable operation of major grids protected. During the grid planning and construction processes, the requirements for environmental protection should be met in all respects, national regulations on environmental protection strictly complied with, environmental management standardised, green construction technology applied, and environmental control on noise, pollutants and waste disposal strengthened, so as to achieve harmony between grid construction and the natural environment. The development of clean energy should be actively supported and the use of green power promoted.

4.3.3.3 Outlook for China's Power Grid Development

The future development of power grids in China should focus on resolving the problem of being 'weak at both ends', namely the UHV and distribution grids, on the basis of

---

[6] The 'Triple Bottom Line' offers a safeguard for the secure and stable operation of China's power grids. The first bottom line refers to the situation when a relay protection device will promptly cut off the faulty component of a power grid to maintain normal power supply in the event that the power grid is likely to have a single fault. The second bottom line moves into action when the grid has encountered the unlikely incident of a single serious failure, in which case a stabilising control device will be activated and such measures as equipment shut-off/load switch-off taken to ensure the stable operation of the power system. The third bottom line is put to work when the grid concurrently experiences a highly unlikely multitude of serious faults that disrupt its stability, in which case the grid is disconnected and load losses minimized through an emergency control device to prevent a system crash.

scientific planning to accelerate the development of the backbone UHV grid, coordinate the construction of power grids at all levels, strengthen the development of distribution grids and improve the urban and rural power grids so as to form an SSG with an appropriate grid structure and powerful resource allocation capabilities.

It is expected that by the end of 2020, the backbone of a UHV synchronous grid covering northern, eastern and central China will be in place, while the power grids in northeastern, northwestern and southern China are linked by DC back-to-back asynchronous interconnections with the synchronous grids in northern, eastern and central China (see Figure 4.14). This will meet the demand for power transmission in the large coal-fired, hydro, nuclear and renewable energy bases and build a strong network platform for receiving electricity on a large scale in the load centres in the eastern and central regions. UHV AC and UHV DC developed concurrently are complementary to each other, forming a strong end-use grid covering the load centres in Beijing, Tianjin, Hebei, Shandong, the eastern four provinces in central China and the Yangtze River Delta Region. Coal-fired electricity in northern China and hydro power in southwestern China transmitted to the end-use grids via multiple UHV AC/DC circuits are able to meet the power transmission needs of the large coal-fired, hydro, nuclear and renewable energy power generation bases. Rapid development of distribution grids should be pursued, the grid structure strengthened, and both power supply capacity and reliability greatly improved. The network functions at different voltage classes should be clearly positioned, structural robustness and coordinated development achieved, key smart technology and equipment widely adopted, and smart technology development covering the whole grid basically accomplished, which should fully meet or even surpass world-leading standards in terms of technological and economic performance and equipment quality.

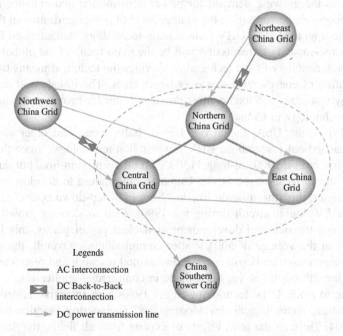

Figure 4.14   Future development pattern of China's power grids.

It is expected that by 2030, the UHV grid framework will pave the way for stronger end-use grids in northern, eastern and central China and strong supply grids in the northeastern and northwestern regions. The UHV power grid has strong load factor capacity, which is able to achieve high capacity, long-range transmission and consumption, while guaranteeing system security and stability. Xinjiang's coal-fired power output and Tibet's hydro power output can be sent out on a large scale by UHV DC transmission. The superconducting power transmission technology can be applied, and the distributed energy systems can become an important supplement to centralised power supply through large power grids. All the large power supply bases, distributed power supply systems and the users and load centres are closely connected by transmission and distribution grids nationwide, with unimpeded transmission capacity to provide a smooth, efficient, safe, and stable foundation and platform for electricity market trading, so that the optimal allocation of energy resources can be fully achieved nationwide. The urban and rural distribution grid structures should be further strengthened and rationalised to raise power supply capacity and reliability to a new height and ensure a supply of safe, reliable, clean and affordable electricity for end-users. The national grid will facilitate 'power transmission from the west to the east', mutually supply between north and south as well as nationwide interconnections, while achieving a balanced supply and demand situation and mutual support countrywide to yield the benefits of nationwide interconnections.

4.3.4   Development of UHV Grids and Grids of All Levels

4.3.4.1   Accelerating Development of Backbone of UHV Grid

In order to meet the growing demand for power transmission and enhance transmission capacity and economic performance, the voltage level of power grids around the world has continued to be upgraded. The UHV transmission technology characterised by ultra long distance and large-capacity transmission will be the main focus of the global competition in transmission technology as the technology provides the technical means to achieve the optimal allocation of energy resources over larger areas. The imbalanced distribution of China's energy resources and load centres provides room for broad applications of UHV transmission technology in China.

Since the 1960s, the USA, the Soviet Union, Italy, Japan and other countries have successively carried out research on UHV transmission technology. Since the 1980s, the Soviet Union has built a 1900 km-long 1150 kV AC transmission line, but due to, among other reasons, the collapse of the Soviet Union and the failure to develop power sources in a timely manner, this transmission line has been in a step-down operation. Japan built a 427 km 1000 kV double-circuit during the 1990s. Due to slowing growth in demand for electricity and the deferred development of nuclear power plants, this line has also been operating at the voltage of 500 kV after commissioning. Overall, the development of UHV in foreign countries is still in the experimental research and practical exploration stages, and a breakthrough has yet to be made in commercial application.

China began to study UHV technology in the 1980s,with intensive efforts in research and demonstration, technological development and engineering practice initiated since the end of 2004. Through the joint efforts of experts from all fields, the development of UHV transmission technology in China has continued to make breakthroughs, overcoming challenges one after another. UHV AC pilot projects and UHV DC demonstration projects developed and commissioned have maintained safe and stable operations. This marks the maturity of China's UHV technology and will lay a solid foundation for further large-scale applications.

UHV Development Having Undergone Scientific Research and Demonstration

On 21 March 2005, Peiyan Zeng, the then Vice Premier of the State Council of the People's Republic of China, chaired a special meeting where the UHV work reports of State Grid Corporation of China were received. The meeting minute states that the development of UHV grids is a concrete manifestation of the power sector carrying out the Scientific Outlook on Development, an important guarantee to meet the growing demand for electricity in China, and an effective way to optimise energy allocations. UHV grid development is conducive to promoting the coordinated development of power grids and power sources, and also greatly improving the security, reliability, flexibility and economic efficiency of the grids. It can help boost the innovative development of proprietary power technology and enhance the capacity of domestic power transmission and distribution equipment manufacturers to upgrade our manufacturing technologies of DC/AC power transmission and distribution equipment and improve our competitive edge in the international market. This highlights the necessity and significance of China's UHV grid development.

Since government permission was given to launch UHV projects in 2005, State Grid Corporation of China has adhered to the guiding principle of 'scientific argumentation, priority demonstration, independent innovation and solid progress' under the strong support of the National Development and Reform Commission, the National Energy Administration and the Ministry of Science and Technology, by rallying various parties to carry out in-depth planning studies, technical demonstration, equipment R&D, project construction and so forth. Based on the foundation of past work and in light of current technological development, more than 300 key technical research studies were carried out on the necessity, feasibility, safety, economic efficiency and environmental impact of UHV development. Expert opinions, especially those from different sources, were collected at UHV international conferences, expert consultation meetings and topical seminars. More than 3000 scientific researchers, engineering and technical specialists, including over 30 academicians, and 11 local and international institutions and organisations, participated in UHV demonstration and over 240 important specialist meetings on demonstration were held. Major local power research and design institutions and 9 universities were involved in UHV research and design. More than 500 construction firms and over 100 000 personnel participated in UHV project construction. More than 200 equipment manufacturers participated in the development and supply of equipment. According to the demonstration results, UHV development is economically viable and technically feasible while safety and environmental well-being can be assured. Development of UHV grids can greatly improve the security, reliability, flexibility and economic efficiency of power grids while promoting structural adjustment, optimising power resource distribution and resolving the issue of energy sustainability. The operating results of the UHV pilot projects show that the environmental benchmarks for the UHV lines are fully compliant with the requirements of national standards. According to the *Advisory Opinions on the Research and Engineering Construction of China's UHV Power Transmission*, proposed by a research group comprised of 27 academicians and 7 experts from four relevant divisions of the Chinese Academy of Engineering, 'as determined by the characteristics of the country's resource and load distributions, UHV transmission technology has important potential for engineering applications and it is necessary to accelerate the pace of research and engineering applications for UHV transmission'.

Based on the result of scientific research and demonstration, UHV development has been included in a number of national programmes. Since 2005, the development of UHV has been incorporated in The *National Outline for Mid-Term & Long-term Scientific and Technical Development (2006~2020)* (2005), *Opinions of the State Council on Speeding up the Revitalization of Equipment Manufacturing Industry* (2006), *The Outline of the Eleventh Five-Year Plan for the National Economic and Social Development of the People's Republic of China* (2006), *Introduction of China's National Climate Change Programme* (2007) and *The Outline of the 12th Five-Year Plan for the National Economic and Social Development of the People's Republic of China* (2011).

UHV Development Having Undergone Verification through Practice
In recent years, China has made important breakthroughs in UHV technology through the joint efforts of different sectors. We have fully mastered the core UHV technology and are able to produce with our proprietary technology a wide range of key equipment, including UHV transformers, reactors, 6-inch thyristors and bulk-capacity converter valves. In addition, a batch of projects as Jindongnan-Nanyang-Jingmen 1000 kV UHV AC Demonstration Project together with its expansion works, Yunnan-Guangdong (5 GW level) and Xiangjiaba-Shanghai (7 GW level) ±800 kV UHV DC demonstration projects have been built successfully. Jingmen Substation as a UHV AC Transmission Demonstration Project and Fengxian Converter Station as a UHV DC Transmission Demonstration Project are shown in Figures 4.15 and 4.16 respectively.

Figure 4.15 Jingmen substation as a UHV AC transmission demonstration project.

Figure 4.16　Fengxian converter station as a UHV DC transmission demonstration project.

The Jindongnan-Nanyang-Jingmen 1000 kV UHV AC Demonstration Project was approved by the state in August 2006, followed by construction in December the same year and commercial operation in January 2009. As of January 2012, the project has been running continuously and safely for three years at the rated transmission capacity, withstanding the test of different weather conditions like freezing, lightning, heavy rain, strong winds and high temperatures. It plays a pivotal role in allocating energy resources over large areas, promoting intensive development of coal-fired power in northern China, fully utilising hydropower resources in central China and easing the power shortage in central China. It has become an important corridor between the north and the south for energy transport. With the commissioning of the extension equipment such as the second transformer cluster in December 2011, the transmission capacity of this line has now reached 5 GW. Having passed the acceptance tests administered by the Ministry of Water Resources, the Ministry of Environmental Protection, the Ministry of Science and Technology and the National Energy Administration in terms of soil and water conservation, environmental protection, technological breakthrough and equipment localisation, the UHV AC transmission demonstration project completed the final national acceptance and handover formalities organised by the National Development and Reform Commission in September 2010. The project has been awarded the 'China Industry Award' and the 'National Gold Prize for Excellent Projects'.

The Xiangjiaba-Shanghai ±800 kV UHV DC transmission demonstration project, 1907 km in length with a maximum transmission capacity of 7 GW, was approved by the state in April 2007, followed by construction in December the same year and commercial operation in July 2010. As of January 2012, it has been operating continuously and safely for more than one and a half years to assure electricity supply for the Shanghai World Expo. Upon full completion, Xiangjiaba Hydropower Station can transmit 35 TWh of clean hydropower to Shanghai annually.

The successful construction and operation of the UHV AC/DC transmission projects has confirmed the technical feasibility, system security, equipment reliability, economic efficiency and environment-friendliness of UHV technology, attesting to China's manufacturing capability and leadership in the global power transmission market. This also shows that China has commanded an internationally leading position in UHV research expertise, equipment manufacturing, as well as construction and operation capabilities, with the capacity for large-scale application of UHV transmission technology. In order to better satisfy the power delivery needs of the hydropower bases in southwestern China and the large coal-fired power bases and wind power bases in Xinjiang, the country is carrying out further research on ± 1100 kV HVDC transmission technology. In the future, China will continue to innovate UHV transmission technology so as to provide technical support for the development of UHV synchronous grids in northern, eastern and central China and realise the optimal allocation of resources nationwide.

The Achievements in UHV Development are Widely Recognised

The achievements of China in the research on UHV technology and engineering practices have commanded widespread electro-technical industry attention in the world. The International Electro-technical Commission, the International Conference on Large High Voltage Electric System (CIGRE) and other international organisations and a number of internationally-renowned experts have spoken highly of China's achievements. They agree that the construction and operation of UHV AC/DC demonstration projects is a milestone in the development of power grid technology in the world and China has become a global leader in UHV transmission technology.

An international conference on UHV transmission technology held in 2006 outlined the latest progress and results in research on UHV technology at home and abroad and fully recognised a number of important breakthroughs achieved in China. At the meeting, a consensus was reached on the feasibility, safety, economic efficiency and technical superiority of UHV transmission technology while the significant advantages and comprehensive benefits of UHV transmission development and the promising prospects for this area of development in China were unanimously agreed upon.

Minutes of 2006 International Conference on UHV Transmission Technology (29 November 2006, Beijing)

From 27th to 29th Nov 2006, representatives from electric utilities, research institutes, consulting companies, associations, universities, financial organisations and equipment manufacturing enterprises from 19 countries and regions attended 2006 International Conference of UHV Transmission Technology in Beijing. The representatives conducted in-depth discussion concerning the challenges faced by electric power industry, studied the goal and orientation of the electric grid development, exchanged the research result of the UHV power transmission technology and reached consensus.

Discovery and application of electricity is one of the most important achievements in human history. As an economical, convenient and clean energy, the consumption level of the electricity is an important symbol of a society's civilisation. Electric power industry is an important base for economic and social development and thus plays an important role in promoting the world economic development and advancement of human being society.

Technology innovation is the drive and source for electric power development. Ever since electricity was invented, generations of electric power engineers have kept making active innovations and continuous efforts to promote the great progress of the electric power development. The scale of electric grid keeps expanding, the voltage grade keeps improving and the economic status and reliability of the electric power system operation keeps improving, which greatly support the economic and social development.

Since the middle of last century, human society began to face the dual burden of resources and environment, which posed new challenges to electric power industry. To meet the needs of continuous and rapid growth of the electric power, some countries in the world have made research on 1000–1600 kV UHV transmission technology, among which the former Soviet Union, Italy and Japan have completed UHV industrial pilot projects.

We realised that the UHV transmission has the advantages of large transmission capacity, long distance power transmission, low line loss, economical investment, efficient utilisation of corridor and strong network connecting ability, etc. It is the important development orientation of the power transmission technology and thus has a significant role in promoting the innovative development of the world electric power industry.

According to the available research result and the construction experience, we believe that the UHV transmission has no insurmountable obstacles in technology; it has stepped into a construction application period and has a broad application prospect.

We have noticed that the Chinese electric power demand keeps growing while the energy resources and electricity load centre are extremely imbalance. It is necessary to build a large scale and long distance AC and DC hybrid electric grid. The construction of the UHV grid could promote the highly efficient development and utilisation of China's energy resources and thus facilitates the sustainable development of China's energy industry.

We appreciate the efforts that State Grid Corporation of China (SGCC) made in the aspects of developing the UHV power transmission technology. In recent years, China actively promoted the research of UHV transmission technology and made breakthrough and thus further improved the UHV power transmission technology. China's construction of the 1000 kV UHV AC and ±800 kV UHV DC pilot projects does not only have an important significance to China, but also to many other countries and regions.

To build the electric grid, develop the electric power and serve the global economic growth and social progress are the historical task we are facing. We propose that technology communication mechanism of UHV transmission should be built to strengthen the technical exchange and cooperation, share the experience and knowledge, promote the international research and application of the UHV transmission technology and deal with the challenges faced by the sustainable development of the electric power industry in the new century.

The 2009 International Conference on UHV Power Transmission was held four months after China successfully built and commissioned its UHV AC demonstration projects. Experts attending the conference spoke highly of the achievements made in UHV transmission technology by State Grid Corporation of China and its cooperative partners at home and abroad. They believed that by making a historic contribution to the development of power grid technology internationally, China has taken a global leading position in this field and reached a major milestone in high voltage technology. It is generally agreed among the experts that based on the earlier research and practices conducted by countries including the Soviet Union and Japan, UHV transmission technology has become mature enough to be applied to engineering projects thanks to the innovation, research, construction and operation of the UHV projects in China, with the Chinese experience providing a benchmark for many countries in the world.

Minutes of 2009 International Conference on UHV Power Transmission (22 May 2009, Beijing)

The 2009 International Conference on UHV Transmission, convened in Beijing on May 21–22, has attracted representatives of government departments, industrial organisations, power utilities, research and design institutes, manufactures, universities and financial originations from 21 countries and regions. The conference has provided an ideal platform of exchange and communication among the participants on the state-of-the-art UHV transmission development to reach a broad consensus.

Energy provides the material foundation for the sustainable development of the human being. Electricity is a clean and efficient secondary energy. To meet the growing demand of economic and social development in the 21st Century, countries around the world are facing new challenges and heavy pressures coming from resources and environment in the process of developing clean energy, high efficiency power transmission while promoting energy conservation and environmental protection. Practice proves that technological innovation is an effective way to address the challenges and also a driving force for energy and power sustainable development.

UHV power transmission shows the way to the future with its advantages in transmitting more power over longer distance with higher efficiency, lower investment and smaller land occupation. Study on UHV power transmission technology started in former Soviet Union, Japan and Italy in 1960s–1970s. In recent years, China together with other countries has been working on this area vigorously. State

Grid Corporation of China (SGCC) has successfully launched the first commercial 1000 kV UHV demonstration project in the world on Jan 6th, 2009, marking a milestone of innovation and breakthroughs in the global power industry. It has validated the technology feasibility, equipment reliability, operation safety and environmental friendliness of UHV power transmission technology, and accumulated positive experience and thus laid a solid foundation for wider application of the technology around the world.

It is inevitable for China to develop UHV power transmission technology with the context of the mismatch between energy resources and energy demand on one hand, and the need for transregional interconnection and the conservation for both energy and environment on the other hand. The development of a strong power grid with UHV network as the backbone is of great significance to the sustainability of the economy. We appreciate the efforts and contribution of SGCC and its partners in UHV power transmission development. The experience of China in this area is an important reference for many other countries in the world.

Smart grid technology is a new global trend. With remarkable attributes in securing energy supply, increasing energy efficiency, improving energy structure, addressing the climate change and improving customer service, smart grid technology has been incorporated into the national strategy in some countries, and are used as an important measure to fight against the global financial crisis in other countries. How to develop the smart grid is a country-specific issue. The result reached by State Grid Corporation of China after extensive research is enlightening to all those who are interested in smart grid. The progress made and experience accumulated so far in smart grid development around the world have paved the way for the roll-out of this promising technology.

We would like to call for wider communication and cooperation, sharing of experience, knowledge and achievement, in UHV power transmission and smart grid technologies, across the global energy community. So we can jointly develop a modern power system with strong infrastructure and smart control technologies to support the harmonious development of the economy, society and the environment.

At the International Forum on Smart Grids held in 2011, participants exchanged views over the latest achievements in the theoretical research on and day-to-day practices of smart grids around the world and looked into the UHV and large power grid technologies. Representatives attending the forum generally agreed that large power grids and smart grids would be the way of the future for development of power grids worldwide, while super large and super long-distance power transmission technologies would command a key role in building competitiveness in the world of transmission technologies. China's achievements in the development of UHV transmission technology also won highly positive comments at the forum.

Through development and practice, China has gained useful experience and insights into the wider application of UHV transmission technology around the world and supported the development of UHV technology globally. Countries like India, Brazil, Russia and South Africa have proposed schemes or plans for long-distance and large-capacity power transmission based on UHV technology while enterprises and organisations in such countries and regions as the USA and the EU have strengthened their cooperation with China in UHV transmission technology.

Committing to Building 'Equally Strong AC and DC Circuits' in UHV Power Grid Development

The functions and characteristics of AC transmission and DC transmission are different. AC transmission has the dual functions of transmission and grid construction similar to an 'expressway network', where unloading at any point of the way is allowed with very flexible grid access and power transmission and consumption to form the basis for the safe operation of the power grid. A higher AC voltage class means a stronger grid structure and greater transmission capacity. DC transmission only serves the function of transmission and cannot form part of a network. It can be compared to a 'direct flight' where getting off midway is not allowed, making it only suitable for large-capacity, long-distance transmission. In light of China's energy situation and load distribution characteristics, UHV AC is positioned for building the main grid framework and facilitating power transmission over wider areas to allow the nearest grid access for the large coal-fired, hydro, wind and nuclear power generation bases covered by a UHV AC power grid. An UHV DC power grid is positioned for long-range, high-capacity power delivery from large energy bases such as the hydropower bases in southwestern China, the coal-fired and wind power bases in northwestern China and Xinjiang as well as cross-border power transmission.

China should place equal emphasis on AC and DC transmission modes to ensure coordinated development in lock step when pursuing UHV projects. Due to its reliance on AC grid commutation principle, high-capacity DC power transmission cannot operate normally unless a stable AC voltage is maintained. This means that it must rely on a strong AC power grid to be functional. The research findings[7] of CIGRE show that insufficient strength of an AC power grid with multiple DC feeds will cause simultaneously multiple DC commutation failures when the AC system breaks down, leading to a voltage collapse and power outage over a large area. A large amount of computational analysis on the domestic power industry further elucidates that the absolute limit of DC transmission capacity fed into an AC grid depends on the structure and scale of the AC grid. In other words, a more powerful grid at the power-receiving end can accept higher DC transmission capacity. The relationship between UHV DC transmission and a UHV AC grid is like that between a 10 000-ton ship and a deep-water port where a larger ship requires a larger port with greater water depth for better manoeuvrability. Therefore, from the perspectives of safety and technology, we cannot solve the problem of sustainable development of power grids in China by relying solely on DC transmission, and we need to develop UHV DC transmission capacity on the basis of a strong UHV AC power grid to form a grid framework where strong AC and strong DC circuits can smoothly work together to complement and support each other. By building a hybrid power grid with 'equally strong AC and DC circuits' and making the best of the functions and advantages of the two transmission modes, we can ensure the security and economy of the power grid.

---

[7] CIGRE Working Group B4-41, Systems with multiple DC feeds, 2008.

The Objectives of Developing UHV Grid Framework[8]

'The 12th Five-Year Plan' period marks an important stage in the development of UHV grids in China, and construction will be accelerated of the UHV synchronous grids in northern, central and eastern China on the basis of UHV AC demonstration projects. It is expected that by 2015 the UHV synchronous grids in northern, central and eastern China will form a framework of 'three longitudinal and three horizontal corridors' where the energy bases located in Xilin Gol League, western Inner Mongolia and Zhangbei will transmit power to northern, central and eastern China through the three longitudinal UHV AC transmission corridors while the coal-fired bases in Shaanxi and Ningxia and the hydropower bases in southwestern China will transmit power to northern, central and eastern China through the three horizontal UHV AC transmission corridors. China will build the three longitudinal power transmission corridors of Xilin Gol League-Nanjing, Zhangbei-Nanchang and western Inner Mongolia-Changsha, as well as the three horizontal power transmission corridors of Northern Shaanxi-Weifang, Jingbian-Lianyungang and Yaan-Shanghai.

At the same time, during the '12th Five-Year Plan' period, China will develop 15 UHV DC transmission projects in line with the development of hydropower bases in southwestern China and the coal-fired and wind power bases in the 'Three North' Region (northeastern, northern and northwestern China).

By 2020, China's UHV power grid will have formed a strong grid platform built on the UHV synchronous grids in northern, central and eastern China to meet the power transmission needs of the large coal-fired power, hydropower, nuclear power and renewable energy bases and the requirement for coordinated AC and DC development.

4.3.4.2   Accelerating Grid Development at All Levels

In the future, as the development of the main UHV grid framework is strengthened, we need to coordinate and advance the development of power grids at all levels by refining the power grid structure and defining rational grid hierarchies and segments to bring about an organic integration of power grids of all voltage classes and coordinated AC-DC development.

Speeding up Construction of 750 kV Grids

Construction of a 750 kV grid in northwestern China, especially Xinjiang, will be expedited and the network extended to northern Shaanxi, western Qinghai and southern Xinjiang to meet the need for leapfrog economic development of northwestern China, in particular Xinjiang, and provide strong support for UHV/HVDC delivery projects in the coal-fired power bases in Hami, Zhundong, Binchang and eastern Gansu, the wind power bases in Hami and Jiuquan, and the solar power bases in the northwestern region. The main load centres in northwestern China will be served by a 750 kV ring grid structure, with significantly strengthened supply reliability to ensure safety and reliability for the consolidation of intraregional power sources, interprovincial power exchanges and cross-regional power delivery.

---

[8] Information on the objectives of UHV grid development is compiled based on the findings of planning studies conducted by State Grid Corporation of China in 2011.

Continued Development of 500 (330) kV Grids

While the UHV backbone is being developed and formed, it is necessary to continuously strengthen and refine the various regional 500 kV grids, optimise and readjust grid structures at all levels by building and upgrading the grids across all voltage classes, opening up the electromagnetic ring grids of all voltage classes in a timely manner, appropriately controlling the short-circuit current level, resolving the problems of excess short-circuit capacity in the regional 500 kV power grids, and enhancing the transmission capacity of all main sections. Attention should be paid to the stability of system voltage and the dynamic reactive support capability of the grids should be enhanced. In the northwestern region, the construction of 330 kV power grids should be strengthened continually for coordinated development with the 750 kV power grids.

Continued Development of 220 kV Grids

The 220 kV power grids in the various provinces will be further strengthened and the coverage of the 220 kV grids gradually extended from the main cities to all the counties to develop dual-circuit power supply and loop power supply around the load centres, forming a relatively independent 220 kV ring grid structure with regional 500 kV substations at the centre. By strengthening the 220 kV grid construction in the key power delivery corridors, the 220 kV grids and the 500 kV/750 kV grids can be better connected and operated while meeting the needs of power delivery.

Strengthened Construction of 110 (66) kV Power Grids

More 110(66) kV substations in regions with fast-growing load demands will be built, with the development of a 110(66) kV grid framework to ensure the availability of at least two trunk circuits and system linkages in the county power grids.

Active Development of DC Transmission Capacity

HVDC development in China started late, but is progressing rapidly. China has now developed three DC transmission voltage classes, including ±500, ±660 and ±800 kV, ranking it first in the world in terms of total transmission capacity and line length.

The large hydropower bases in the southwest and the large coal-fired, wind and solar power bases in the western and northern regions are far away from the load centres in eastern and central China, necessitating large-scale outward power delivery. The development of DC transmission is an important technology option to achieve safe and efficient delivery of electricity. The importation of power resources from neighbouring countries by DC transmission can better avoid grid security and stability issues brought about by different construction standards and operational conventions.

In light of the delivery demands of the energy bases and the trend of technological development in China, efforts will be made to further increase the rated transmission capacity of DC transmission, simplify the series of DC voltage classes, achieve a dynamic synthesis of standardisation and serialisation, facilitate the attainment of cluster effects and economies of scale, and improve the economic viability of DC transmission. Based on different capacities and transmission distances, the DC transmission projects in China should adopt one of the four voltage series of ±500, ±660, ±800, and ±1100 kV, with rated capacity at 3.5 GW, 4.6 GW, 8 GW and 10 GW respectively, and economic transmission distances of <900 km, 700~1350 km, 1100~2400 km and 2200~4500 km respectively.

4.3.4.3 Accelerated Development of Urban and Rural Power Grids

China's distribution network is usually divided into two parts: urban grids and rural grids. Certain differences exist in such aspects as technical standards and equipment and management levels. In the future, to meet the needs of socioeconomic development, it is important to continue to increase investment in urban and rural power grid construction and promote the unified and coordinated development of urban and rural power grids. In the long term, with advances in grid technology and accelerating development of the rural economy and the rural power grids, the difference between urban and rural grids will become narrower and eventually move toward integration, while the boundary between urban grids and rural grids will also gradually blur and even disappear. With the continued development of urban and rural power grids, it is expected that by 2020 the levels of equipment technology and supply reliability of urban and rural power grids will be greatly improved, and the difference between the rural and urban grids will gradually narrow. From a medium to long term perspective, urban and rural grids will gradually be merged after 2030, bringing about the integration and unified standards of the two and marking a new era in the construction and operation of distribution grids. By then a distribution grid with structural rationality, operational reliability, economic efficiency and significantly enhanced smart capabilities will come in place.

Urban Power Grids
With the fast economic growth of China's urban communities and accelerating urbanisation, the urban population will increase sharply, the size of cities will expand rapidly, and the load density of population centres will also rise. With the ever-higher safety and reliability standards for urban power supply that impose a more stringent requirement in the development of urban power grids, urban power grid construction must be aligned with urban development, with the development of urban grids of different voltage classes strengthened, the grid structure constantly optimised, and the voltage classes appropriately simplified. The grid supply capacity should also be increased, power quality improved, system losses reduced and supply reliability enhanced to ensure that electricity can be 'obtained and used' to meet the higher requirements from urban socioeconomic development and people's rising living standards. In order to ensure harmonious development of the urban power grids and the urban landscape, appropriate measures should be adopted in light of local conditions by installing underground transmission cables, indoor substations, underground (semi-underground) substations, fully buried (semi-buried) distribution stations and landscaped and prefabricated distribution stations in urban centres and scenic areas based on the investment capacity in the construction process.

Rural Power Grids
Since 1998, after the first two phases of grid upgrading, power supply quality in China's rural areas has greatly improved as power supply reliability and rural users' access to supply services have been significantly enhanced, with a fairly rapid reduction in overall

line losses. However, the upgraded coverage and the upgrade standards remain low. With the overall progress in the socialist reconstruction for the new countryside, rural socioeconomic development is bound to usher in a period of faster development and greater requirements for the development of rural power grids. In January 2011, at the State Council's standing committee meeting, the decision was made to implement a new round of modification and upgrade projects for rural power grids, with the proposed construction of new rural power grids that are 'safe, reliable, energy-saving, environment-friendly, technologically advanced and operating under standardized management'. During the '12th Five-Year Plan' period, State Grid Corporation of China will invest more than RMB460 billion to resolve the weakness in the rural power grids, aiming to put in place a new form of rural power grid that is 'safe, reliable, energy-efficient, environment-friendly, technologically advanced and operating under standardized management'.

For a long time to come, efforts are needed to continuously strengthen the development of rural power grids by scientifically organising and implementing upgrade projects, promoting standardisation in grid construction and optimising grid structures to achieve an increasingly rational grid system, higher technology levels of the equipment used, and enhanced supply capacity and reliability. At the same time as development and construction of the rural distribution grids is strengthened, vigorous efforts are also required to develop 220 kV rural power grids, optimise 110 kV (66 kV, 35 kV) voltage grids, build up robust 10 kV power grids to form a grid structure with a solid framework and a rational layout to meet the growing needs of rural power loads of all categories. Priority should be given to solving the weakness in the interconnection between the county grids and the main power grids and the problem of a single supply source for key accounts, while ensuring an uninterrupted domestic power supply in rural areas during peak demand periods. Appropriate measures attuned to the local conditions should be adopted to resolve the problem of difficult access to power for certain rural households.

4.3.4.4 Speeding up the Implementation of Cross-border Interconnection Projects

The development of China's UHV and DC transmission technology has created favourable conditions for importing electric power from other countries. In view of her participation in the joint development of energy resources with neighbouring countries, China should further expand interconnection projects with neighbouring countries like Russia, Mongolia, Kazakhstan and Myanmar and actively import foreign electric power. In particular, electricity from Russia, Mongolia and Kazakhstan will be imported and fed into the power grids in northeastern, northern and central China through UHV DC grid interconnections.

As the conventions adopted by Russia, Mongolia and Kazakhstan for power grid operations are different from those observed in China, long-distance and high-capacity DC or DC back-to-back systems should be used for power transmission. Based on the economic delivery distance of DC transmission circuits of different voltage classes, the DC back-to-back system of ±500, ±660 kV and ±800 kV DC transmission schemes should be used for importing electricity from Russia's energy bases; ±500, ±660 kV and ±800 kV DC transmission schemes are suited to the Mongolian energy bases; and ±800 kV and ±1100 kV DC transmission schemes should be adopted for the energy bases in Kazakhstan.

During the '12th Five-Year Plan' period, China needs to intensify efforts on cross-border interconnection projects like the China-Russia DC back-to-back project. During the '13th Five-Year Plan' period from 2020 to 2030, in consideration of the progress of work on bringing power from neighbouring countries to China and also the load development needs of northeastern and northern China, work should start on the DC transmission projects to support power transmission from Russia and Mongolia to Liaoning, Tianjin, Shandong and other regions.

4.3.5 R&D and Application of Grid Technology

Technology innovation is essential to the successful scientific development of power grids in China and technical progress is a key factor to ensure the smooth implementation of grid development strategies. To achieve the development goal of building strong and smart grid networks, China must develop its own capacity for independent innovation and, riding on the new wave of energy technology revolution, increase investment and achieve breakthroughs in certain advanced technology fields in order to provide a solid technological foundation for future power grid development.

4.3.5.1 New Power Electronics Technology

The application and development of power electronics technology represents a major change in the traditional power industry. Power electronics technology can not only improve transmission efficiency, but also enhance the controllability of power systems to realise rapid, continuous and flexible grid control.

The application of power electronics technology in power systems mainly encompasses flexible AC transmission technology (FACTS) and DC transmission technology. Combining power electronics technology with modern control technology, flexible AC transmission is capable of continuously readjusting and controlling power system parameters. Currently, the more widely used flexible AC transmission technologies include: Static Var Compensator (SVC), Static Synchronous Compensator (STATCOM), Thyristor Controlled Series Compensator (TCSC), and Controllable Shunt Reactor (CSR). Given the trend towards higher voltage classes and transmission capacity in the conventional HVDC transmission technology of thyristor clusters, the capacity of the $\pm 1100\,kV$ DC transmission design still under research and development is at the 10 GW class. Thanks to the technologies of multiterminal DC power transmission, capacitor commutated DC and tripolar DC transmission, the scope of applying DC transmission is widening. As the performance of the new generation of power electronics devices based on switchable appliances continues to improve, the new power electronics technology will play a more important role in future power grids. The Unified Power Flow Controller (UPFC) and Voltage Sourced Converter-High Voltage Direct Current (VSC-HVDC) will provide strong support for improving the reliability, economic efficiency, environment-friendliness and flexibility of power grids.

Significant progress has been made in China's power electronics technology industry. 6-inch silicon thyristors independently developed in this country have been successfully applied in the UHV DC transmission field. The flexible AC transmission technology based on thyristor half-controlled devices has matured and commanded a world-leading position in some aspects. Significant technical breakthroughs have also been achieved with the STATCOM based on controlled devices. However, the R&D of high-end power electronic devices still lags far behind internationally advanced levels. Research on high-capacity and high-voltage flexible AC transmission equipment needs to be improved. Multi-Terminal Direct Current (MTDC) and compact converter stations have yet to catch up with world advanced levels. In the future, more investment will be made in R&D of basic components manufacturing, analysis simulators for power electronics technology systems and other technologies in order to narrow the gap with the leading countries.

4.3.5.2 Superconducting Power Technology

Superconductivity is a phenomenon of zero electrical resistance occurring in certain materials when cooled below a certain critical temperature. The application of superconducting materials to the manufacture of electric power equipment can reduce not only losses but also equipment size and weight to save infrastructure investment and make large-capacity and high-density power transmission possible.

Currently, superconducting power technology is still in the research and experimental stage and experimental devices have been developed for various types of superconducting power equipment. A 610 m long 138 kV superconducting cable commenced operations in April 2008 in Long Island, New York, USA. It operates at a record high voltage level with the longest transmission distance in the world. In April 2004, the Puji substation in the Chinese city of Kunming put into operation a 34 m long 35 kV superconducting cable. In April 2011, China's first superconducting substation was officially commissioned in Baiyin City, Gansu Province. The substation features a range of superconducting power devices, including high-temperature superconducting current limiters, high-temperature superconducting energy storage systems, high-temperature superconducting transformers and high-temperature superconducting cables. The facility represents the latest result of China's R&D efforts in superconducting power technology over the past 10 years.

Although some progress has been made in the research and application of superconducting technology, technical bottlenecks are yet to be overcome. More efforts should be directed towards research on the technology of interconnection among high-temperature superconducting components and further improvement of the insulation level and transmission capacity. Since no superconducting properties will emerge unless the superconducting material is kept at very low temperatures, superconducting materials are costly with stringent operational and maintenance requirements. Before major breakthroughs are made in superconducting materials, superconducting technology can only be applied to transmission over very short distance or under very special settings. There is still a long way to go before superconducting technology can be applied to long-distance transmission or the development of large superconducting grids.

4.3.5.3 Control Technology to Ensure Security and Stability of Large Power Grids

With the construction of the UHV power grids and grid-tie projects nationwide, China's power grid system will have the world's highest voltage class and system capacity and also the strongest resource optimisation capabilities. Therefore, research on control technology to ensure the security and stability of ultra large hybrid AC and DC grids is essential for China to enhance the operational standards of its power systems and protect grid transmission capability.

At home and abroad, extensive research on the safety and stability mechanisms and properties for large power grids has been carried out. In the field of technical research into the safety analysis technology for large power grids, research has been conducted in overseas countries on online analysis and protection systems for large power grids, with the results applied in the energy management systems. But large-scale application is yet to be implemented. China has developed an on-line dynamic safety assessment and early warning system for large power grids and promoted its application in some dispatching centres.

In order to enhance significantly the operational safety and stability of China's existing power grid systems while speeding up the construction of power grids, it is necessary to carry out research into grid safety assessment, hybrid AC and DC control systems, power grid failure recovery, and protection and safety control. It is necessary to establish a theoretical system for stability and control of complex AC and DC hybrid systems and online safety analysis, assessment and decision-making methodology to achieve real-time, dynamic and intelligent grid-wide safety assessment. It is also necessary to promote Wide Area Measurement System (WAMS) technology and develop coordination control technology based on multiple data sources to further improve the coordination of the 'Triple Bottom Lines' in China's power grids.

4.3.5.4 Micro-grid Technology

To address the adverse impacts of distributed energy sources on power grids and to coordinate the development of large power grids and distributed energy, the concept of micro-grid was proposed and generated much interest. A micro-grid makes use of a large number of modern power electronics technologies, combining the power source, load, energy storage and control device to form a single controllable and relatively independent power supply system, which can achieve self-control, protection and management. Micro-grid technology can improve the quality of distributed energy supply and expand the range of optional resources for system operation and adjustment. By connecting some of the distributed energy sources to large power grids in the form of micro-grids for grid-tie operation, we can combine the competitive advantages of distributed energy systems and eliminate the impacts on the power grids. This is one of the effective ways to realise the performance and value of distributed energy systems.

As a new technology adapting to the requirements of distributed energy development and intelligent distribution networks, micro-grid technology is closely linked to the development of power electronics and energy storage technologies. The progress and maturity of these two technologies determine the extent to which micro-grid technology can be applied and promoted. Generally speaking, the application of micro-grid technology in the engineering field is still in its infancy, and the EU, the USA and Japan have all started working on micro-grid demonstration projects with the major objective of proving the concept, verifying the control programmes and studying the operational features. Many Chinese research institutions and enterprises have conducted research into relevant technologies of the micro-grid, and State Grid Corporation of China has developed a number of pilot projects on micro-grid operation and control in, among other locations, Henan, Zhejiang and Shanxi. As an effective tool to complement larger power grids, a micro-grid has great potential for reducing losses in energy transmission and improving the operational reliability and flexibility of power systems. In order to promote the healthy development of micro-grid technology, specialised research must be conducted into the micro-grid structure, simulation analysis, operational control, protection devices and energy management in the future to accelerate the application process of micro-grid technology.

4.3.5.5 Energy Storage Technology

The instantaneous nature of the production, transmission and consumption of electricity necessitates the maintenance of a real-time balance between power production and consumption. A breakthrough in energy storage technology can achieve large-capacity storage of electric energy with a significant impact on the traditional way of electricity production and consumption.

Energy storage comes in a variety of ways, including physical storage (such as pumped storage, compressed air energy storage, flywheel energy storage, etc.), chemical energy storage (such as lead-acid batteries, sodium-sulphur batteries, flow batteries, lithium-ion batteries, etc.) and electromagnetic energy storage (such as superconducting magnetic energy storage, super capacitors, etc.). As energy storage technologies vary in terms of energy density and power density, it is difficult to find an energy storage technology that is capable of meeting all the requirements for different applications in the power system.

At present, different energy storage technologies are developing rapidly. Among the large-capacity energy storage technologies, the development of pumped storage technology is relatively mature, but is limited by the geographical distribution of resources. Superconducting energy storage technology is in the experimental and research stage and large-scale application is still a long way off. Compressed air energy storage technology, flywheel storage energy technology and super capacitors are in the early phase of commercial application. Among the chemical energy storage technologies, lead-acid battery technology is relatively mature and widely used, and the sodium-sulphur battery, flow battery and lithium-ion battery are expected to enjoy promising prospects as one of the potential technologies of choice for large-capacity energy storage power stations in the future.

In the future, it is necessary to conduct comprehensive and in-depth analysis of the possible applications of energy storage technology in power systems. Amid intensive efforts to build pumped storage stations, the research on various types of energy storage technology must be strengthened and pilot projects actively pursued. Commercial applications of pumped storage should be accelerated as the technology matures.

4.4   Construction of UHV Synchronous Grids in Northern, Eastern and Central China

Building UHV synchronous power grids in northern, eastern and central China has become a tough issue facing the development of the nation's electric power industry. International experience has shown that grid voltages will be continually upgraded in line with ever-larger grids, while the development of transmission technology at higher voltage levels will in turn drive the expansion of grid scale. To develop UHV synchronous power grids in northern, eastern and central China by applying UHV technology is a necessary and inevitable option for the sustainable development of energy in China.

4.4.1 Development of Large Synchronous Grids in Overseas Countries

4.4.1.1 Typical Large Synchronous Power Grids in Overseas Countries

The development of the world's electric power industry throughout history has been characterised by ever-expanding grid scale. Wide area synchronous grids have emerged in some regions around the world, with the more typical examples being the synchronous power grid serving eastern North America and Canada, the IPS/UPS synchronous power grid covering Russia and Eastern Europe, and the ENTSO-E synchronous grid in Continental Europe. In addition, there are international interconnections of synchronous grids in southern Africa and the Middle East.

The Synchronous Grid Serving Eastern North America

The synchronous grid serves a region that includes eastern and central USA and five Canadian provinces, covering an area of about 5.2 million km^2, with the highest voltage of 750 kV and total installed capacity of more than 700 GW.

The maximum load in the USA comes in summer, while the maximum load in Canada occurs in winter. Hydropower forms the basis of Canada's electric power structure. Canada's rich hydropower resources and the thermal power in the USA can well complement each other. There are over 100 north–south grid-tie lines between the two countries, with a power exchange capacity of approximately 20 GW. The level of cross-border power exchange stood at 72.6 TWH in 2009.

Russia–Eastern Europe Synchronous Power Grid

The Russia–Eastern Europe synchronous power grid has evolved from the synchronous power grid in the Soviet Union. The Soviet Union synchronous power grid crosses the Eurasian continent, with the total installed capacity once amounting to 460 GW. It covers an area of close to 20 million km^2, stretching 6000 km from east to west and 3000 km from south to north. After the collapse of the Soviet Union, Russia and some East European countries saw their grids connected to a synchronous power grid for operation at the highest voltage of 750 kV. It covers an area of around 10 million km^2, with a total installed capacity of more than 300 GW.

Synchronous Power Grid of Continental Europe

As one of the world's largest interconnected synchronous grids, the synchronous power grid of Continental Europe covers 24 European countries. In 2008, its total installed capacity was about 670 GW, with a maximum power load of approximately 400 GW and annual electricity consumption of about 2600 TWh.

The power grids of the countries in Continental Europe are closely linked. There are nearly 300 transnational interconnection circuits, with a power exchange capacity of over

100 GW. These European countries have taken full advantage of power generation and transmission equipment and promoted international power trading through multinational grid interconnections, laying the physical basis for the establishment of a uniform European power market.

Europe also plans to build a Mediterranean grid to link together the power grids of over 20 countries in North Africa and the Middle East, complementing each other and making the best of the abundant wind and solar power resources in North Africa.

4.4.1.2 Lessons from the Development of Large Synchronous Power Grids in Overseas Countries

Optimal allocation of energy resources over large areas is an important driving force for the development of wide area synchronous grids. As the power grid is an important platform for the optimal allocation of energy resources, the grid topology is largely determined by the distribution as well as the supply and demand situation of energy resources. The USA and Canada have yet to form their respective nationwide synchronous grids. They are committed instead to developing Canada–USA south–north power grid interconnections, so as to take full advantage of the cheap hydropower resources in Canada. The synchronous power grid development in Europe is closely related to the development of the European power market and changes in the energy structure. After the Fukushima nuclear leakage incident in Japan, Germany announced that it would gradually close its domestic nuclear power plants. Italy also rejected plans for restarting its nuclear development programme, with the majority of the resulting supply shortfall to be met by imports. As a result, the demand for reinforcing transnational interconnections and improving transnational power transmission capacity will become more acute in the future.

Wide area synchronous grids have significant interconnection benefits. The development of wide area synchronous grids has its advantages. The larger a synchronous grid is, the more power sources can be connected to the grid and the stronger ability it will have to withstand the impact of disturbance and failure, with a better adaptability to the supply structure, load distribution and power current changes. Synchronous power grids can fully enjoy interconnection benefits such as peak load shifting, peak-shaving, complementing hydropower with thermal power, interregional compensation, mutual back-up and balanced supply and demand. An expanded synchronous power grid will deliver significantly higher capacity for long-distance and large-capacity power imports.

There is no absolute correlation between the size of synchronous power grids and the incidence of major power failure. The development of large synchronous power grids internationally has undergone a period of transforming from initially weak interconnections to a gradually stronger structure. With enhanced grid connection and improved power grid construction, the transmission capacity, security and stability of synchronous power grids have also experienced an initially weak and continuously improving process. However, there is no definite connection between the incidence of major power failure and the size of synchronous power grids. Analysing the mechanism and the development process of the world's major power blackouts in recent years, we can find that an important cause of such incidents is a lack of unified planning, scheduling and coordination control for power grids. It can also be attributed to inadequate information, communication and emergency response systems that stand in the way of establishing a reliable line of defence for grids. With a centrally coordinated and managed system, major power failures could be avoided with the implementation of decisive and correct control measures to avert vicious chain reactions.

4.4.2 The Necessity of Building UHV Synchronous Power Grids in Northern, Eastern and Central China

The choice of grid topology is determined by national conditions and the technological factors in grid development. The uneven distribution of energy sources and power loads in China will definitely result in interregional, long-distance and large-scale flows of energy and power. The future development of China's power grids will be characterised by power flowing from west to east and from north to south on a large scale, with scattered landing points along the way and concentrated development of large coal-fired, hydro, nuclear and wind power bases to produce power for consumption over expansive areas. The power load centres in China are concentrated in the northern, eastern and central regions. The answer to China's requirements for energy and power development lies in linking up the grids in these three regions to form a strong UHV synchronous grid network that also connects the coal-fired power bases in the north, the hydropower bases in the southwest and the renewable energy bases in the western and northern regions, with the load centres in northern, central and eastern China.

Building UHV synchronous power grids in northern, eastern and central China has obvious advantages in terms of interconnection benefits, optimal allocation of resources and electricity market development. This superiority compared with independent regional power grids is mainly manifested in the following ways.

4.4.2.1 Conducive to Yielding Interconnection Benefits

Building UHV synchronous power grids in northern, eastern and central China will enable the on-grid power sources to form strong connections with the loads, which can maximise the role of various grids in various periods, optimise grid operations in a more convenient and more economical manner, and allocate resources on a wider scale. This will generate substantial benefits like power backup and sharing, a balance between hydropower and thermal power, lower capital investment, stronger incident support, improved reliability and lower operating costs. It is expected that by 2020, with centralised dispatch and maintenance programmes, UHV synchronous power grids in northern, eastern and central China can reduce more than 30 GW of installed capacity. The UHV synchronous power grids in northern, eastern and central China will expand the uptake of hydropower and greatly improve the economic effectiveness of run-off-river power plants with poor adjustability, which can reduce about 34.3 TWh of electricity loss through water abandoning.

4.4.2.2 Conductive to Receiving Externally-sourced Power on a Large Scale

While meeting local requirements, power from the large-scale generating bases in the western and northern regions is also delivered to the eastern and central regions via UHV for consumption. The eastern and central regions can only achieve balanced supply and demand of power and secure a supply of electricity by accepting a great deal of externally-sourced power. The capacity of a ±800 kV UHV DC transmission line can reach 8 GW, which places very demanding requirements on the carrying capacity of the end-use grids. To better receive the externally-sourced power on a massive scale and ensure the security of system operations, China really needs to build strong UHV synchronous power grids in the northern, eastern and central regions as end-use markets, so that power from large power generating bases in the western and northern regions can be delivered and received.

4.4.2.3 Conductive to Massive Consumption of Clean Energy

The UHV synchronous power grids in northern, eastern and central China offer such benefits as peak load shifting, reduced peak-valley difference and the adjustability of hydropower to help improve the grid-wide uptake capacity of wind power and accelerate the development of clean energy. Relevant research shows that without the development of UHV synchronous power grids in northern, eastern and central China, the grid-wide uptake capacity of wind power in these regions would total about 60 GW by 2020. But with the development of UHV synchronous power grids in these regions, the grid-wide uptake capacity of wind power would increase by 43 GW to about 103 GW.

4.4.2.4 Conductive to Developing a Unified Electricity Market

The northern, eastern, and central regions of China are in close proximity to each other, with complementary power structures. Building a strong and flexible UHV synchronous grid and leveraging its market functions helps to eliminate grid congestion while promoting electricity trading in northern, eastern and central China and also the electricity trading between these and other regions to optimise the allocation of power resources over larger areas for establishing a platform for building a unified electricity market in China.

4.4.3 Safety of UHV Synchronous Grids in Northern, Eastern and Central China

4.4.3.1 Safety Analysis of UHV Synchronous Power Grids in Northern, Eastern and Central China

The safety of a power grid depends on structural soundness, the robustness of security and stability control as well as scientific dispatching and management. In a synchronous grid, all generators run concurrently at the same frequency, with the inherent capacity for automatic recovery of synchronous operations after experiencing a disturbance and fault. The larger the grid and the more generators are connected, the stronger ability it will have to withstand disturbances and failures and the higher supply reliability there will be.

It is expected that by 2020, UHV synchronous power grids in northern, eastern and central China will cover an area of 3.2 million km², with a total installed capacity of 1000 GW and a 1000 kV interconnection. The power grids are closely connected, with equivalent electrical distance roughly equal to only a quarter of that of a 500 kV circuit. The grid morphology will develop from an original 'long chain' pattern covering northeastern, northern and central China (formed in September 2003 to subsequently become two synchronous power grids in northern and central China in November 2008 after readjustment for the Gaoling Back-to-Back Converter Station Project) into a stronger 'cluster' structure that covers northern, eastern and central China, forming a DC connection with three synchronous power grids in northwestern, northeastern and southern China.

In order to demonstrate the safety of the UHV synchronous power grids in northern, eastern and central China, State Grid Corporation of China has conducted a lot of simulation calculations of current flow, static security, transient stability, dynamic stability, serious failure checking for each level year and different planning options and operation modes. The results show that the UHV synchronous power grids in northern, eastern and central China have met the requirements set out in the *Guide on Safety and Stability for Power Systems*, and their safety can be fully assured.

4.4.3.2   Safety Comparison of Different Project Proposals for UHV Synchronous Power Grids in Northern, Eastern and Central China

To further demonstrate the safety of UHV synchronous power grids in northern, eastern and central China, State Grid Corporation of China has selected for comparison three proposals for power grid development in northern, eastern and central China according to different technical conditions. Proposal 1: UHV synchronisation projects in northern, eastern and central China. The main framework for synchronous power grids is to be built covering the three regions based on a UHV AC interconnection, while four DC-interconnected synchronous grids will be built nationwide. Proposal 2: 500 kV asynchronous projects in northern, eastern and central China with the main grid framework adopting 500 kV AC and linking up with each other via DC asynchronous interconnections to form six nationwide synchronous power grids. Proposal 3: UHV asynchronous projects in northern, eastern and central China, with the main grid framework adopting UHV AC and the grids in these regions linking up with each other via DC asynchronous interconnections to form six synchronous power grids nationwide.

Safety analysis shows that the 500 kV asynchronous projects in northern, eastern and central China will result in excessive short-circuit current over a large area, and there is no long-term solution to this problem. A massive amount of power is transmitted to the load centres in eastern and central China in DC mode, giving rise to a typical grid structure characterised by 'strong DC system and weak AC system'. A failure of the AC system or DC system will trigger chain reactions like flow transfer and voltage collapse, leading to large-scale blackouts. Even the installation of SVC/STATCOM cannot meet the Level I Safety and Stability Standards stipulated in the *Guide on Safety and Stability for Power Systems* or ensure the safety of power grids.

The proposed UHV asynchronous projects for northern, eastern and central China can solve the problem of excessive short-circuit current over large parts of the power grids in these regions, especially eastern China. They can also solve the problem of the AC system's insufficient power transfer capability in the event of a monopolar or bipolar DC failure. However, given the same power shortfall under fault conditions, the grid frequency will fluctuate more significantly compared with the case of developing UHV synchronous power grids in the same regions. When the DC power transmitted to power grids in eastern China exceeds the handling capacity of the AC grids, it will give rise to the serious problem of voltage instability on the AC circuit. Even if a large number of SVC/STATCOM is installed, it cannot meet the 'N-1' standard requirement in the *Guide on Safety and Stability for Power Systems* and widespread blackouts are still a potential risk.

The proposed development of UHV synchronous grids in northern, eastern and central China is conducive to opening up a more secure and reliable UHV AC grid corridor for the load centres in eastern and central China to receive externally sourced power, which is transmitted via UHV AC and UHV DC modes. Such a pattern with 'equally strong DC and AC systems' and combined AC and DC transmission can not only fundamentally solve the problem of large-area excessive short-circuit current, but also completely eliminate the likely instability of the 500 kV power grids in eastern China caused by AC-DC system interactions. In terms of safety, stability, operational reliability and the ability to withstand different incidents, the UHV synchronisation project, proposed for northern, eastern and central China, compares much more favourably with the proposed 500 kV asynchronous and UHV asynchronous projects and fully meets the safety and stability standards stipulated in the *Guide on Safety and Stability for Power Systems*.

Marking a stage in the development of UHV synchronous power grids in northern, eastern and central China, power grids in northern and central China are currently synchronously connected through the Southeastern Shanxi–Nanyang–Jingmen UHV project and are in turn interconnected by asynchronous DC mode with the power grids in eastern China. In the future, UHV development plans should be accelerated in accordance with the principle of 'equally strong DC and AC systems', and a strong UHV grid structure constructed as soon as possible. It should be noted that in the early stage of UHV grid development when a synchronous power grid is not yet fully formed, the operation of a UHV project will be constrained by equipment and technology conditions. A UHV grid will undergo a process of continued enhancement in terms of safety, stability and operational capability. With the continuous improvement of the UHV AC grid framework structure, the interconnectability of the synchronous power grids will be constantly enhanced, smart grid control and other advanced technologies will be widely used, and the reliability and management levels of grid equipment will continue to be improved. This development will result in safer and more reliable UHV synchronous power grids in northern, eastern and central China.

4.5 Smart Grid Development

The pursuit of smart technology is becoming a new trend in today's world. The so-called smart development refers to efforts to integrate information into industrial applications, continuously create new areas of economic growth, new markets and new employment patterns, and raise the operational efficiency of society, through which interconnection and communication, information sharing, intelligent processing and collaboration can be delivered.[9] Smart grids reflect the development trend of smart technology in the field of energy resources, serve the inherent need for technical upgrading of energy systems, and symbolise a deep push into the growing development of smart technology around the world. It is also an important feature of a new energy reform now underway. Triggering a global wave of smart grid development, the major developed nations in the world have

[9] Remarks made by President Hu Jintao at the 15th General Assembly of the China Academy of Sciences and the 10th General Assembly of Academicians of the China Academy of Engineering in June 2010.

made developing smart grids an important strategic measure to command a high position in the future low-carbon economy. In 2009, State Grid Corporation of China took the lead in proposing an SSG development strategy, drawing widespread community attention. Research and practical work on smart grid development have since been comprehensively initiated and proceeding smoothly. Having been incorporated into the '12th Five-Year Plan' as an emerging industry of strategic importance, smart grids have entered a new stage of development in China.

4.5.1 The Essence and Features of Smart Grids

The term 'smart grid' has yet to be given an internationally established definition. Different countries place emphasis on different aspects of smart grid development because they differ in socioeconomic development, the status of grid development and the allocation of resources. However, they have the same overall objectives and fundamental requirements as they all need to resolve the problems of energy security and environmental protection, combat climate change, ensure a safe, reliable, quality and efficient supply of power, and satisfy the diverse demands for electric power brought about by socioeconomic growth.

In China, the term 'smart grid' covers all voltage classes and grid accesses, as well as power transmission, transformation, distribution, consumption and dispatching, with UHV grids as the backbone framework and power grids at different levels developing in a coordinated manner. As a new modern power grid with the capacity for intelligent responses and system self-recovery, a smart grid can facilitate the connection and disconnection of various power sources and facilities, interact dynamically with the users and dramatically improve the safety, reliability and operational efficiency of the power system, through the use of modern communication and information technologies, automatic control technology, decision-support technology and advanced power technology.

Smart grids are information-based, automated and interactive. Being information-based means that smart grids are supported by a communication and information platform where information, power and business flows are integrated seamlessly for pooling and sharing of real-time and nonreal-time information. Being automated means that the automation of grid production, operation, dispatching and management can be improved comprehensively with the adoption of advanced automatic control measures. And being interactive means that smart grids can support two-way information flows between the power grid and the power source and between the power facilities and the users, with the capability to adapt to interactions and adjustments and achieve the power system's optimal operating condition.

4.5.2 Strategic Significance of Smart Grids

The development of smart grid technology has resulted in expanding network functions to promote the optimisation of resource allocation, ensure the safe and stable operation of the power system, provide diverse and open power services, and enhance the development of emerging industries with strategic importance. As a significant platform for resource transportation and allocation, smart grids generate substantial benefits for the national economy, for the modes of energy production and utilisation and for the protection of the environment throughout the funding, construction, production and operation processes.

4.5.2.1 Realising Highly Efficient Operation of Power Systems

Due to the growing size and complexity of the modern power system, more and more contributing factors have to be taken into account at the operational level. It is also important to timely identify and eliminate the potential risks caused by various factors, improve the anti-interference capability of the power system and ensure its safe, efficient and coordinated operation. Promoting the development of smart grids can benefit the comprehensive monitoring and flexible control of the power system and improve operational stability and supply reliability. By promoting the utilisation of various advanced technologies and facilities, the production and utilisation efficiency of the different processes, including power development, transmission, transformation, distribution, consumption and dispatching, can also be improved.

4.5.2.2 Satisfying the Users' Increasingly Diverse Demands for Electricity

With the continued progress of society and the widespread application of smart equipment, the users have come to expect ever-higher service quality and richer product offerings. They demand that the energy system should provide energy marked by greater safety, reliability, economy, efficiency, user-friendliness, interaction, transparency, openness, cleanliness and environment-friendliness. With the development of smart grids, the ability of the power system can be greatly strengthened to optimise resource allocation and resist interference while providing the users with a sufficient and quality power supply. Besides, not only obtaining timely information on power usage and electricity tariffs, users can also actively participate in the management of power consumption and the formulation of operation strategies for power facilities. This will help achieve more sophisticated control of power consumption and ensure greater access to more satisfactory power services.

4.5.2.3 Promoting Development of Clean Energy

By developing smart grids and making use of advanced automation, coordination control and energy storage technologies, accurate forecasts and reasonable control can be achieved of various energy resources, including wind power and solar energy, while improving the output characteristics of new-energy power generation to effectively address the technical problems arising from large-scale development of renewables such as wind and solar power, and expand the market consumption of renewable energy. This will help promote the optimisation and readjustment of the energy structure and reduce the over-reliance on conventional fossil fuels.

4.5.2.4 Enhancing the Development of Electricity Industry and its Associated Sectors

The development of smart grids will make the electricity industry more smart-oriented, marking a new step forward in the development of this industry. It will also boost the development of smart grid-related high-tech and new-tech sectors such as new energy and new materials and contribute to the advancement of such new emerging industries as electric cars. This may exert a multiplier effect on the stimulation of consumption and economic growth. As initially estimated based on input-output analyses, compared with conventional grids, an incremental investment of RMB50 billion in a smart grid every year between 2010 and 2020 will translate into RMB194 billion more in economic output, create 140,000 new jobs, and improve the population's income by RMB3.5 billion.

4.5.3 The Priorities and Practices of Smart Grid Development

Judging by the driving force and priorities of smart grid development around the world, different countries and regions have adopted different approaches to development and practice. The USA is focused on upgrading and updating its existing grid infrastructure, raising supply reliability, maximising information and communication technologies, integrating smart grids with conventional grids to promote power grid modernisation, actively developing clean energy, popularising plug-in hybrid vehicles, and achieving the integrated operation of distributed energies. The countries in Europe instead place emphasis on the research and resolution of issues like the grid uptake of wind power (especially large-scale offshore wind power), grid-tie integration of distributed energy sources and stronger demand side management. Equally important is the improvement of supply reliability and power quality as well as the enhancement of value-added services for the public. Though different countries stress different aspects of smart grid development, they share the same overall objectives and fundamental requirements, given the common need to resolve the problems of energy security and environmental protection, combat climate change, ensure a safe, reliable, quality and efficient supply of energy, and satisfy the diverse demands for electric power brought about by socioeconomic growth.

The smart grid is still in its early stage of development both at home and abroad. Despite an early start in research on smart grids conducted by the developed countries around the world, no substantial investment or construction was pursued by governments until 2008. So in terms of smart grid development, China is at the same starting line as the developed countries. The development of the Strong and Smart Grid has presented China with a good leapfrog opportunity to secure a commanding position in international grid technologies.

4.5.3.1   The Priorities of Smart Grid Development in China

The development of smart grids in China must be pursued in a coordinated fashion based on systematic research and scientific planning while maintaining a strong focus.

Coordinated Promotion of Different Areas of Smart Grid Development
Power Sources Access
We should optimise the structure of power sources, strengthen the coordination between the generating plants and the grids, and enhance the operational safety of power systems. We should make the grids more adaptable to the development of different types of clean energy, advance the development and uptake of clean energy, make the different energy resources complement and balance each other, promote highly efficient utilisation of clean energy, and improve the overall efficiency of energy utilisation in light of the intermittent

and uncertain nature of clean energy, including wind and solar energy. The emphasis on the adoption of smart technology for power access should mainly involve the R&D on and application of key technologies for coordination between conventional power plants and grids as well as R&D on advanced technologies to facilitate the analysis, power prediction and grid integration operation of wind and solar power generation. Other focus areas include the R&D on and application of high-capacity energy storage equipment, formulation of the relevant standards, and the R&D on and popularisation of critical conventional power source equipment, critical equipment for large-scale renewables-based power generation and large-capacity energy storage.

Power Transmission

We should maximise our existing grid resources, improve the capacity of transmission lines and reduce the operating costs of transmission. The stability of the power system should be enhanced. We should also implement condition-based monitoring of major transmission equipment, fully promote the application of smart technology for line inspection, extensively promote the condition evaluation, condition-based maintenance and risk early warning for transmission lines, in order to ensure that the operation of transmission lines is controllable and under control. The emphasis on smart technology applications in power transmission should be placed on implementing digitised surveys, modularised design, operation conditioning, information standardisation and network-based application through the integrated application of new technologies, new materials and new processes. We should actively adopt flexible AC transmission systems to improve transmission capacity and the flexibility of voltage and flow control. In-depth studies for developing analytical, evaluation and diagnostic capabilities and support technologies for decision-making should be conducted with the support of communication, information and control technologies and by means of GPS, intelligent monitoring and advanced inspection technologies to realise intelligent condition-based evaluation of transmission lines. We should also strengthen the R&D on and application of the technologies for condition-based maintenance, full lifecycle management and intelligent disaster prevention for transmission lines, with a view to applying smart transmission technology at an advanced level.

Power Transformation

The adoption of smart technology in power transformation can help significantly improve grid stability and reliability, transmission capacity and the health of equipment. We should strengthen the role of smart equipment in providing information support for optimised grid dispatches and operational management, while providing a base for optimisation and decision-making in smart grid dispatch and also operational management at the equipment level. We should enhance the standard of asset management and the operation level of substations. The adoption of smart technology for power transformation should focus on the key technologies and equipment for the automation of smart substations, integrated on-line monitoring and self-diagnosis at the equipment level, the R&D on and application of the key technologies and requirements for smart applications in primary transformation facilities, as well as the testing and assessment, technical standard development, operating environment monitoring and integrated operation, maintenance and management of the monitoring and automation equipment for smart substations.

Power Distribution

The adopting of smart technology for power distribution will contribute to improved supply reliability, system operational efficiency and power quality at the end-user level. It can help realise not only distributed power generation as well as integrated, coordinated and optimised operation of energy storage equipment and micro-grids, but also efficient and interactive demand side management. It can support an integrated and optimised system for power distribution asset operation and management by employing advanced modern management concepts. The emphasis on adopting smart technology for power distribution should be placed on stepping up the construction of automatic power distribution and support systems on the basis of a stronger power distribution network, so as to achieve flexible adjustments and optimised operation of power distribution networks and improve supply reliability and power quality. It can strengthen the development of information systems for grid production and control as well as operational maintenance and management in order to provide full support for the planning, operation, maintenance and management of power distribution networks and achieve an organic integration of the various application systems and the two-way interaction between power dispatch and power utilisation. It can promote the research and widespread application of integration technologies as well as uniform and coordinated control of the distributed power generation, energy storage and micro-grid systems, which can help maximise the role of these technologies in ensuring supply reliability and system peak-shaving to enhance supply reliability.

Power Consumption

We should build up a smart power service system to achieve modernised operation of marketing management and the application of smart technology in marketing activities. We should fully develop two-way interactive power services to enable two-way interactions between the grids and the end-users, improve service quality for the users and meet their diverse requirements. We should encourage technological innovation in the area of smart power utilisation to drive the development of the associated industries while transforming the mode of energy utilisation at the end-user level so as to improve the efficiency of power utilisation. The adoption of smart technology for power utilisation should be oriented towards developing and perfecting a smart two-way interactive service platform and the support platforms of the associated technologies, which may bring about two-way interactions among energy flow, information flow and business flow at the power user level and fully improve the capacity of the two-way interactive power services. More specifically, efforts should be made to promote the smart meters and build up a system for collecting information on power consumption that covers all power consumers and networks, which will facilitate on-line monitoring and real-time collection of important data on power load factors, power usage and voltage for providing technological support for power services. We should also build a smart system of energy services together with customer-side distributed power sources and energy storage at a community/building level to enable real-time interactive response between the smart grids and the power users. The systems are designed to improve a grid's integrated service capability, satisfy the user demands for interactive services, enhance service quality, provide guidance on the use of electricity in a scientific manner, reduce power consumption, and improve end-user energy efficiency and grid operational efficiency. A smart charging service network should be developed based on scientific and rational planning so as to facilitate two-way energy exchanges between the electric automobiles and the grids, hence meeting the demand for new power services such as that from electric automobiles.

Power Dispatch
As an integral part of a smart grid, dispatch is closely related to other aspects and an important safeguard for the safe, quality and economical operation of the Strong and Smart Grid. With the development of smart applications for dispatch operations, we can make power dispatching information-based, automated and interactive while achieving scientific organisation, accurate control, forward-looking guidance and efficient coordination of power production in order to refine the planning arrangements for power dispatching and improve the levels of basic automation and practical applications. We should also pursue integrated development of the support systems and dispatch technologies, strengthen the safety protection for the secondary systems, improve the ability of power dispatch to regulate wide area power grids and fully enhance the ability of power dispatching to optimise resource allocation. In the process we can prevent and reduce risks, make rational decisions and carry out scientific management, conduct flexible and efficient adjustments and market resource allocations based on the principle of equality and user-friendliness while comprehensively improving the security and economical operation of the grids. The development of smart dispatch capability mainly concerns setting up a defence line covering annual or monthly grid analyses, checks on daily schedules, and real-time dispatch operations. We should also implement network-based data transmission, panoramic monitoring of operations, dynamic safety evaluation, fine-tuned dispatch decision-making, automatic operational control and optimised coordination between the grids and the power plants, forming a support system for smart dispatch technology and establishing an integrated smart dispatch mechanism to support the centralised and distributed grid-tie integration of renewable energy resources on a massive scale.

4.5.3.2   Pushing Forward Development of Communication and Information Platform

As an important means to support the development of Strong and Smart Grids, a communication and information platform will help to realise the comprehensive collection, smooth transmission and efficient handling of information on grid planning, design, construction, production and operation, in addition to improving the automation levels of equipment and business processes. It will facilitate formation of an information system covering all business processes, enhance the level of modern business management, realise the optimal allocation and efficient utilisation of resources in the entire grid network, and exercise comprehensive risk control. We can build a collaboration and interoperability platform that can transparently share information, integrate processes and protocols, and incorporate powerful functions to raise the level of interaction between the human operators and the application systems, between the various grid processes, between the different business activities, and among all the interested parties. We can maximise the potential value of the diverse and massive information generated by the Strong and Smart Grids to enhance the intelligent analyses and scientific decision-making for the grids. We can also put the various power supply and powered equipment together as an integrated whole through the smart grids, while providing a communication carrier to and expanding the application of the Internet of Things.

The development of a communication and information platform is focused on maximising modern telecommunication and information technologies to continuously expand data collection, transmission, storage and utilisation for power generation, transmission, transformation, distribution, consumption and scheduling based on the development of digitalised and automated grids. Through this development, we can implement digitalised data collection, automated production processes, interactive business transactions, information-based business operation and management, and scientific strategic decision-making. By focusing on the requirements of smart grid development, we can promote the coordinated development of communication networks at all levels, improve the deployment of communication resources, enhance the carrying capacity of various communication services, and strengthen the capability to withstand various natural disasters and damage from external forces. We can step up the regulatory control of information security and the development of operation and maintenance systems to form a defence-in-depth system for information protection across different classes.

4.5.3.3 Promoting the Development of Intelligent Public Service Platform

Relying on the smart grid and its associated intelligent information and communication systems, we can expand the functionality of network services and promote integration and amalgamation with public service resources, which will be conductive to the close integration and comprehensive application of energy and information flows and the creation of a new, smart grid-based intelligent public service platform.

The smart grid-based Fibre To The Home (FTTH) approach can be adopted to push forward 'Three-Network Integration'.[10] FTTH is a generic term for any broadband network architecture using Optical Fibre Composite Low Voltage Cable (OPLC) to replace the traditional low-voltage cable that links the backbone network with the end user, so as to realise FTTH at the same time as Power Cable To The Home is achieved. FTTH enables broadband network access for thousands of households based on a power optical fibre cable, providing an important physical channel for the integration of the three networks so as to overcome the difficulties with network channel integration and resolve the problem of terminal access to the information highway. Promoting the concept of Three-Network Integration with the smart grid and power optical fibre as the carrier can achieve better economic benefits and at the same time realise multiple uses for a single cable and shared network infrastructure to reduce duplicate development, improve grid integration efficiency, and optimise social resource utilisation. As a key option for providing user access to the smart grid, FTTH has the network resources and service capacity for providing the community with public services. It offers a network application platform to the community featuring high bandwidth, large capacity, wide coverage, energy efficiency, and the potential for more value-added services for Three-Network Integration.

The smart grid will help promote the creation of an intelligent community and support the development of smart cities. An intelligent community changes the traditional mode of one-way power supply to achieve two-way interactions between the power suppliers

---

[10] 'Three-Network Integration' refers to the integration of the telecom, cable and Internet sectors.

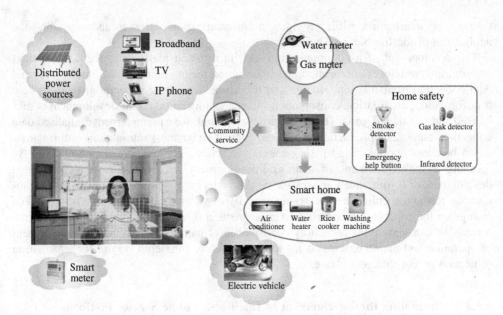

Figure 4.17 Main functions of an intelligent community.

and the users in terms of power flow, information flow and service flow (see Figure 4.17). Power companies may gain real-time information on power usage and consumption patterns so that they may initiate flexible and sophisticated demand response according to the characteristics of power consumption and guide the users to change their behaviour and use electricity rationally. Users may also be encouraged to participate in power load balancing and grid operations, support grid access to a variety of user-side distributed power sources (including solar photovoltaic, wind energy, electric vehicles and energy storage), and promote the use of clean new energy. Smart grids can better ensure electricity supply to cities and encourage the cities to use green energy more. At the same time, as an important public service platform in the cities, a smart grid will provide full support for development of intelligent cities in the future.

4.5.3.4 The Development Practices of China's Smart Grids

In recent years, China has performed a lot of work on the planning study, technical research, equipment development, standards formulation and pilot projects for smart grids, with fruitful results (see Table 4.4). Our smart grids have produced a significantly greater impact on the international community, giving China a leading position in global smart grid development.

The integrated demonstration project of smart grids in Shanghai World Expo (see Figure 4.18) is the first project of its kind in China. It includes 9 subprojects for new-energy grid access, energy storage, smart substation systems, distribution automation, emergency repair management system, power quality monitoring, systems for collecting information on power consumption, smart buildings and smart home appliances, and

Table 4.4   State Grid Corporation of China's practical experience in smart grids.

Type of work	Progress and results
Initiating studies of the theory and planning of smart grids	• Completing medium- and long-term planning and the '12th Five-Year Plan' for the application of smart technology in State Grid Corporation of China's power grids and its affiliated power grids in different provinces (autonomous regions and principal municipalities) • Releasing the *R&D Plan for Critical Smart Grid Equipment (System)*, and taking the international lead in proposing the standard systems for smart grid technology • Releasing the *Standard Systems Planning Proposal for Smart Grid Technology*, and taking the international lead in proposing the standard systems for smart grid technology • Formulating the *Comprehensive Development Action Plan for Smart Grid* • Formulating the *Summary Report of Strategic Research on Strong and Smart Grids*
Carrying out smart grid technology research, equipment development and standards formulation	• Setting up national-level technology R&D and testing organisations for smart grids, such as the National Energy and Smart Grid Technology R&D Centre whose integrated experimentation and testing capabilities have reached a world-leading level. • A series of technological breakthroughs in smart meters, smart grid scheduling technology support systems, smart transformers, smart switching equipment, optical-fibre composite low-voltage cables, charging equipment, and control systems for electric vehicles and micro-grid operations • Publishing 166 enterprise-class standards, with 42 national and industry standards being developed and amended under contract

Carrying out smart grid pilot projects	• Arranging a total of 287 pilot projects in 29 categories, and completing and commissioning 238 projects by the end of 2011
	  • Phase I of the Wind/Solar PV Energy Storage and Transmission Demonstration Project completed and put into operation
	  • The Integrated Demonstration Project of Smart Grids in Shanghai World Expo and Sino-Singapore Tianjin Eco-City completed and put into operation
	  • Building 65 smart substations of between 110 kV and 750 kV
	  • Constructing smart power distribution networks in the core areas of 23 cities
	  • Completing and commissioning 243 charging stations and over 13 000 AC charging poles for electric vehicles covering 26 provinces (autonomous regions and principal municipalities), and building smart charging service networks in Qingdao, Hangzhou and other cities.

(continued overleaf)

Table 4.4 (*continued*)

Type of work	Progress and results
	• Installing a total of 51.62 million smart meters, and collecting electricity consumption data from 76.45 million users
	• Completing and commissioning the support system for smart grid scheduling technology with panoramic monitoring of grid systems, dynamic analyses and real-time early warning functions
	• Deploying condition-based monitoring systems for transmission equipment in main stations in 15 provinces
	• Completing 28 intelligent communities and smart buildings, with a service platform catering to 35 000 users

Note: As of the end of 2011.

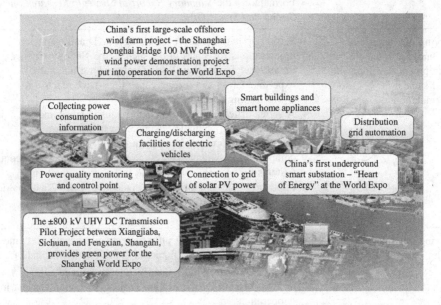

Figure 4.18 The Integrated Demonstration Project of Smart Grids in Shanghai World Expo.

charging facilities for electric vehicles. There are also 4 demonstration projects covering UHV transmission displays, support systems for smart grid scheduling technology, information platforms and visual display systems.

The integrated demonstration project of smart grids in Sino-Singapore Tianjin Eco-City is one with the most comprehensive functions at home and abroad, comprised of 11 subprojects covering grid access to distributed power sources and energy storage systems, condition-based monitoring systems for equipment, smart substations, distribution grid automation, power quality monitoring, systems for collecting power consumption information, intelligent communities, charging facilities for electric vehicles, communication networks, visibility platforms for power grid operations and interactive online business service centres, based on the requirements for 'Feasibility, Replicability and Marketability'.

The wind and solar PV energy storage and transmission demonstration project is a key element of China's 'Golden Sun' programme and a major project under the national science support programme. It is also one of the first smart grid pilot projects established by State Grid Corporation of China. The project is the first of its kind in the world combining wind and photovoltaic power generation, energy storage and smart transmission. Phase I of the project was commissioned on 25 December 2011, providing 100 MW of wind power, 40 MW of photovoltaic power, 14 MW of energy storage, a joint control centre for wind and PV energy storage and transmission, and a 220 kV smart substation.

At the International Smart Grid Forum held in September 2011, representatives spoke highly of the practical results of China's smart grid development and the concept of 'building a Strong and Smart Grid' proposed by State Grid Corporation of China was well recognised. At the forum, delegates expressed appreciation of the work and breakthroughs of State Grid Corporation of China in such fields as the theoretical research on Strong and Smart Grids, standards formulation, key equipment development and pilot project operations. As affirmed by Chuck Adams, the President of the Standardization Association of the Institute of Electrical and Electronics Engineers (IEEE), 'China has commanded a leading position in the field of smart grids'.

Minutes of the 2011 International Smart Grid Forum (29 September 2011, Beijing)

From 28 to 29 September 2011, the State Grid Corporation of China and the Institute of Electrical and Electronics Engineers (IEEE) cohosted the 2011 Smart Grid International Forum in Beijing on the theme of 'Strong & Smart Grid—The Driving Force for Energy Development in the 21st Century'. A total of more than 400 people, including representatives from such international organisations as the International Electrotechnical Commission and International Telecommunication Union and other representatives from 16 national government departments, industry organisations, power companies, research institutions, design firms, manufacturing enterprises and universities in Asia, Europe, North and South America attended the forum.

Energy is an important foundation and safeguard of socioeconomic development. Currently, various countries in the world are vigorously pursuing innovative development in the field of energy, and the energy transformation characterised by the use of clean energy and development of smart grid are in the ascendant. The Strong and Smart Grid is an extension of the smart grid concept, where the close integration of the 'strong' and 'smart' aspects has become an inherent requirement and inevitable trend of the development of modern power grids.

The Forum believes that the development of smart grid can play an important role in improving the energy structure, optimise the allocation of resources and improve the efficiency of energy production and utilisation, and will play an important role in safeguarding the security of energy supply, combating the global climate change, improving the people's quality of living and nurturing strategic and emerging industries. Participants gave a comprehensive review of the latest progress made by various countries in the field of smart grids, and carried out extensive exchanges and discussions in the relevant issues in the field of smart grids, and offered comments and suggestions on strengthening cooperation in the area of standards formulation for smart grids. The Forum is of milestone significance in promoting the development of smart grids in the world.

IEEE has more than 400,000 members around the world, consisting of 45 associations and professional committees, and its interdisciplinary experts have played a role in the development of smart grids. Through taking advantage of its solid foundation and broad collaboration, the IEEE develops standards for the smart grid, and shares the best practice experience, as well as publishes research results in the area of energy conversion. IEEE also provides relevant educational products and services to promote the development of the smart grid. The Forum fully acknowledged the benchmark role of its experience for the various countries in the world.

The Forum fully affirmed and highly appraised the work and achievements made by State Grid Corporation of China in the field of smart grids. With the support of the Chinese government, State Grid Corporation of China has vigorously promoted and actively developed the Strong and Smart Grid characterised by information application, automation and interaction with the UHV power grid acting as the backbone and coordinated power grid development in all voltage classes, and SGCC has carried out fruitful work in such areas as the theoretical research, standards formulation, key equipment development, pilot project operation and building up experimentation and testing capabilities, achieving breakthroughs in core technologies in a number of areas and accumulating a wealth of experience. It has made important contributions to promoting the development of smart grids in the world.

The Forum has proposed that the various countries should make joint efforts and strengthen exchanges and cooperation as well as share experience and achievements in accordance with the principle of mutual benefit, win-win relationship, friendship and openness to jointly promote the innovative development of smart grids in the world.

4.5.4 The Development Principles of Smart Grids

The majority of developed countries in the world attach importance to promoting the smart grid and regard it as an important direction for optimising and adjusting the energy system. Through strong policy support and guidance, China should adjust its existing management systems, price mechanisms, pricing models and investment and financing mechanisms to promote the synergistic relationship and coordinated planning among the energy, communication and information sectors, and incorporate smart grid development into the nation's economic and social development plans and the strategic working system of energy development for the purpose of giving it a greater role in energy, economic and social progress.

4.5.4.1 Committing to Unified Planning and Phased Implementation

The smart grid development in China must be built on the nation's overall energy development strategy, with the adaptation to and promotion of clean energy (such as wind power and solar power) development and utilisation as one of the basic objectives, which will be met by the provision of a strong grid. China must enhance grid operational efficiency and improve the potential of a grid for energy conservation and emission reduction while promoting energy efficiency, energy conservation and emission reduction on the user demand side to achieve the sustainable development of China's energy and power industry. Therefore, the development of China's smart grids must be built upon the physical grid infrastructure and aligned with the overall plans for national grids, power distribution grids and communication services. We must see to it that planning authorities at a lower level shall submit to planning authorities at a higher level, with power generation, transmission, transformation, distribution, utilisation and dispatch as well as telecom and information platforms coordinated and guided by the overall plans for state grids in order to realise the coordinated development of the entire power supply chain.

4.5.4.2 Committing to Independent Innovation and Leadership in Setting Standards

In view of the size of the smart grid market, governments around the world are competing to promote technological progress in smart grids and the development of standards to secure a commanding position in the global smart grid industry. At present, China's smart grid construction is basically at the same level as the world's major economies, with relatively well-developed power and information industries. Therefore, we should seize the historic opportunity in the energy sector's wave of technological innovation, with a commitment to independent innovation and leadership in setting standards to improve the overall level of technological competence and systematic capacity for innovation in China's power and related sectors. Efforts should be focused on key business segments and support technologies to constantly improve the research framework for key smart grid technology. This will help develop products, technologies and brand names with proprietary intellectual property rights to make China an important leader in the development of smart grid technology, with a competitive edge in the competitive international marketplace. By developing a framework of standards for the smart grid, we should promote and protect the rapid and orderly development of the smart grid industry and have a greater say in the development of international standards for smart grid technology.

4.5.4.3 Committing to Overall Planning and Coordination while Making Progress through Joint Efforts

We should strengthen the coordinated development of the smart grid with the new energy industry, energy conservation and environmental protection, electric vehicles, Internet of Things and the high-end equipment manufacturing industries while strengthening R&D in core technology and realising the organic integration of the leapfrog development in various technology fields with China's power industry programme. Smart grid construction is a complex systems engineering project, constituting an integrated whole that covers the entire power industry and all voltage classes. It cannot succeed without the active participation and efforts of all parties concerned. We should therefore coordinate the priorities of different industry segments, project scale and construction schedules, as well as appropriately allocating different resources to achieve the goal of coordinated development between the different segments. We also need to bring together the expertise from materials R&D, equipment manufacturing, operation management and energy services to establish and improve a coordinated, interactive, harmonious and win-win mechanism for cooperation, while actively coordinating the resources of various parties at home and abroad to create a mutually beneficial and win-win situation for smart grid development.

4.6   Oil and Gas Pipeline Networks

China's oil and gas pipeline networks have undergone rapid development in recent years. However, problems remain such as relatively low total mileage and coverage, inadequate interconnectivity and insufficient load shifting capacity, which call for continued improvements to accelerate development of transmission and distribution networks, strengthen overall development planning, increase development efforts in load shifting facilities, and continue to pursue technological improvements.

4.6.1   Present Situation of Oil and Gas Pipeline Networks

Pipeline transmission is a mode of transport accomplished by the use of compressor equipment to pressurise and transport the fluid inside the pipeline to a destination. This has the advantage of high throughput capacity, small footprint, low pollution, cost effectiveness, safety, reliability and all-weather operability. It is particularly suitable for transporting liquefied or gaseous forms of commodities such as natural gas. In developed countries, the long-distance transport of natural gas and refined petroleum products is accomplished mainly by means of piping.

China's oil and gas pipeline networks have been constantly growing and expanding alongside the development of the oil industry. The first long-distance crude oil pipeline (connecting the Karamay oilfields to the Maytag Refinery in Xinjiang) was built and commissioned in 1959. In the 1970s, as the oil production base was being developed in the northeast, the first wave of oil and gas pipeline development was initiated, culminating in the completion of transmission pipeline networks connecting to northeastern and northern China, among other areas. Since the 1990s, in response to the development of the oilfields in the western region and the importation of oil and gas resources from central Asia, oil and gas development in China entered a period of rapid growth, with such important gas pipelines as the Shaanxi–Beijing pipeline and the West–East Pipeline Project (WEPP) completed and commissioned. By the end of 2010, China had built 22 000 km of crude oil pipelines, 18 000 km of refined petroleum product pipelines and 39 000 km of natural gas pipelines, giving a combined total mileage of approximately 79 000 km of oil and gas pipelines, or 9.5 times the total mileage in 1978 (see Figure 4.19). Currently, China has largely completed a backbone pipeline network covering the whole country.

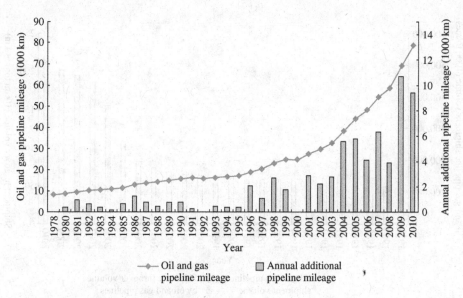

Figure 4.19 China's oil and gas pipeline mileage.
Source: *China Statistical Yearbook*, National Bureau of Statistics of China.

At the same time as the continued growth in total mileage, total shipment volume and shipment turnover transiting through the oil and gas pipelines have also experienced significant growth. In 2010, bulk shipment volume transited through oil and gas pipelines in this country reached 499 720 000 tons, or 4.8 times the figures for 1978. Shipment turnover reached 219.7 billion tonnes·km, or 5.1 times the figure for 1978 (see Figure 4.20).

The crude oil pipelines are mainly used for transporting crude oil from the production regions to the refining bases. The refinery bases in China are mostly distributed around the Yangtze Delta, the Pearl River Delta, the Bohai-rim, along the Yangtze River, and the western regions. The locations of the oil pipelines are mostly laid out according to the distribution of the refining and processing bases forming a series of regional crude oil pipeline networks in certain parts of the country such as the northeastern, northern, central south, eastern and northwestern regions, with the bulk of the land-transported crude oil largely delivered through pipelines. Among the main trunk crude oil pipelines are the Daqing–Tieling pipeline, Tieling–Dalian pipeline, Tieling–Fushun pipeline, Tieling–Qinhuangdao pipeline, Qinhuangdao–Beijing pipeline, Alataw Pass–Maytagh/Karamay pipeline, western region crude oil pipeline, Dongying–Huangdao pipeline, Dongying–Linyi pipeline, Shandong–Nanjing pipeline, Zhejiang–Shanghai-Nanjing pipeline, and Yizheng–Changling pipeline.

Development of pipelines for transporting refined petroleum products has experienced significant growth in recent years with a continued steady increase year after year in the ratio of pipeline transportation of refined petroleum products, as a number of sizable backbone pipelines for refined petroleum products have been built and commissioned in the Pearl River Delta, the northwestern and southwestern regions while other smaller regional networks of refined petroleum product pipelines are being built in eastern, northern and northeastern China. The trunk pipelines for refined petroleum products include

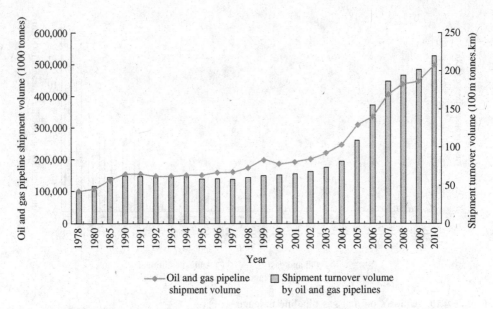

Figure 4.20 Shipment volume and shipment turnover by oil and gas pipelines in China.
Sources: China Statistical Yearbook, National Bureau of Statistics of China.

the Western Region refined products pipeline, Lanzhou-Chengdu-Chongqing pipeline, Lanzhou-Zhengzhou-Changling pipeline, Maoming-Kunming pipeline, Shandong-Anhui pipeline and Zhanjiang-Huizhou pipeline.

In line with the rapid growth of the natural gas industry, the pace of natural gas pipeline development in China is continuing to accelerate, as comparatively sound regional natural gas pipeline networks have been developed around the municipality of Chongqing in Sichuan province, northern China and the Yangtze River Delta Region, while the central south and the Pearl River Delta have also developed the main framework for their own regional gas pipelines, with the formation of a basic topology of nationwide trunk pipeline networks consisting mainly of supply pipelines from the four major gas-producing regions, receiving pipelines connected to undersea pipelines and imported LNG gas sources. The main trunk gas pipelines are WEPP I and II, Shaanxi–Beijing Gas Pipelines I and II, Zhong County–Wuhan pipeline, Nibei–Xining–Lanzhou pipeline, Changqing–Ningxia pipeline, Lanzhou–Yinchuan pipeline, Huaiyang–Wuhan pipeline, Ji–Ning pipeline and eastbound Sichuan pipeline.

4.6.2 The Main Problems of Oil and Gas Pipeline Networks

Significant progress has been achieved with a succession of breakthroughs in the development of oil and gas pipeline networks in recent years. However, due to constrains in terms of resources, capital, market regulation and technology, problems and adequacies remain in pipeline transmission capacity, the scale of interconnection, the means of capacity allocation and technology levels.

4.6.2.1   Total Pipeline Network Mileage and Coverage are Inadequate to Meet Demands

China's pipeline networks account for less than 3% of the global total in length, with total pipeline mileage and coverage both lagging far behind the developed countries in Europe and America. The total mileage of US oil pipelines was 236 000 km in 2008, or 6.4 times the total figure for China in 2010. This compares with Europe's 156 000 km for natural gas pipelines, or 4.7 times the total figure for China in 2010. In 2010, China had an average pipeline coverage of 0.0034 km/km^2, compared with the competitive figures in 2008 for France and German of respectively 0.0670 km/km^2 and 0.1064 km/km^2, roughly equal to 20 times and 31 times the figure for China.

4.6.2.2   Comparatively Low Level of Interconnection

As oil and gas pipeline networks are still in the early stage of development in China, with an emphasis on development of trunk oil and gas pipelines at the expense of branch lines and linking lines, the result is a relatively low level of interconnection in the oil and gas pipeline networks and an absence of pipeline connection in different regions. This situation is not connective to the flexible allocation of transmission capacity and has hampered the expansion in the scope and size of the terminal consumption markets for natural gas.

4.6.2.3   Low Proportion of Piped Refined Petroleum Products

In China, most refined petroleum products are still transported by rail and waterway while pipeline delivery accounts for less than 30% of the total volume carried. In contrast, in the developed countries in Europe and America, pipeline delivery is the main means of transport for refined petroleum products, with this means of transport accounting for over 80% of total volume in some countries. In China, pipeline transportation of refined petroleum products is markedly lagging behind that of the developed countries, making it difficult to fully leverage the benefits of pipeline transportation in terms of economy and efficiency. An over-reliance on rail transport has created a situation of untimely and inflexible transport and allocation of refined products, and the bottleneck resulting from inadequate rail transport capacity is also one of the main factors causing a 'shortage' of oil.

4.6.2.4   Development of Natural Gas Load Shifting Facilities is Lagging Behind

Development of gas storage reservoirs and other infrastructure facilities has suffered from an insufficient level of investment in China where the overall capacity of load shifting reserves and contingency response capability are relatively weak. In the USA and European countries where the development of gas pipeline networks is relatively well-advanced, the reserve capacity of storage facilities generally accounts for over 20% of total annual consumption volume. In contrast, the total usable gas reserves in this country can only meet around 3% of annual demands. Among the various trunk natural gas pipelines in China, only WEPP and Shaanxi–Beijing pipelines have gas storage reservoirs built along the routes while all the other pipelines have no load shifting facilities. To avoid affecting the normal upstream operations of gas fields, load shifting can only be effected by the trunk pipelines. As a result, the transmission capacity of these trunk pipelines cannot be fully utilised.

4.6.2.5 The Problem of Severe Ageing in Some Crude Oil Pipelines

The crude oil pipeline networks in the northeast have been in continuous operation for over 30 years. They are now plagued by problems such as equipment ageing and usage beyond service life. They have entered an accident-prone period, posing potential safety risk to the transmission of crude oil. An oil leak will cause damage to the ecological environment. In addition, the facilities for communication and scheduling in these old pipeline networks are obsolete with a very low degree of automation, making it more and more difficult to meet current demands for crude oil transmission. So there is a need for upgrades and replacements on a step-by-step basis.

4.6.2.6 Trailing Behind Developed Countries in Pipeline Transport Technology

As the development of the oil and gas pipelines in China has a relatively short history, a gap remains with the developed countries in the field of pipeline transmission technology. While the pipelines for refined petroleum products in overseas countries are moving towards the direction of 'large gauge, large throughput and multibatches', the transmission facilities in China are only capable of handling a single product or sequentially handling a few products, which trail far behind the developed countries in multibatch, multiproduct and multidestination pipeline delivery. Moreover, China relies on imports for the supply of some of the key pipeline equipment and components, with a low level of domestic content.

4.6.3 The Basic Thinking Behind the Development of Oil and Gas Pipeline Networks

4.6.3.1 Accelerating Development of Oil and Gas Pipeline Networks, Implementing Interconnection·in Transmission and Distribution

In the future, along with the rising proportion of natural gas and other energy resources in China's energy consumption, demands for oil and gas pipeline transportation will continue to grow. Therefore, there is a need to continue to expand the efforts in developing the oil and gas pipeline networks, focusing on the development of the outward transmission facilities from the major oil production bases, while aiming to develop interregional and long-distance trunk oil and gas transportation pipelines with reference to the distinctive distribution patterns of oil and gas resources in China. The branch line coverage of refined petroleum products and natural gas should be increased on the base of expanded and improved trunk oil and gas pipeline networks nationwide. Linking lines should also be expanded to build an interconnected topology of oil and gas pipeline networks to form a comparatively robust transmission and distribution network for refined products and natural gas while enhancing the flexibility, economy and reliability of the pipeline networks for oil and gas. Efforts should be stepped up in developing the transmission networks and infrastructure facilities for importing oil and gas, and to ensure an effective link-up with the existing domestic oil and gas infrastructure facilities.

4.6.3.2 Strengthening Overall Development Planning for Oil and Gas Pipeline Networks, Ensuring Coordinated Development between Upstream and Downstream Projects

Plans for delivering oil and gas pipeline networks should be incorporated for consideration as part of the network programme for integrated transport systems to ensure shared use and development and full utilisation of pipeline facilities of the same category, thereby avoiding waste in pipeline resources. The oil and gas industry features a high degree of integration between the upstream and downstream sectors. As the pipeline is the key element linking the upstream and downstream sectors of the oil and gas industry, the distribution of the upstream oil and gas fields, the geographical spread of refining capacity and the downstream consumption demand should be taken into full consideration in the planning process for oil and gas pipeline networks. At the same time as planning for the trunk oil and gas pipelines, consideration should also be given to the support infrastructure for branch lines to ensure the trunk lines and the support branch lines are planned and developed as an integrated system to achieve the optimal integration of the oil and gas pipeline network. At the very early stage of design and planning for refinery product and natural gas pipelines, the likely prospects of fast-growing oil and gas demands should be considered and construction planned ahead of time to expand design transmission capacity to an appropriate level.

4.6.3.3 Strengthening Efforts in Building Up Load Shifting Capacity for Natural Gas to Increase Throughput Volume in Pipeline Networks

Load shifting facilities are the principal means to ensure supply security and are key to coping with seasonal load fluctuations in the oil and gas pipeline networks. They play a significant role in ensuring the transmission capacity of pipeline networks and the optimal allocation of operational capacity. Therefore, during the development process of natural gas pipeline networks, due attention should be given to developing load shifting facilities. Adjacent to the load centres, work on site selection should be expedited for building underground gas reservoirs in depleted oil and gas fields, rocky caves and a variety of other terrains. To the extent permitted by site conditions, priority should be given to the use of underground gas reservoirs as the means for load shifting in the natural gas pipeline networks. For locations with restrictive conditions for building underground reservoirs, consideration should be given to building LNG load shifting stations/ Liquefied Petroleum Gas (LPG) load shifting stations, or using LNG receiving stations for load shifting.

4.6.3.4   Expanding Technological Input to Further Raise the Technology Level of Oil and Gas Pipeline Networks

The government should resort to taxation policy, state subsidy and other incentives to drive corporate investment in R&D on pipeline network technology and fully leverage the major roles of companies in innovation. Government measures should be strengthened to coordinate R&D on pipeline technology and studies conducted to raise the domestic contents of key pipeline equipment, with a focus on boosting research on the weak link of operation automation to eventually achieve centralised monitoring and control for all pipeline networks and to catch up with the leading international industry players as soon as possible. The pace of commercialisation of research findings should be quickened and the use of domestic equipment in pipeline construction encouraged.

5

Terminal Energy Consumption

To transform its energy development mode, China should attach importance not only to energy exploitation and conversion and its transportation and allocation, but also to terminal energy consumption. Establishing a model of green energy consumption by starting with end-use consumption has important implications for optimising and adjusting the energy consumption structure and utilisation mode, promoting energy conservation and emission reduction, improving energy efficiency, and achieving sustainable energy development. The key to establishing a green energy model lies in implementing energy conservation as a strategic priority, which focuses on terminal energy conservation, terminal energy substitution strategy targeting electrification improvement, and green transport strategy based on electric vehicle development.

5.1 Model of Green Energy Consumption

From the perspective of consumption, China is facing many challenges for sustainable energy development and urgently needs to innovate its energy consumption concepts and establish a model of green energy consumption in line with the transformation of its energy development mode and energy strategy.

5.1.1 Challenges for Energy Consumption

Since the implementation of its reform and opening up policy, China has seen substantial growth of its integrated national power, with its GDP rising from RMB364.5 billion in 1978 to RMB40 trillion in 2010, representing an average annual growth rate of about 10%. With its rapid economic growth, China's total energy consumption has kept surging and the imbalance between energy supply and demand has become increasingly evident. In 2010, China saw the total consumption of primary energy hit 3.25 billion tonnes of standard coal, which was five times the level recorded at the beginning of the reform and opening up programme. Sustained growth in energy consumption has put great pressure on energy security and environmental care.

Electric Power and Energy in China, First Edition. Zhenya Liu.

5.1.1.1 Sustained Growth in Energy Demand

As China is in the middle stage of industrialisation, energy demand will continue to rise in the long term. As a major developing country, China needs basic raw materials and heavy industrial products such as equipment to support its economic development. China cannot depend on massive imports to meet its need for these products but should instead turn to domestic manufacturing. Moreover, given a large population and insufficient energy resources in China, without a fundamental change to the extensive mode of economic growth, the sustained growth in energy demand will pose great challenges to China's energy supply and have a significant impact on the environment.

5.1.1.2 Irrational Energy Consumption Structure

Since the launch of its reform and opening up programme, China's industrial restructuring has been characterised by a decreasing proportion of the primary industry, a rising proportion of the tertiary industry and no marked change in the secondary industry. The proportion of industry has remained high. Industrial energy consumption has accounted for a whopping 70% of total terminal energy consumption. In 2010, terminal energy consumption by the industrial sector amounted to 1.56 billion tonnes of standard coal, up 800 million tonnes from 2000, representing an average annual growth rate of 8.7%. In addition, constant expansion of China's energy-intensive industries and massive exports of energy-intensive products (especially rough-wrought products) in recent years has resulted in substantial indirect energy exports. In 2010, China exported steel, copper and aluminium products and cement totalling 42.56 million tonnes, 508 000 tonnes, 2.18 million tonnes and 16.16 million tonnes respectively. According to the World Bank's estimates, more than 40% of energy consumed in China ultimately goes into the production of commodities for export.

5.1.1.3 Extensive and Inefficient Energy Utilisation

In 2010, the energy consumption per unit of GDP in China reached 1.03 tonnes of standard coal/RMB10 000 (with GDP calculated at comparable prices in 2005), 2.5 times the world average. The total efficiency of energy processing and conversion was 72%, 10%~20% lower than the world average. The average energy consumption per unit of output value of the major energy-intensive industries was around 15% higher than the world's advanced level.[1] Some energy-consuming equipment were marked by low efficiency of energy utilisation, examples being industrial coal-fired boilers with an average utilisation efficiency of about 15% lower than the world's advanced level, small and medium electric motors, blowers and water pumps with a system operation efficiency 10%~15% lower than the world's advanced level, and motor vehicles with a level of fuel economy about 20% lower than the comparative level in Europe.

---

[1] Estimated based on energy consumption per unit of GDP in China's major energy-intensive products.

5.1.1.4 Low proportion of High-quality Energy in Terminal Energy Consumption

In China, large quantities of coal go into terminal energy consumption while high-quality energy like oil, gas and electricity is utilised insufficiently. In 2010, the terminal coal consumption in China aggregated 1 billion tonnes of standard coal, accounting for 44.0% of the total terminal energy consumption, somewhat lower than the level (69.0%) recorded in 1980 but still high as compared with the world average of less than 10%. Electric energy's share of in China's terminal energy consumption was 3~6 percentage points lower than the comparative levels in Japan and France. A high proportion of coal consumption and an irrational terminal energy consumption structure are to blame for the severe environmental pollution in China and spell the need for an optimised energy structure.

5.1.1.5 Increased Pressure on Energy Supply due to Upgrading of Domestic Consumption Structure

Based on international experiences, when per capita GDP exceeds US$1000, the domestic consumption structure will gradually transform and upgrade from one mainly involving food and clothing to one oriented more towards housing and transport. With accelerated urbanisation and the population's increasing income, urban infrastructure and housing construction will boost consumption of such energy-intensive products as iron and steel, nonferrous metal and building materials. Given the rapidly increasing popularity of urban vehicles and the inevitable tendency of motorised transport, energy demand will post rapid growth. Since 2000, China has registered an average annual growth rate of 15% in private car ownership and a doubling of transport fuel consumption. In 2030, China is expected to see a car ownership rate of 80% among urban families, and a growing share of oil in total energy consumption, which will exert enormous pressures on oil supply and the urban environment.

5.1.2 Establishment of a Green Energy Consumption Model

Given the current situation of energy consumption in China, the current model of irrational and inefficient energy consumption is unsustainable. China must switch to a new concept by establishing scientifically-based ways of production and living and a model of green energy consumption featuring energy conservation and efficient and clean energy consumption, so as to better promote transformation of the nation's energy development mode and energy strategy and support sustainable socioeconomic development.

A model of green energy consumption is one built on the sustainability concept with the focus on rational consumption and overall reduction. The aim is to achieve sustainable energy utilisation through pursuing maximum efficiency with minimum energy input by saving energy, improving utilisation efficiency and promoting substitution of high-quality energy for low-quality energy. A green energy consumption model emphasises a dynamic balance between energy consumption on one hand and socioeconomic development and environmental requirements on the other hand. It is eco-friendly, with the basic characteristics of energy conservation, environmental protection and sustainable growth to aim at achieving coordinated development between man and nature and between man and society.

The establishment of a green energy consumption model can promote social production, construction, circulation, consumption and other areas. It is also conducive to the scientific and reasonable utilisation of energy resources in various aspects of socioeconomic development, the improvement of energy utilisation efficiency and quality and a rational energy consumption structure, with important implications for adjusting and optimising industry structure, transforming the mode of economic growth, and building a conservation-conscious and environment-friendly society. The key strategies for establishing a green energy consumption model are as follows:

5.1.2.1 Establishing Terminal Energy Conservation as the Focus of a Strategic Priority in Energy Conservation

For the benefit of socioeconomic development, China should adhere to scientifically-based ways of production and living, reverse the over-reliance on energy investment for economic growth, limit unreasonable demands for energy consumption, advocate rational, economical and moderate consumption, realise higher economic efficiency of energy with a more intensive and efficient mode of energy utilisation, and boost sustainable energy utilisation. Energy conservation at the end-user level features huge amplification effects. Every 1% increase in the relative efficiency of terminal equipment (including blowers, pumps and compressors) will translate into a 4%~5% rise in the relative efficiency of energy production. 1 kWh of electricity saved means saving approximately three times that level of primary energy.

5.1.2.2 Establishing Electrification Improvement as the Focus of a Substitution Strategy for End-use Energy

It is the objective requirement of energy development to advance energy substitution at the end-user level and optimise energy structure. China's energy resources mainly come from coal. A shortage of oil and gas resources and a high dependence on oil imports suggest that a notable rise in the proportion of oil and gas in terminal energy consumption is unlikely, and it is more attuned to China's basic national conditions to press ahead with energy substitution at the end-use level by focusing on electrification improvement and the substitution of electric energy. In the face of China's ever-higher total energy consumption in the future, improving electrification in the energy-consuming industrial, transportation, business and domestic sectors is of strategic importance to optimising energy consumption structure and realising sustainable energy development in China.

5.1.2.3 Green Transport Strategy Based on Electric Vehicle Development

To actively develop, promote and use green transport like electric vehicles is a modern transport development trend. It is also an objective requirement for and a major step in establishing a model of green energy consumption. The development of electric vehicles is very important for promoting transformation of energy consumption in the transportation field, optimising the energy consumption mode and structure, alleviating pressure on energy supply and environmental protection, and realising green energy consumption.

5.2 Energy Conservation as a Strategic Priority

Energy conservation plays a key strategic role in China. According to the Chinese Government's *Energy Conservation Law* revised in October 2007, energy conservation means 'stronger energy utilisation administration, with the implementation of measures which are technologically feasible, economically rational and bearable to the environment and society for lowering energy consumption, loss and waste discharges, preventing wastage and achieving efficient and rational utilisation of energy resources from energy production to consumption'. The Central Committee and the State Council attach great importance to energy conservation. The 5th plenary session of the 16th Central Committee of the Chinese Communist Party ('CPC') incorporated resource conservation into China's basic state policy, and the report delivered at the 17th National Congress of the CPC stressed the necessity of enhancing energy resource conservation and taking a new road of industrialisation with Chinese characteristics. Energy consumption per unit of GDP and total emissions of major pollutants were included as binding targets in the outlines of China's '11th Five-Year Plan' and the '12th Five-Year Plan' respectively. During the '11th Five-Year Plan' period, energy consumption per unit of GDP in China declined 19.1% and sulphur dioxide emissions and chemical oxygen demand fell 14.29% and 12.45% respectively. The outline of the '12th Five-Year Plan' embodies the binding targets of lowering energy consumption and carbon dioxide emission per unit of GDP by 16% and 17% respectively for the Plan period. The focus is placed on the need to 'persist in building a resource-efficient and environment-friendly society as an important step to accelerate the pace of transforming the current economic development model. Efforts should be devoted to conserve energy, reduce the greenhouse effect, develop a circular economy, promote low-carbon technology, actively address global climate change, promote the coordination between socio-economic development on one side and population growth, resources and the environment on the other, and embark on the path of sustainable development.'

Although China has achieved some results in energy conservation and emission reduction, it is still facing a grim situation at the moment. There is a tendency of emphasising development and speed but ignoring conservation and effectiveness. The operability of laws and regulations on energy conservation needs to be improved and there is no effective incentive policy on energy conservation and no new energy conservation mechanism that complies with the requirements of the market economic system. The development, spread and application of energy conservation technology should be strengthened. The idea of prioritising energy conservation must be further affirmed and efforts should be increased in energy conservation.

5.2.1   Thinking behind Energy Conservation as a Strategic Priority

Prioritising energy conservation means adhering to the guideline of 'giving consideration to conservation and development simultaneously, and placing top priority on conservation' in energy development, while focusing on conservation in energy exploitation, production, transportation and utilisation with a view to improving energy efficiency. A commitment to energy conservation as a priority is a must for building a resource-efficient and environment-friendly society, an important prerequisite for safeguarding China's energy security and promoting the building of green culture, an inevitable choice to fall in line with international energy trends and respond to global climate change and also one of the core tasks of national energy strategy. The basic idea of committing to the priority of energy conservation concerns the following.

5.2.1.1   Committing to Incorporating Energy Conservation in the Overall Socioeconomic Development Strategy and Energy Strategy

As energy conservation is not only a fundamental national policy, but also an important part of China's national strategy, we should, from a strategic and all-round perspective, fully appreciate the important implications of energy conservation for relieving energy constraints, safeguarding national energy security, improving the quality of economic growth, and protecting the environment. China should incorporate energy conservation in the overall strategy of national socioeconomic development, as well as in the national macroeconomic, industry and trading policies. Meanwhile, energy conservation as the $(N+1)^{th}$ energy should be incorporated into the energy planning and management system, overall planning conducted by integrating the various energy resources on the supply side and the demand side, and energy conservation adopted as a primary approach to meeting emerging energy demands. China should efficiently, economically and reasonably make use of the potential supply-side and demand-side resources, constantly improve integrated energy efficiency, and promote rational and effective energy utilisation, so as to support the sustainable socioeconomic growth with a minimum input of energy resources.

5.2.1.2   Committing to Combining Government Regulation with Market Allocation

To promote conscious and scientific energy conservation across the community, the impact of the government's macro-controls must be maximised, policy and information guidance strengthened by formulating energy conservation-related laws and regulations, a policy and institutional environment conducive to energy conservation created, incentives and restrictions for energy conservation in line with the rules of market economy developed. At the same time, China should attach importance to the role of the market mechanism, specify the dominant position of enterprises and other social entities in energy conservation and emission reduction, improve the means and capability for market-oriented energy conservation, promote the development of the energy efficiency sector, and create a market environment favourable for energy conservation.

5.2.1.3   Committing to Energy Conservation at the Structural, Technology and Management Levels

By adjusting the industry and product structures and reasonably planning industry allocations, China can increase industry concentration and economies of scale and eliminate backward energy-intensive enterprises so as to promote the optimisation and upgrading of industry structures. It is estimated that every 1% rise in the proportion of the tertiary industry with a 1% drop in the proportion of the secondary industry will lead to a reduction of about 1% in energy consumption per unit of GDP. China should improve the efficiency of energy utilisation through technological progress by developing, spreading and applying advanced and efficient energy conservation and substitution technologies, comprehensive energy utilisation technologies and utilisation technologies of new energy and renewable energy. Production tools, operating equipment and work processes should be changed, operation processes and skills improved, and mature energy conservation technology adopted for equipment or system renovation. Management should also be strengthened to reduce energy wastage, dripping and leaking in the production, circulation and consumption processes, a green consumption concept of health, civilisation and conservation built, and publicity and training in energy conservation improved so as to enhance community awareness of energy conservation.

5.2.1.4   Committing to Combining Direct and Indirect Means of Energy Conservation

China should strengthen rational energy utilisation across the energy system, reform inefficient production processes and improve energy utilisation efficiency via new equipment and technology in a comprehensive manner, in order to reduce the energy consumption per unit of product (workload). By building on direct energy conservation, indirect energy conservation should be given a bigger role. By conserving regular consumables such as raw materials and daily consumables, economies of scale and product yield and quality can be improved, product structure reasonably adjusted and human resources saved to realise indirect energy conservation and increase GDP per unit of energy consumption.

5.2.2   Focus Areas of Energy Conservation as Strategic Priority

Energy conservation, which involves every aspect of production, living and the whole community, is a complex systems engineering project. Emphasis should be placed on what is truly essential, with specific guidance to push forward energy conservation in an all-round way. Currently, China should strengthen energy conservation in such key areas as industry, building and transportation so as to promote energy efficiency on a nationwide basis.

5.2.2.1   Strengthening Energy Conservation in Industrial Sector

Industry consumes the most energy in China. Over the past decade, China's industry has experienced unprecedented growth in terms of scale and speed. In 2010, terminal energy consumption in the industrial sector increased by 840 million tonnes of standard coal over 2000. But the energy consumption per unit of industrial product in China is still significantly higher than the world's advanced level. The country should focus energy conservation on the industrial sector, speed up industrial restructuring, eliminate obsolete industrial capacity and reduce energy consumption for industrial products, so as to improve the efficiency of energy utilisation in the industrial sector as soon as possible.

Adjusting and Optimising the Industrial Structure

In recent years, China's heavy industry has maintained relatively fast growth in added value, especially in energy-intensive industries like iron and steel, nonferrous metals, petrochemicals and building materials, which have registered average annual growth of over 20% in added value. Curbing the excessive growth of energy-intensive industries has become an important task in effecting macroeconomic controls and optimising the structure of industrial energy consumption. It is estimated that changes in industry and product structures play an important role in reducing the energy consumption per output value in the industrial sector. Given the current structure of the industrial sector, every 1% increase in high-tech industries and every 1% decrease in energy-intensive industries (including metallurgy, building materials and chemicals) in terms of the proportion of value added produced could translate into approximately 1.3% lower energy consumption per unit of GDP. Therefore, China should vigorously adjust the industrial structure, stick to the road of new industrialisation, step up the development of high-tech industries, promote the upgrading of traditional industries and push forward the restructuring of key industries such as equipment manufacturing, shipbuilding, automobiles, metallurgy, building materials, petrochemicals, and light and textiles industries. The share of emerging manufacturing industry in the industrial sector should be increased, the development of energy-intensive industries such as iron and steel, nonferrous metals, building materials and petrochemicals reasonably controlled, and the proportion of energy-intensive heavy chemical industries in industrial energy consumption lowered. The structure of industrial products should also be optimised, with the focus shifting from resource-intensive products to technology-intensive products.

Eliminating Obsolete Industrial Capacity

Many small and medium-sized enterprises (SMEs) in energy-intensive industries in China are still employing backward production processes and energy-intensive technologies, with low-level technology, equipment and management. According to a survey, SMEs consume 30%~60% more energy per unit of output value than large enterprises. During the '11th Five-Year Plan' period, China achieved remarkable success in eliminating obsolete capacity, with small thermal power generating units of 72 GW shut down and obsolete capacities of 121.72 million tonnes of iron, 69.69 million tonnes of steel and 330 million tonnes of cement eliminated. For some time to come, it is still an important means of promoting energy conservation in China's industrial sector by eliminating backward energy-intensive industrial products, equipment and production processes, and transforming the mode of industrial development. It is necessary to develop and implement the objective of elimination, stronger performance-based initiatives, improved technical standards as well as policy measures governing the energy consumption limits for major energy-consuming products to maximise the role of the market mechanism and effectively control the unreasonable growth of obsolete capacity in energy-intensive industries.

Improving Industrial Technology and Reducing Energy Consumption for Industrial Products

In recent years, China's major energy-intensive industries have seen major advances in overall technological levels. From 2005 to 2010, the share of thermal power units of over 300 MW level in the thermal power industry's total installed capacity had increased to 73% from 47% whereas the percentage share of large blast furnaces of over 1000 m³ level in the iron and steel industry had risen from 21% to 52%. As a share of total output, large pre-baked anodes in the electrolytic aluminium industry grew to 90% from 80% while the output proportion of new dry-process cement clinker in the building materials industry increased from 39% to 81%. But on the whole, problems remain in major energy-incentive industries, such as different technology and equipment levels and backward production processes and equipment in some enterprises, leading to low efficiency of energy utilisation and high energy consumption per unit of output value. Therefore, China needs to develop policy on energy-saving technology for major energy-intensive industries, guide the key energy-intensive industries in optimising the production process by stepping up research and promotion of new energy-efficiency technology, new processes and new materials, drive the progress of energy efficiency technology in energy-intensive industries and the transformation of energy efficiency technology at the corporate level. All this will have the benefits of improving the level of production technology in key sectors, including ferrous metals, nonferrous metals, electric power, chemicals, building materials and machinery, and reducing energy consumption by major energy-intensive products such as iron and steel, electrolytic aluminium, cement, ethylene, synthetic ammonia, caustic soda and calcium carbide. By 2020, according to the targets (see Table 5.1) for energy consumption per unit of China's major energy-intensive product specified in the *Medium & Long-Term Energy Conservation Plan*, the energy consumption per unit of major product in China's key industries will further decline, and the overall energy efficiency will improve.

China's power industry operates on relatively high overall technical standards, with some reaching or approaching the world's advanced level. Future energy conservation should focus on vigorously developing (ultra) super-critical units of 600 MW level or above and large combined cycle units, adopting efficient and clean power generation technology to renovate the operating thermal generating units and improve the generation efficiency of the units, raising single-unit capacity by 'replacing small units with larger ones', 'encouraging large projects and discouraging small inefficient power plants' and eliminating small units. In addition, multi-cogeneration technologies including heat and power cogeneration, heat, power and cool cogeneration, and heat, power and gas cogeneration should be developed. The focus should also be placed on promoting the development of ultra-high voltage and smart power grids and implementing the economical operation of power grids, adopting advanced power transmission, transformation and distribution technologies and equipment to phase out old energy-intensive equipment and reduce losses from power transmission, transformation and distribution, while strengthening management to reduce power consumption at the power plant level.

Table 5.1 Target energy consumption per unit of China's major energy-intensive product in 2020.

Category	Unit	2020
Coal for thermal power generation	g SEC/kWh	320
Integrated energy consumption per tonne of steel	kg SEC/tonne	700
Integrated energy consumption per tonne of steel	kg SEC/tonne	640
Integrated energy consumption for 10 different nonferrous metals	tonnes of standard coal/ tonne	4.45
Integrated energy consumption of aluminium	tonnes of standard coal/ tonne	9.22
Integrated energy consumption of copper	tonnes of standard coal/ tonne	4
Oil refining unit energy factor consumption	kg standard oil/tonne factor	10
Integrated energy consumption of ethylene	kg standard oil/ tonne	600
Integrated energy consumption of large-scale synthetic ammonia	kg SEC/tonne	1000
Integrated energy consumption of caustic soda	kg SEC/tonne	1300
Integrated energy consumption of cement	kg SEC/tonne	129
Integrated energy consumption of flat glass	kg SEC/weight case	20
Integrated energy consumption of building ceramic	kgce/m^2	7.2

Source: Data in the table from the Medium & Long-Term Energy Conservation Plan issued by the National Development and Reform Commission in 2004.

5.2.2.2 Strengthening Energy Conservation in Buildings[2]

Building and construction-related energy consumption accounts for about 30% of the global total. In particular, this area of energy consumption in developing countries and regions (such as India, Brazil and Africa) accounts for 20%~25% of total energy consumption, compared with 30%~40% in developed countries (such as the USA, Canada and Japan). The high proportion of building-related energy consumption in total energy consumption has driven many countries to practise energy conservation. In recent years, China has witnessed ever-expanding city size and ushered in the peak period of housing development with over 2 billion m^2 of newly-built space every year, and has leaped to world No. 1 in terms of existing urban and rural building stock (over 40 billion m^2 in total). In addition, with the progress of society and changing ideas of lifestyle, people also have more diversified demand for building-related services, with higher new requirements for building-related energy consumption. It is expected that by the end of 2020, the floor area developed in China will increase 25~30 billion m^2. At the current rate of building-related energy consumption, each year will see 1200 TWh of electricity and 410 million tonnes of standard coal consumed, or nearly three times the national total of building-related energy consumption recorded at the beginning of the '10th Five-Year Plan' period. With the continued improvement in living standards and the advancement of the urbanisation process in the future, China will continue to see mounting building stock and a rising proportion of building-related energy consumption in community-wide energy use, indicating a huge potential for energy conservation in the building sector.

---

[2] According to internationally adopted energy statistical methods, residential and commercial energy consumption is included in energy consumption in building operations.

Placing Emphasis on Energy Efficiency Plans for the Construction Field

Building-related energy conservation touches on wide-ranging concerns and represents a huge challenge. Energy conservation should be taken into full consideration in the early stage of formulating urban plans and building programmes. Therefore, the key lies in the level of commitment and effort committed by local governments, especially municipal governments. To promote energy conservation from the planning stage and based on the goal to build a harmonious and energy-conscious society, efforts should be made to formulate urban plans in a scientific and reasonable and moderately advanced manner. The direction for new developments should also be determined to avoid constructing any 'short-lived buildings' due to a lack of thoughtfulness in the blind pursuit of style at the expense of functionality. Meanwhile, attention should also be paid to maintenance, repair and reasonable use of existing buildings. By doing this, China can lower the demand for floor space while ensuring building functionality to ease the pressure on energy supply by reducing the massive requirements of steel, cement and other energy-intensive products brought about by rapidly growing construction scale.

Strengthening Energy Conservation in New Buildings

The key is to fully implement energy efficiency design standards for new buildings. Relevant design standards and technological requirements for building-related energy efficiency should be enforced in the building design, construction, management, acceptance and other processes. A higher level of energy efficiency design should be mandated particularly for large and energy-intensive public buildings. Efforts should be made to strengthen supervision and inspection of energy efficiency design standards for buildings and to raise the bar when appropriate. Currently, the energy efficiency design standards for buildings in China date back to 1996 with a stipulated efficiency rate of 50% and some provinces and cities subsequently raising the efficiency ratio to 65%. In line with the increasingly stringent requirement for building-related energy efficiency nationwide, energy efficiency design standards based on an efficiency ratio of 65–75% should be applied to new buildings in around 2020, with the bar to be raised to 85% beyond 2030.[3]

Promoting Energy Efficiency Retrofits on Existing Buildings

In accordance with the features and energy consumption modes of different buildings, energy efficiency retrofits should be carried out on protection structures, heat supply systems, heating and cooling systems, lighting equipment and hot water supply facilities

---

[3] Phase I of *The Energy Efficiency Design Standards for Civilian Buildings (Heated Residential Areas)* has been implemented in China since 1 August 1986 with a target efficiency rate of 30%. A 30% energy efficiency rate means a saving of ~30% in energy consumption against the design energy consumption levels for residential buildings in 1980–1981. Phase 2 of *The Energy Efficiency Design Standards for Civilian Buildings (Heated Residential Areas)* has been implemented since 1 July 1996 with a target efficiency rate of 50%. A 50% energy efficiency rate means a saving of 50% in energy consumption against the design energy consumption levels for residential buildings in 1980–1981. In line with the ever-higher requirements for building energy efficiency nationwide, some provinces and cities have implemented Phase 3 of *The Energy Efficiency Design Standards for Civilian Buildings (Heated Residential Areas)* with a target energy efficiency rate of 65%. A 65% energy efficiency rate means a saving of ~65% in energy consumption against the design energy consumption levels for residential buildings effective since 1980.

of existing buildings that do not comply with the mandatory energy efficiency design standards for buildings. Energy efficiency retrofits should be mandatorily enforced on existing buildings through the setting of proportions and deadlines. With the rigorous implementation of energy efficiency design standards for new buildings and constant promotion of energy-efficiency retrofits on existing buildings, it is expected that by 2020 about 420 TWh of electricity and 260 million tonnes of standard coal can be saved and 846 million tonnes of carbon dioxide and other greenhouse gases emissions reduced.[4]

Reinforcing Management of Energy Consumption in Building Operations

Energy consumption in building operations is the focus of energy efficiency management in buildings. During the life-cycle of buildings, energy consumed in building material production and building construction generally accounts for about 20% of the total building-related energy consumption, with the remaining 80% attributable to building operations. In particular, heating and cooling take up half of the total energy consumption in building operations. On a per-floor-area basis, energy used for heating and cooling in China is about 3 times that in developed countries on the same latitude with similar climatic conditions. To reduce energy consumption in buildings in China, the heating structure must be optimised, with wider applications of quality power and renewable energy in heating systems and the development of solar-powered and geothermal heat pumps in light of local conditions to reduce the proportion of decentralised heat supply fuelled by coal. In addition, various energy-efficient products, including energy-saving air-conditioners, lights and refrigerators, should be promoted to maximise the positive impact of energy-efficiency technology on enhancing building energy efficiency. In the meantime, energy consumption in buildings should be better managed through developing advanced energy management systems. It is expected that by 2020, energy consumed in residential and public buildings in China may approach or reach the level of moderately developed countries, if China should strive to implement various building energy efficiency standards and measures.

5.2.2.3 Enhancing Energy Efficiency in the Transportation Field

China has seen rapidly growing energy consumption in transportation in recent years. According to the data of *The China Energy Statistical Yearbook*, terminal energy consumption in the transportation sector in 2010 accounted for ~10.6% of China's total terminal energy consumption, an increase of 140 million tonnes of standard coal from 2000, representing average annual growth of 9%. The transportation sector is a major oil consumer, led by road transport (excluding private transport) with a 60% share of the total energy consumption in transportation, compared with 15% each for waterway and railway transport and 9% for civil aviation. On the back of a rapidly rising economy, accelerating urbanisation and the development of the tertiary industry, ever-growing flows of people and cargo in China will drive rapid growth in transport demands in the future. Therefore, the transportation field is a key area where China should strengthen energy efficiency, with the focus on reducing oil consumption.

---

[4] Remarks by Wang Guangtao at 2nd International Conference on Smart, Green and Energy-efficient Buildings & New Technologies and Products Expo in 2006.

Boosting Construction of Modern Integrated Transport System
Currently, structural irrationality stands in the way of fully leveraging the comparative advantages of China's modes of railway, road, waterway, civil aviation and pipeline transport. In the future, based on the technological and economic features of different transportation modes and taking into account China's economic geography and national conditions, efforts should be made to further the structural improvement and upgrading of the transportation field, maximise the competitive edges of various transportation modes, optimise the allocation of transport resources, bring into play the combined efficiency of integrated transportation, and realise the coordinated development between different transportation modes and between different segments of each transportation mode. Meanwhile, China should maximise the vital roles of power grids in energy transportation, with a commitment to parallel development of coal transportation and electricity transmission, faster expansion of electricity transmission capacity and the building of a modern integrated energy transport system closely linked to the construction of an integrated transport system in order to improve the overall energy efficiency in the transportation sector.

Strengthening R&D on Energy Efficiency Technology Utilisation and Energy Efficiency Management in the Transportation Field
In highway transport, the focus is on improving motor vehicle fuel efficiency, implementing mandatory fuel efficiency standards, encouraging the development of low-emission vehicles and new-energy automobiles, and the building of a fast public transport system. In rail transport, efforts should be made mainly to develop electrified railway operations and enhance energy efficiency management of locomotives. In civil aviation, emphasis is placed on adopting fuel-saving aircraft and increasing load factor and passenger load factor. In waterborne transport, attention is focused on building ultra-large vessels, improving vessel power design and optimising vessel power structure.

5.2.3   Implementing Measures to Ensure Strategic Priority of Energy Efficiency

5.2.3.1   Improving and Implementing Laws and Regulations and Incentive Policy Governing Energy Efficiency

A new social order for energy efficiency should be established through legal and economic means. On the one hand, laws and regulations and supporting standards in respect of energy efficiency should be formulated and improved to move energy efficiency gradually towards institutionalisation and standardisation. The establishment and improvement of a legal system for energy efficiency based on the provisions of *Energy Conservation Law* should be sped up and coordinated, with supporting regulations and standards to intensify the supervision and management of energy efficiency work and the implementation of energy efficiency regulations. On the other hand, the energy pricing system should be improved and the tax and financial policies governing energy efficiency promoted. The reform of energy product pricing should be stepped up and a more scientific and rational energy pricing system established so that energy product prices can fully reflect the scarcity of resources, supply/demand situation, environmental costs and other external factors, thereby maximising the role of prices in guiding energy efficiency efforts.

Meanwhile, financial, tax, credit and other economic policies to encourage energy efficiency should be improved, enterprises and other sectors of the community should be guided and encouraged to practise energy efficiency. Financial support for energy efficiency initiatives should be strengthened and a special energy efficiency development fund set up. Tax preferential policies conducive to promoting energy efficiency in the production and consumer areas should be studied and implemented. Investment and financing policies should be readjusted to provide loans with subsidised interest for energy efficiency projects, with commercial banks guided to make investment in the energy efficiency field.

5.2.3.2   Speeding up R&D, Demonstration and Promotion of Energy Efficiency Technology

Efforts should be devoted to organise R&D on generic, key and cutting-edge energy efficiency technologies and carry out major energy efficiency demonstration projects aimed at commercialising energy efficiency technology. An enterprise-based system for energy efficiency technology innovation should also be set up and the transformation of technological achievements expedited. China should introduce and absorb advanced overseas energy efficiency technology. A campaign should be organised to promote and apply advanced and maturing new technology, processes, equipment and materials for energy efficiency while energy end-use equipment (products) listed in the energy efficiency equipment (products) catalogue should be accorded priority for promotion. Plans for development, demonstration and promotion of energy efficiency technology should be developed with well-defined milestones and key support policies implemented in phases. More funds should be invested to establish generic energy efficiency technology and a base (platform) for R&D on general equipment. China should improve the technology and production capacity of energy equipment manufacturers, strengthen policy guidance and incentive, and encourage wide adoption of efficient equipment and processes in energy-intensive industries in order to reach or approach world-leading levels.

5.2.3.3   Implementing a New Market-oriented Energy Efficiency Mechanism including Contract Energy Management and Power Demand Side Management

There should be proactive exploration and promotion of a new market-oriented, China-specific energy efficiency mechanism as an important means to promote energy efficiency. China should actively implement contract energy management and encourage the development of energy efficiency service companies to provide one-stop service from diagnosis and design through financing and retrofit to operation and management for enterprises engaged in energy efficiency retrofits. China should formulate and enforce investment, tax, credit and other policies favourable for the energy efficiency service industry, guide and encourage energy efficiency service providers to expand the market and scope of their services while speeding up the commercialisation of energy efficiency services. It should also enhance power demand side management, give full play to the initiative of power grid operators and launch relevant incentive policies.

5.2.3.4 Implementing Energy Efficiency Standards, Labels and a Certification System for Energy Efficiency Products

China should push for the implementation of a management system for energy-efficient product certification and energy efficiency labels and help continuously improve product energy efficiency standards and energy efficiency technology. It should strengthen the energy-efficient product certification system, enlarge the scope of energy efficiency standards and labels, enhance the mandatory nature and the supervision and management of energy-efficient product certification, and maximise the role of energy-efficient product certification in guiding energy efficiency efforts. Users and consumers should be guided to purchase energy-efficient products through a market mechanism. For example, the 'Energy Star' certification system effective since 1992 in the USA currently covers nearly 4 000 product types of electric home appliances, consumer electronics and buildings. It helps reduce electricity demand by over 5% annually.

5.2.3.5 Launching Energy Efficiency Campaign Nationwide

It is a huge systems engineering project to implement the strategic priority of energy efficiency, which entails forming a government-led, enterprise-based work structure with concerted efforts of all sectors. China should fully leverage the leadership position of government agencies, with fundamental support from families, communities and schools, especially enterprises, in energy efficiency. The focus should be placed on energy efficiency in energy-intensive enterprises, reinforcing government supervision and management of energy efficiency in key energy-consuming industries, helping enterprises speed up energy efficiency retrofits, enhancing energy efficiency management and improving energy utilisation. In the meantime, China should unleash the initiative of enterprises, help them fulfil social responsibilities and pursue energy efficiency spontaneously and voluntarily through effective incentives and restrictions. Targeting the corporate sector, the 'Ten Key Projects for Energy Efficiency'[5] and 'Energy Efficiency Action Plan for Top 1000 Enterprises'[6] achieved satisfactory results during the 11th Five-Year Plan period. As the largest state-owned public utility enterprise, State Grid Corporation of China has strengthened its own energy efficiency initiatives while taking advantage of the competitive edge of power grids to promote energy efficiency and emission reduction in the electricity industry and the community at large. In 2010, State Grid Corporation saved

[5] The 'Ten Key Projects for Energy Efficiency' refer to a coal-fired industrial boiler (kiln) retrofit project, a district cogeneration project, a residual heat and pressure utilisation project, a petroleum saving and substituting project, an electrical system energy efficiency project, an energy system optimisation project, a building energy efficiency project, a green lighting project, an energy efficiency imitative among government agencies, and an energy efficiency monitoring and technology service system project. Implementation of the Ten Projects helped save energy totalling 340 million tonnes of standard coal during the 11th Five-Year Plan period.

[6] The 'Top 1000 Enterprises' refer to 998 independently audited enterprises in nine key energy-consuming industries, namely iron and steel, nonferrous metals, coal, electricity, petroleum and petrochemicals, chemicals, building materials, papermaking and textiles with annual consumption of 180 000-plus tonnes of standard coal. The 'Top 1000 Enterprises' saved energy equivalent to 150 million tonnes of standard coal during the 11th Five-Year Plan period.

electricity equivalent to over 1.3 million tonnes of standard coal through reduced line loss rates, raised power output by the equivalent of 6 million tonnes of standard coal through optimised power grid dispatch and water utilisation efficiency, saved 12.66 million tonnes of standard coal by promoting trading of generation rights, and saved up to 2.58 TWh of electricity by carrying out 88 000 demand side management projects that covered green lighting, high-efficiency electric motors, reactive-load compensation equipment and energy-efficient transformers.

5.3   Electrification in Socioeconomic Development

In the future, China's energy consumption will continue to rise, with the constraints on energy resources and environmental capacity posing great challenges to the fast-growing economy. In response, improving electrification in all areas, including industry, transportation, commerce, the urban sector and the countryside, and raising electricity's share of terminal energy consumption can increase energy efficiency, relieve pressure on oil imports, help upgrade and optimise energy consumption structure, improve people's livelihood and environmental quality, and promote sustainable development in economic, societal, energy and environmental terms.

5.3.1   Substitution of Electric Energy in Terminal Energy Consumption

5.3.1.1   End-use Energy Types are Intersubstitutable to a Certain Extent

Despite their different properties, all energy products can be used to generate motive power or heat. Different energy products are intersubstitutable to a certain extent, some common examples being the substitutability of electric cars for petrol vehicles, gas-fired boilers for coal-fired boilers, and electric water heaters for gas water heaters.

Terminal energy substitution refers to the intersubstitutability among different fuel options for end use, like coal, oil, natural gas and electricity. It should be noted that such fossil fuels as coal, oil and natural gas have a more notable role as substitutes when used as industrial raw materials, and electric energy can be substituted for them in many other important sectors.

China's energy resource endowments have to be taken into full account when it comes to energy substitution. Provided that scientific guidance, technical feasibility and economic viability are in place without affecting the cost efficiency of terminal use, fuels with abundant reserves can substitute for scarce fuels, low-carbon energy for high-carbon ones, high-efficiency energy for low-efficiency ones, and safe and convenient energy for dangerous ones, in order to complete the adjustment of energy consumption structure. Improving energy substitution and optimising energy consumption structure by all possible means and measures are significant initiatives to drive the transformation of China's energy strategy and promote sustainable socioeconomic development.

5.3.1.2   Substitution of Electric Energy is an Important Trend in Terminal Energy Substitution

Energy substitutions have occurred throughout the course of socioeconomic development at home and abroad, reflecting the technological progress in the energy field and the demand for reasonable use of energy resources. In human history, coal had substituted for firewood before it was replaced by oil and natural gas. At present, new and renewable energy products (chiefly electricity for terminal consumption) substitute for traditional fossil energy. From the perspective of development trends, practical needs and technical conditions, the substitution of electric energy is an important trend in terminal energy substitution.

Raising Electricity's Share of Energy Consumption is a Worldwide Trend

In 2009, electric energy accounted for 17.3% of the global terminal energy consumption, up 6.4 percentage points from 1980, while coal and oil declined by 3.2 percentage points and 3.7 percentage points respectively. Despite the widely different resource endowments of different countries, major countries in the world are seeing a rising electrification trend. In developed countries such as the USA, Japan, France and Germany, electric energy's share of terminal energy consumption has increased over 10 percentage points from the 1970s. Developing countries like India and Brazil have also seen significantly higher electrification levels and the proportion of electric energy in Brazil's terminal energy consumption has increased 12 percentage points from the 1970s. See Figure 5.1 for the mix of global terminal energy consumption.

Electricity's share of China's terminal energy consumption reached 21.3% in 2010, higher than the world average level but about 5 percentage points lower than Japan's 2008 level. This compares with coal's relatively high share of terminal energy consumption at 44.0% in 2010 (down 25 percentage points from 1980 but still higher than the world average in the same period), indicating growth potential for electricity's share of total energy consumption. See Figure 5.2 for the mix of China's terminal energy consumption.

Energy Security Calls for Substitution of Electric Energy

In the future, fossil fuels like coal, oil and natural gas will be exhausted and new and renewable energy will be widely exploited, so the substitution of electric energy is an inevitable choice in developing and utilising new and renewable energy. With oil reserves being far less than coal reserves, China can develop electricity as a substitute for oil to reduce its reliance on oil imports and better ensure energy security.

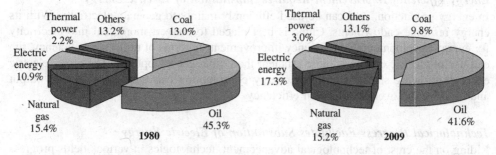

Figure 5.1 The mix of global terminal energy consumption in 1980 and 2009.
Source: IEA.

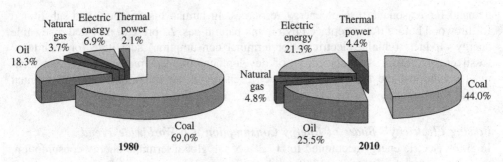

Figure 5.2 The mix of China's terminal energy consumption in 1980 and 2010.
Source: *China Energy Statistical Yearbook*, National Bureau of Statistics.

Substitution of Electric Energy Required for Overcoming Environmental Constraints

As energy is widely explored and utilised, ecological and environmental constraints will become increasingly prominent in the future. Coal's considerable share of terminal energy consumption is one of the major causes of China's environmental problems. Developed countries put nearly all or most of their coal into electricity generation and little into end-use consumption. With advanced combustion technologies and well-developed devices for dust elimination, desulphurisation and denitrification, large modern thermal power plants can reduce the air-borne pollution caused by dust and sulphur dioxide emissions and effectively lower the total volume and density of pollutants discharged, to facilitate concentrated, large-scale treatment of pollutants. Therefore, converting coal into electricity for end-use consumption is one of the best ways to reduce coal's impact on the environment. In the meantime, China's power source structure has been increasingly optimised as the proportion of clean energy like hydropower in total installed generating capacity continues to rise. The optimisation of resource allocations on a larger scale via power grids makes electricity consumption more eco-friendly and cleaner. With the increasingly intensive substitution of electric energy in cities, the reduced levels of direct coal combustion and the promotion of electric vehicles, urban air quality and the environment can also be improved.

Energy Efficiency Improvement Requires Substitution of Electric Energy

In energy production, coal can be used efficiently mainly to generate electricity. With its energy resource endowments, China is best placed to convert more coal into electricity for energy development and efficiency improvement. In terms of utilisation efficiency, the end use of electric energy is more efficient than other energy options. Therefore, electric energy's growing share of terminal energy consumption can increase economic output and improve the overall utilisation efficiency.

Technological Progress Facilitates Substitution of Electric Energy

Riding on the crest of technological advancement, technologies in various fields progress at a rapid pace. From a supply side perspective, the efficiency of coal-fired power generation technology is gradually improving, and power generation technologies in renewables, including wind, solar and biomass energy, are maturing and gradually applied on a large scale. Technologies and equipment for electricity consumption have been developed and applied on a large scale, power batteries and related technologies have broken fresh ground and have been commercialised, contributing to the rapid growth of energy storage technology. All this will play a positive role in supporting the substitution of electric energy and community-level electrification improvements.

5.3.2 Electrification in the Industrial Sector

With continuous industrialisation, energy consumption for industrial purposes will continue to increase. Improving electrification in industry can help change and upgrade the industrial structure, improve industrial automation levels, contribute to the efficient use of industrial energy and improve the overall efficiency of industry to reduce environmental pollution and support China's new-style industrialisation.

5.3.2.1 Promoting Electrification with Industrial Informationisation

Informationisation and electrification are complementary, as electrification serves as the base of informationisation, which in turns drives electrification. Thanks to the rising contribution of the information industry to China's national economy, ever-growing use of information science in traditional industries and the increasing development of information infrastructure, the information industry will play a more and more dominant role in the national economy to propel the electrification of industrial informationisation to a new height.

5.3.2.2 Major Contributions to Higher Levels of Industrial Electrification

The major contributors are motive power and heating. In motive power, electric energy shows a competitive edge; in heating, electric heating is represented by an energy-efficient and environment-friendly induction heating technology capable of heating metallic materials most effectively and efficiently. In comparison with heating by common fuels, electric heating can produce higher temperature (e.g. the temperature of electric arc heating can reach over 3000 °C) and make automatic and remote temperature control easier, with higher heat efficiency and the capability to produce higher temperature faster. In line with the technological requirements of heating, it can make even heating or partial heating possible and easily achieve vacuum heating and controlled heating. Electric heating generates less exhaust gas, residue and soot, which can keep heated objects clean and is friendly to the environment. Therefore, electric heating can be widely used for production, scientific research, experimentation and other purposes.

5.3.2.3 Major Sectors Conducive to Improving Industrial Electrification

Among the major segments of industry, the manufacturing sector, including steel and iron, building materials, general equipment and electrical and electronic equipment manufacturing, rubber, pharmaceuticals, food and textiles, offers the potential for substitution of electric energy. Specifically, substitution of electric energy in the steel and iron, building materials and light industries is the focus of the electrification drive. In the steel and iron sector, electrification is pursued mainly through the intensive development of steel for electric furnace production, the structural adjustment of the steelmaking process to bring out the distinct advantages of efficiency and positive environmental impact, and optimise the sector's energy consumption structure. In the building materials sector, the promotion and use of electrically heated kilns in the ceramics and glass-making fields is good for the environment. It can also improve the safety and stability of technical operations and the quality of products. For light industries including textiles, papermaking and food, the substitution of electric boilers with thermal storage capability for coal-fired and oil-fired furnaces in the drying, heating, steaming and boiling processes ensures energy is consumed in a safe, economical and environment-friendly way.

5.3.3 Electrification in the Transport Sector

Transportation is an energy-intensive tertiary industry.[7] In 2010, the total terminal energy consumption in China's transport sector reached the equivalent of 240 million tonnes of standard coal, with 89% attributed to oil products. In the future, energy consumed in the transport sector will continue to increase with the rapid industry development, exerting an important influence on China's energy security and supply–demand balance. Promoting the substitution of electricity for oil to improve the transport sector's electrification level can help lower the amount of oil consumed, adjust energy consumption structure, contribute to efficient use of transport fuels, and reduce environmental and pollution problems to build a green integrated transport system. The key aspects of electrification in this sector are as follows.

5.3.3.1 Developing Electric Rail System

Electric rail has its technological and economic advantages in the form of huge transport capacity, high speed, low energy consumption, low costs and good operation conditions. As an electric locomotive provides an average total traction of nearly 500 tonnes higher than a diesel locomotive but only consumes about 1/3 of the energy required by a diesel locomotive, China will focus on electric traction in developing railway traction power. Under the *Medium and Long Term Plans for Development of the Multimodal Transport Network* released by the National Development and Reform Commission, it is predicted that the mileage of China's railways in operation will reach more than 120 000 km in 2020, with electric rail accounting for 60% of the total.

5.3.3.2 Developing Urban Rail Transport

Urban rail transport comes in various forms, including urban railways, subways, light rails, trams, magnetic suspension lines, etc. They are driven mainly by electricity with

---

[7] To maintain consistency and comparability, the data on the energy consumption and structure of China's transport industry are extracted from the *China Energy Statistical Yearbook* of previous years. The oil consumed in transportation only includes the oil consumed by public transportation and excludes the oil consumed in other segments and by private vehicles.

large carrying capacity, with the benefits of being fast, safe and punctual. To ease the congestion of surface transport in modern cities, urban rail transport is more widely used and has become an important part of the transport network in large and medium-sized cities. China has now become the fastest moving country on urban rail in the world, with its subway investment increasing by RMB200 billion year-on-year. It is predicted that by 2020, the cumulative mileage of urban rail transport in operation in China will hit 7 395 km. In particular, the subway mileage of Beijing is planned to exceed 1000 km in 2020, with the average road length reaching 1.9 km within the fourth ring, equivalent to the density of the traffic networks of Tokyo and New York. It is predicted that by 2030, the cumulative mileage in urban railway operations in China will soar to 10 000 km.

5.3.3.3 Developing Electric Vehicles

As an important category of new-energy vehicles, electric vehicles are the development focus of China's automobile industry (including pure electric vehicles, fuel cell cars and plug-in hybrid vehicles) in the future. Developing electric vehicles is a significant step in realising the 'substitution of electricity for oil' in the transport industry. With a variety of maturing technologies, lower production costs, better support policies and facilities, and rising market acceptance, the advantages of electric vehicles in terms of low operation costs and high environmental performance will be fully felt. It is forecast that in 2020, China's electric vehicle industry will move into a growth period, with EV's share of total car ownership coming at 5%~8%. In 2030, China's electric vehicle industry will enter a period of popularity, during which all kinds of electric passenger vehicles will be developed, with the performance, cost effectiveness and output of electric vehicles improved remarkably and electric cars accounting for 10%~15% of total car ownership. (The strategic significance of developing electric vehicles will be further discussed in the following section.)

5.3.3.4 Developing Electric Bicycles

Known for their low energy consumption, space efficiency, freedom from parking worries and user-friendliness, electric bicycles offer the feasibility of substituting for motorbikes on the grounds of technology, cost, environmental protection and safety. In the future, China should gradually increase the share of electric bicycles in daily short-haul trips. As the largest producer of electric bicycles in the world since 2003, China's electric vehicle ownership exceeded 120 million units in 2010 and has been increasing by 30% annually, reflecting tremendous market potential. China should perfect the support measures for electric bicycles as soon as possible to ensure orderly industry development.

5.3.4 Electrification for Businesses and Urban Population

Reflecting fast-growing energy usage, the terminal energy consumption by businesses[8] and urban residents in China amounted to 42.85 million and 150 million tonnes of standard

---

[8] According to the classification standards of China's national economy and electricity industry, the term 'businesses' mainly refers to the wholesale, retail and F&B industries.

coal in 2010, or 2% and 6.5% of the total, respectively. Originally coal-oriented, energy consumption by businesses and urban residents is similar in structure and type. The proportion of coal had subsequently fallen sharply, in contrast with the gradual rise of electricity. In 2010, coal accounted for 35% of the energy consumed by businesses and 13% of the energy consumed by the urban population, against the comparative shares of electricity at 37% and 25% respectively.

Looking ahead, against a backdrop of economic restructuring and growing urbanisation, the living standards in China will see constant improvement and energy consumption by the tertiary industry and urban residents will increase more significantly. By promoting the substitution of electric energy among businesses and urban residents and increasing the percentage share of electricity and other clean energy, we can adjust and optimise energy consumption structure, better support urban development, improve living conditions and quality of life, and promote clean, environmental-friendly and efficient urban energy consumption. The electrification of energy consumption for businesses and urban residents can be promoted primarily in the following ways.

5.3.4.1 Developing Electric Heating and Electric Water Heaters

Electric heating takes the form of a heating system powered by electric energy. It comes in various devices and equipment like electric heaters, air-conditioners, electric boilers with thermal storage, heating cables and thermoelectric film. With the advantages of being clean, pollution-free, silent, safe, convenient and capable of individual metering, electric heating satisfies the requirements of various environments and conditions and promises huge potential for promotion in commercial and residential areas. It can supplement central heating and replace coal-fired boilers by using electricity during off-peak periods. The electric water heater is cost effective, eco-friendly, convenient and safe. Tailored to local conditions, it can be used to supply hot water to businesses and residents. With the fast-rising living standards, domestic consumption of hot water will increase gradually, promising great growth potential for electric water heaters and other energy-efficient appliances.

5.3.4.2 Promoting and Applying Heat Pump Technology

Heat pumps transfer heat from low-grade heat sources to high-grade ones. Based on heat sources, they may be categorised into water-source, ground-source, air-source and double-source (water-source combined with air-source) heat pumps. As an energy-efficient air-conditioning system capable of both heating and cooling with shallow underground geothermal heat resources, ground-source heat pumps have made decent headway and been widely used in North America and Europe in the last decade. The number of ground-source heat pumps in the USA increases by 10% every year, with more than 400 000 sets installed so far. Compared with traditional air-conditioners and heat supply systems, heat pump technology enjoys great potential for application given its recyclability, low operation cost, small footprint, water conservation, eco-friendliness, high automation, safety and reliability. In recent years, China has started working on ground-source heat pump technology, giving it a strong growth momentum. By 2010, the total floor area of premises installed with heat pumps in China had exceeded 30 million m^2 and a stronger push should follow.

5.3.4.3 Advancing Household Electrification

Along with the growing economy and higher living standards, China is seeing ever-rising expectations of quality of life. As an integral part of domestic energy consumption, electricity offers increasingly prominent benefits such as cleanliness, efficiency, security, controllability and convenience. The electrification of kitchens, bathrooms and the whole residential unit has become an important development trend. In the USA, 65% of energy consumed in kitchens is electricity, and the terminal utilisation efficiency of induction cookers and other electric cooking appliances has reached 90% or more. Compared with coal-fired and gas cookers, the induction cooker features higher thermal efficiency, lower pollution and greater safety and reliability, with easy maintenance and control. Promoting the substitution of electric energy in the residential sector can optimise energy consumption structure by raising the level of electricity in the domestic consumption of energy.

5.3.5 Rural Electrification

Energy problems in the rural sector affect the energy consumption and the improvement of quality of life among more than half the Chinese population. The development of rural energy is an important part of an optimised national energy system. Promoting the substitution of electric energy in the rural sector and raising the level of rural electrification can facilitate the improvement of quality of life and better support socioeconomic development in rural areas to serve the construction of a new countryside. Quality clean energy like electricity in place of direct burning of coal, firewood and straw may also reduce environmental pollution and ecological damage in rural areas. Rural electrification can be improved mainly in the following ways.

5.3.5.1 Promoting Electrification of Agricultural Production

Driven by the increasing population density and improving living standards, demand for grains in China will continue to grow strongly in the next two to three decades. By 2020, the nation's arable land is expected to be no less than 1.8 billion *mu*, and the effective irrigation areas will exceed 900 million *mu*. There is great potential for the substitution of electric energy in farmland irrigation and crop production. Farmland irrigation will be based on water pumps driven by electric motors instead of diesel engines. Electric submersible pump technology is used extensively for farmland irrigation, sprinkler irrigation, garden spray irrigation, water tower delivery and the breeding industry's water supply and drainage to reduce equipment investment and operating costs and improve efficiency. In agricultural and sideline production, China will popularise rural electricity and develop new technology for electricity usage to facilitate intensive crop processing and enhance the level of automation so as to better improve crop output and quality.

5.3.5.2 Propelling Electrification of Rural Life

Energy for China's rural life needs to be improved structurally. Traditional biomass energy, mainly straw and firewood, accounts for a large proportion of energy consumed in rural China, while the percentage of electric and other commercial energy has been low, so electrification of energy consumption in rural areas has huge development potential. By increasing sales of electric home appliances to the countryside, we can improve the level of home appliance ownership in the rural population and contribute to higher electrification of rural life. The efficiency of energy utilisation can be improved by replacing direct burning of coal, straw and firewood with electric cooking, which can also contribute to better living and sanitary conditions and significantly enhance the health and quality of life of rural residents.

5.3.5.3 Strengthening Infrastructure Development for Rural Electrification

To enhance power supply security more effectively for rural areas and advance rural electrification and the 'Electricity for Every Household' project, China should actively pursue rural power grid construction and upgrading, and implement electrification work for a new countryside, small hydropower projects and a new round of rural grid upgrading. In light of the characteristics of rural grids and electricity demand in the rural sector, prior consideration should be given to the expansion of large power grids to meet the power requirements of areas without electricity supply and the development of small hydropower stations and distributed generation in areas not served by large power grids. At the same time, the application of advanced and practical new technologies and new equipment should be promoted, coupled with continued efforts to improve grid equipment and automation levels for the rural sector and expedite the construction of rural smart grids to level the playing field between urban and rural residents in terms of supply grids and tariffs. During the '11th Five-Year Plan' period, State Grid vigorously implemented the rural power development strategy of 'New Countryside, New Power, New Services' to successfully achieve the objective of 'Electricity for Every Household', benefiting 5.09 million people in 1.34 million households and the economy of the newly electrified areas. By fully implementing the electrification strategy for a new countryside, State Grid in the '11th Five-Year Plan' period electrified 407 counties, 4991 towns and 90 053 villages. Compared with the period before executing these projects, the ownership of electric home appliances among rural households had grown significantly. Ownership of TVs, refrigerators, washing machines, electric fans, air-conditioners and electric cookers increased 18.9%, 32.4%, 28.3%, 19.4%, 64.5% and 42.2% respectively, contributing to rural China's all-round economic development.

5.4 Development of Electric Vehicles

As an important trend of the global automotive industry in the 21st century, the development of electric vehicles reflects a profound change in the way of energy consumption. Many countries have developed an electric vehicle development strategy, with a package of support policies. Globally, R&D investment in and demonstration of electric vehicles have gathered unprecedented momentum, marking a new stage of marketing and promotion. The USA, France, Japan, Germany, Britain and Switzerland have taken the lead in electric vehicle industrialisation and commercialisation. In view of the tremendous market potential of electric vehicles, China has set its sights on electric vehicles as the strategic focus of new-energy automobile development. The development of electric vehicles is of great significance to optimising the way and structure of energy consumption, easing the pressures on oil supply and the environment, developing a green energy consumption model, and promoting automobile industry upgrading and technological leapfrogging.

5.4.1 Important Implications of Electric Vehicle Development

5.4.1.1 Electric Vehicle Development is a Significant Strategic Initiative to Ensure China's Energy Security

Though a resource-rich country, China lacks quality fossil energy. The energy security situation is grim, with increasing crude oil imports to meet 56.5% of its oil requirements in 2011. Growing oil consumption is mainly attributable to the transport industry. It is forecast that by 2020, car ownership in China will reach 200 million units, so controlling oil consumption growth in the transport industry will be an important part of energy restructuring. By promoting the wider use of electric vehicles, 'the substitution of electricity for oil' may be realised and national energy security achieved by reducing the proportion of oil consumption and the reliance on oil imports. Through intensive efforts in electric vehicle development, the oil saved per year in the USA is estimated to be 15% of its annual demands by 2030, and the increased demand for electricity will be 5%~6% higher if compared with the scenario of no electric vehicle development. According to the *Plans for Development of Energy-efficient and New Energy Automobile Industry (2011–2020)* (exposure draft) of the Ministry of Industry and Information Technology, the number of all-electric vehicles and plug-in hybrid vehicles on the market will reach 5 million by 2020. China may see electric vehicles increasing at a faster pace to hit 5~10 million units by 2020, with 10~20 million tonnes of oil saved every year. Along with the maturity of battery and other key technologies, the number of electric vehicles is expected to reach 30 million in 2030, with oil consumption reduced by more than 60 million tonnes per annum.

5.4.1.2 Developing Electric Vehicles is a Realistic Option for Promoting Energy Efficiency and Emission Reduction

Automobile exhaust is a major contributor to urban pollution at present. Electric vehicles have significant advantages over traditional oil-fuelled vehicles in terms of energy efficiency, emission reduction and environmental protection, so the development of electric cars may help effectively address greenhouse gas emissions and the need for environmental protection. In the event of large-scale substitution of electric vehicle for traditional ones, the technology of interactivity between electric vehicles and power grids may directly reduce energy consumption and also carbon dioxide and other pollutants in the service life of vehicles, and promote the development of wind, solar and other clean energy. It is estimated that by 2015 and 2020, an electric passenger car may reduce carbon dioxide emissions by 6 kg and 7.6 kg per 100 km respectively, compared with regular petrol passenger cars.

5.4.1.3 Electric Vehicle Development Provides a Historic Opportunity for China's Automobile Industry to Reach and Surpass World-leading Standards

Though ranked as the world's largest automobile producer and consumer, China is still battling with the long-standing problems in its car industry, including low R&D capability and a lack of core competences. The development of electric vehicles provides a rare historic opportunity for China's automobile industry to catch up technologically. Technically, China has developed a certain level of R&D capability with the right conditions in place to accelerate the progress of industrialisation in three key electric vehicle technologies. With the large number of patents it owns, China is fully capable of taking the historic opportunity of automobile power technology reforms to reverse the over-reliance on foreign technologies and achieve breakthroughs in the core technologies of electric vehicles, so as to take the lead in realising the industrialisation of electric vehicles, securing a pre-emptive position in the competitive electric vehicle market, and reshaping the global automobile industry.

5.4.2 Key Areas of Electric Vehicle Development

After years of hard work, China has made significant breakthroughs in independent innovation in electric vehicles, with the development conditions improving and a solid foundation laid for industrialisation to offer favourable conditions and comparative advantages for accelerated growth. The government attaches great importance to the development of the electric vehicle industry, by launching a package of relevant policies and kick-starting a drive to demonstrate and promote energy-efficient new-energy vehicles and a pilot scheme for subsidising purchasers of new-energy vehicles. Electric vehicle development has also been incorporated as one of the strategic emerging industries and an important strategic initiative to advance China's economic restructuring during the '12th Five-Year Plan' period. With the goal of electric vehicle development and the roadmap for developing all-electric cars initially established, China has made important EV-related technological breakthroughs in vehicle integration, power system integration, power assembly and key parts through close cooperation among enterprises, universities and research institutes. Now that electric vehicle patents, standards and market access systems are in place, China has formulated over 40 EV-related standards with the development of testing, assessment and product certification capabilities covering finished vehicles, power storage batteries, driving motors, etc. Moreover, institutions engaged in electric vehicle R&D and production have obtained about 1800 patents and developed a variety of electric vehicle prototypes. All these developments have laid a solid foundation for the industrialisation of electric vehicles with Chinese intellectual property rights.

The road to electric vehicle development is still beset with obstacles and difficulties, such as immature technology, high costs and poor support facilities. In particular, power battery technology and charging and battery swapping facilities are two crucial factors in the realisation of leapfrog development of electric vehicles.

5.4.2.1 Overcoming Bottlenecks in Power Battery Technology

As the core component of an electric vehicle and the power source of the automobile driving system, the battery's performance indicators and economic costs determine the industrialisation process of electric vehicles. China is the world's second largest producer of lithium batteries with a massive industry scale and a solid foundation. With increasing input in recent years, China's power battery industry has begun to enter the industrialisation stage characterised by rapid development. An abundant supply of lithium is also conducive to battery development and accelerated commercialisation of the power battery industry. However, there are technological bottlenecks in such areas as safety and battery life. China must be committed to independent innovation to master as soon as possible key EV battery technologies to help establish a competitive EV industrial system.

The focus of electric vehicle marketisation now lies in making breakthroughs in the core power battery technologies, increasing energy density, service life, safety and reliability while lowering the cost of batteries. To increase energy density, the focus should be on developing new cathode and anode materials with high specific capacity. Given the importance of consistency performance and assembly technology for battery life, the structural design, production process and assembly technology of batteries should be studied. As the safety and reliability of power batteries are of utmost importance, high-performance diaphragms and flame-retardant electrolytes should be developed, the technology for assessing battery operation conditions studied, and effective and reliable battery management system developed to make the battery safer and more reliable. Research should also be carried out on cost-effective mass production technology for key battery materials and the regeneration technology to reduce the cost of using power batteries.

China should step up efforts to develop the domestic capacity for manufacturing key materials and equipment for power battery production, with a focus on supporting enterprises with the right technological foundation and development potential to engage in proprietary R&D and manufacturing of key materials such as cathode and anode materials for lithium ion batteries, diaphragms and electrolytes. In the meantime, qualified equipment manufacturers should also be encouraged and supported to independently develop production, control and testing equipment for power batteries and key materials, which will help to break up foreign monopoly in this area. China should, on the strength of a state-level research and test base of power batteries, establish a system for power battery technology development, carry forward prospective studies of new materials and new systems for a new generation of power batteries with high energy density, and conduct research on technologies for applying new structures and processes, so as to obtain core intellectual property rights. China should strive to improve the energy density of the power battery system to 200w·h/kg and reduce costs to RMB1.5/(w·h) in 2020.

5.4.2.2 Scientific Planning for Charging and Battery Swapping Facilities

Charging and Battery Swapping Facilities are Important for Electric Vehicle Industry

Charging and battery swapping facilities provide power for electric vehicles. So an efficient and robust charging and battery swapping service is the prerequisite for the wide adoption of electric vehicles. Efforts should be devoted to establish charging and battery swapping service networks with standard services to meet EV users' requirements for free intercity and cross-regional driving to help promote the wider use of electric vehicles.

Electric vehicles feature new characteristics of power consumption. The first is mobility where, as opposed to common electricity customers, the demand for EV charging changes with the parking location. The second is diversity where electric vehicles (such as electric buses, sanitation trucks, taxis and private cars) differ widely in power consumption. The third is guidability where through smart grid technology and certain incentives, EV users may be guided to do their charging during off-peak periods at night to help realise peak load shaving.

To meet the EV requirements for mobility and diversity in terms of power supply, it is necessary to establish support charging and battery swapping service networks, with a rational allocation of service outlets and modes so as to provide network-based and standard charging and battery swapping services and drive the market acceptance of electric vehicles and achieve industrialisation as soon as possible.

Making EV Charging and Battery Swapping Facilities More Network-based, Intelligent and Standardised

To prepare for the major development of electric vehicle in the future, China needs to develop a charging and battery swapping service network based on a uniform operation and management platform. The construction of charging and battery swapping service networks can meet EV requirements for mobility and diversity in terms of power supply, help boost the standardised and orderly development of the charging and battery swapping service industry, leverage economies of scale, and cut down operation and maintenance costs.

Data communication between electric vehicles and smart grids is an inevitable development trend in EV charging and battery swapping facilities. Advanced smart grid technology should be employed to direct electric vehicle charging to low load periods and carry out load management and peak-load regulation, thereby improving the overall operation efficiency of the power system. Given the energy storage function of power batteries, an electric vehicle can also serve as a mobile energy storage unit to provide a fundamental platform for maximising the benefits of electric vehicles. State Grid has formulated the Plan for Development of Smart Charging and Battery Swapping Service Networks for Electric Vehicles during the '12th Five-Year Plan' period. According to the Plan and in light of the distribution of EV charging and battery swapping demand nationwide, State Grid has chosen to develop a regional smart charging and battery swapping service network in the Bohai Rim and Yangtze River Delta Region, with a focus on the state-level pilot cities for demonstrating and promoting electric vehicles. The Company also strives to achieve interconnection of the charging and battery swapping networks between the two regions. Refer to Figure 5.3 for State Grid's EV charging and battery swapping facilities.

Figure 5.3 State Grid's EV charging and battery swapping facilities.

Charging and battery swapping infrastructure facilities need to be constructed and operated based on unified standards in order to provide standard and consistent charging and battery swapping services for various electric vehicles. To speed up the development of electric vehicles, it is necessary to formulate relevant standards governing electric vehicles and charging and battery swapping facilities to support the development of a standard and consistent charging and battery swapping system and effectively improve the standard service provided through EV charging and battery swapping facilities.

Coordinating and Promoting Construction of Charging and Battery Swapping Networks and Upgrading of Distribution Networks

As the development of electric vehicles is closely related to electricity and power grids, large-scale adoption of electric vehicles requires coordination with the planning, construction and operation of power grids in many aspects. On the one hand, the planning for charging and battery swapping networks must be closely linked with that for distribution networks. As EV charging load may come at random in time and space, scientific control and guidance are needed to help reduce the peak load of power grids and avoid harmonic pollution, increased peak-valley differences in power grids and grid equipment overload caused by large-scale charging. Development of charging and battery swapping networks entails not only construction of charging and battery swapping stations and poles, but also upgrading of distribution networks so as to ease the pressure from massive EV charging demand. On the other hand, greater benefits can be achieved by combining the construction and management of charging and battery swapping networks and smart grids. While building a charging and battery swapping network, it is important to take into consideration its integration with the advanced metering system and various energy management systems in order to develop an interactive platform that connects power grids with EV users and make the charging and battery swapping network an integral part of the smart grid. In respect of operation management, interaction between the smart grid management platform and the management platform for charging and battery swapping network operations can be achieved to provide smart charging and battery swapping services based on the level of power grid load, real-time tariffs and other information. Large centralised charging stations in the charging and battery swapping network can also serve as power storage stations and emergency power sources.

5.4.3 EV Energy Supply Model

As energy supply is an important part of the EV industry chain, China should establish a scientific and rational EV energy supply model to promote commercialisation of electric vehicles.

5.4.3.1 EV Energy Supply Model

Charging Model

The charging model can be divided into conventional charging and fast charging.

The conventional charging model, also known as common charging or slow charging, involves the use of low current at constant voltage or constant current to provide charging time usually of 5–8 hours. The conventional charging model requires that electric vehicles be parked for a long time and the driving distance sustained by battery power be as long as possible. The conventional charging model can make full use of electricity in off-peak periods for charging due to the long charging time involved, but it cannot meet the urgent demand for EV operation. Conventional charging stations are mainly located in residential areas or large parking lots near large office sites.

The fast charging model, also known as emergency charging, involves the use of high current in a short time (between 20 minutes and 2 hours of EV parking time, with the exact charging times to be determined by the capacity of EV power batteries). The fast charging model refers to quick make-up charging between rides to meet EV operation needs in special circumstances. As high current demand may be detrimental to public utility grids, the fast charging model requires dedicated charging stations, which may be built mainly in public places including airports, railway stations, hospitals and shopping malls at higher installation costs. It is also necessary to unify EV charging equipment, popularise fast charging station technology and the metering method. The key to major development of fast charging stations lies in developing 'large-capacity, low-cost, fast-charging and long-life' battery products to meet user demand for convenience.

Battery-swap Mode

The battery-swap mode charges up electric vehicles by directly replacing the used battery pack with a new one. Professional assistance is required to quickly replace, charge and maintain batteries with special machinery, as replacement work requires specialist skills given the weight of a battery bank. As for the battery-swap model, users can rent a fully charged storage battery and recharge the replaced storage battery intensively in off-peak periods, which is conducive to improving the battery's service life and reducing the charging cost. Application of this mode requires the standardisation of batteries and electric vehicles, the coordination of EV design improvements, the construction and management of charging stations and the management of battery flows.

5.4.3.2   Research for an EV Energy Supply Model Suitable for China's National Conditions

Internationally, experimentations and applications have been conducted on EV energy supply models. The USA has initiated the 'EV Project' (a large charging infrastructure readiness project) with plans to build 11 210 charging points in three years. The 'Charge Point American' project funded by the US Department of Energy envisages the development of about 5000 charging stations in 9 pilot areas. In the area of EV demonstration, following active efforts to explore the feasibility of developing charging infrastructure, battery rental service and other support services, 200 public charging stations had been built as at 2008 in France. In the UK, the government implemented an EV promotional programme and built 500 charging stations. In Germany, plans are underway to build at least 550 EV charging stations in downtown Berlin, while about 1300 EV charging stations are expected to be developed in Portugal. Tokyo Electric Power Company has been actively involved in EV-related infrastructure construction, with more than 250 units already in place by end 2010 based on plans to build more than 1000 charging stations by 2013. In early 2011, Better Place in partnership with Renault opened in Denmark their first operating centre in Europe, providing a one-stop battery-swap service solution, including public battery-swap and private charging pole services.

China has also actively carried out construction of charging facilities. In 2008, a 5000 m^2 electric bus battery-swap station was built in Beijing Olympic Central Zone. In 2010, the world's largest electric bus battery-swap station was built in Shanghai Expo Park Zone. By end 2011, State Grid had built and put into operation 243 standard charging and battery swapping stations and 13 000 AC charging poles, making China the world's largest operator of charging and battery swapping facilities. Zhejiang Electric Power Company has built the first intercity smart charging and battery swapping service network in China. In 2011, an EV charging and battery swapping station was established and commissioned in Xuejia Dao, Qingdao, marking China's first one-stop smart operation with integrated battery charging, replacement and storage capabilities. An EV charging and battery swapping station was officially commissioned in Gaoantun, Beijing, in 2012. Combining battery charging, replacement and distribution capabilities, the station is the world's largest and most powerful facility of its kind.

With China now in the crucial period of rapid EV development, an energy supply model should be selected in light of the trend of technology development and realistic conditions, the difficulty level and cost of energy supply network development, the economics of energy usage and the business model of energy suppliers.

Firstly, the difficulty level of energy supply network development and user accessibility are the most crucial factors in determining network universality and user-friendliness. Cities are the focus area for EV development. However, unlike developed countries where living in detached houses with garages or dedicated parking spaces is the norm, China is densely populated with most people living in high-rise buildings and parking spaces are in extreme short supply. In the long run, it is difficult to achieve the goal of making fixed parking spaces available for most vehicles. Building charging stations, transformers, lines, meters and other grid facilities and devices extensively in residential areas and public places also faces the problem of revamp and upgrading and is a capital-intensive and time-consuming complicated systems engineering project. It is also hard to build charging facilities in communities. Given its impact on battery life, the fast charging mode has yet to gain popularity pending further technological breakthroughs. The battery-swap mode is not subject to housing or office conditions. Intensive charging is available, without any specific requirements for usage. Coupled with efficient distribution, a battery-swap service network can achieve extensive coverage, making its service easily accessible. Like fuelling, EV users can charge their batteries in service outlets spread over the city and transport nodes. However, the battery-swap mode requires organisation and coordination efforts from government due to the difficulties and obstacles in standardising battery boxes, especially in terms of dimension and interface.

Secondly, the economics of usage is related to the competitiveness of energy supply modes. The self-charging mode can take full advantage of concessionary tariffs for off-peak periods. However, the cost of initial acquisition and subsequent replacement is high (30%~50% of the total cost of EV ownership). The battery-swap mode can reduce the cost of initial acquisition but the day-to-day operating expenses may be higher compared with the self-charging mode.

Thirdly, the sustainability of energy supply networks is essentially a function of profitable business models. As a new mode of energy logistics, the battery-swap mode can significantly lower total operating costs by facilitating large-scale and standardised battery production and large-scale purchases and intensive management of energy supplies. In the battery-swap mode, energy suppliers play the role of independent intermediate operators, which will better enable the government to implement more targeted supportive and preferential policies, such as tariff policy and subsidy for battery purchases, and establish a clear profit model. But what should also be taken into consideration are the interests of users, electric vehicle manufacturers, battery manufacturers, energy suppliers and other stakeholders. The relatively simple self-charging mode allows all the stakeholders involved to balance their interests subject to negotiation on a fair, reasonable and equitable basis.

Generally speaking, different energy supply modes are unique in their own ways with different scopes of application and restriction. Development of a single mode will not benefit the healthy development of the electric vehicle industry. In view of the different and varied requirements of EV users, a 'combined charging and swap' mode with coordinated self-charging and battery swapping capabilities will be an ideal choice for the development of EV energy supply services in China. Based on the models, uses and development trends of electric vehicles, China can build different charging and battery swapping facilities in light of local conditions so as to meet the needs of different EV users.

Provided that fast charging technology is safe and mature, quick charging of EV can be achieved with the fast charging mode to meet EV users' demands for a shorter charging time. Quick replacement of batteries will also be possible through a battery-swap station to meet users' demand for a quick energy refill. By building centralised battery-swap stations, centralised load management can be realised and the demand for EV energy supply can be satisfied by revamping the distribution network to a certain extent. On the strength of smart grid technology, efficient management of smart charging can be achieved for centralised charging and battery swapping stations so as to improve the overall efficiency of system operations. Centralised charging and battery swapping stations can also serve as emergency and standby power sources for cities to enhance the flexibility and reliability of grid operations in urban areas. With the rapid development of electric vehicles and charging technology, equipment and management systems, coupled with the gradual improvement of relevant policies, laws, standards and regulatory system, China will see increasingly significant expansion of charging and battery swapping facilities for electric vehicles.

5.4.4 Policies Supporting the Development of Electric Vehicles

In the preliminary stage of the electric vehicle industry, the support and guidance of state policies are needed.

5.4.4.1 Establishing an Effective Mechanism for Coordination to Promote Win-win Development for All Parties

As the formation of the electric vehicle industry chain and the establishment of battery-swap service networks carry wide-ranging social and economic implications, an effective mechanism for coordination under the government's coordinated guidance has to be developed to build a platform for sustainable development of electric vehicles by bringing together the relevant departments, vehicle manufacturers, energy suppliers and research institutes to share strong initiative to promote the healthy development of electric vehicles and relevant industries.

5.4.4.2 Reinforcing R&D on Key Technologies and Support for Sales of Electric Vehicles

China's electric vehicle enterprises, research organisations and academic institutes should be given support to step up R&D efforts to overcome key technological bottlenecks in batteries, electric motors and control systems. Technology support should be strengthened to establish important research facilities and bases, key laboratories and technology (research) centres focusing on electric vehicles. To speed up the transformation of hi-tech achievements related to EV, the government should provide special funds, preferential tax policy and stronger support for relevant industries, such as the production and marketing of batteries and key EV components.

5.4.4.3 Strengthening Policy Support for Charging and Battery Swapping Service Networks

As a crucial link of the electric vehicle industry chain, charging and battery swapping service networks shall be incorporated into urban and rural development and land-use planning as a key component of urban infrastructure. In the initial stage, the government should, based on the principles of flexibility, practicality and applicability, provide developers and operators of EV charging and battery swapping service networks with a certain level of financial subsidy, tax incentive and policy support for land-use rights approval, investment and financing, so as to accelerate the cultivation of relevant industries and markets.

5.4.4.4 Establishing a System for Effective Convergence and Integration among the Relevant Technical Standards of the Electric Vehicle Industry Chain

Uniform technical standards are essential for the leapfrog development of the electric vehicle industry. The standardisation work on electric vehicles in China has fallen short of the current requirements of EV development. Therefore, a system of uniform standards should be established for electric vehicles, especially the standards governing batteries, charging and battery swapping facilities and interfaces between batteries and charging and battery swapping facilities in order to avoid inconsistencies in local systems, ineffective convergence of different industry segments and disorderly development. This will guide and regulate the standard and large-scale development of the electric vehicle industry. At the same time, convergence with the relevant standards of America and Europe should also be considered in order to lay a foundation for electric vehicle exports in the future.

6

Energy Market

As a socialist market economic system has been established and improved following the introduction of the reform and opening up policy, market reforms in China's energy sector are gradually taking place, resulting in significantly growing vitality in the energy industry and the gradual evolvement of the energy market's fundamental role in energy resource allocation. In order to further the transformation of its energy development mode for the future, China needs to further strengthen the building of the energy market, guide the sound development of the coal market, and improve the pricing mechanism for oil and gas. It also needs to steadily advance the electricity marketisation process with a commitment to improvement and innovation for better regulatory control to create a good institutional environment for the sustainable development of the energy market in China.

6.1 Overview and Development Ideas in Respect of the Energy Market

China's reforms in the energy field began in the early 1980s with the introduction of policies to utilise foreign capital, raise funding through various channels for electricity operations and carry out oil production on a contract basis, followed by the gradual liberalisation of market access, adjustment of the management systems, separation of government functions from enterprise management, relaxation of price controls and introduction of a competitive mechanism. Although the reform process in different energy departments is different and each department has its own specific circumstances, in general the market assumes a growing guiding role in energy development, thus creating a diverse competitive situation and gradually deepening the link between the domestic and the international markets.

In the energy-market reform process, a number of issues and contradictions have emerged and remain to be resolved, such as the inadequate capacity for resource allocation, irrational price relativity, and the failure of the regulatory system to keep up with the actual development of the energy sector. It is therefore necessary to continue to improve the energy market system and the operational mechanism in line with the socialist market economic system, and further define the ideas on the building of the energy market and a regulatory regime.

Electric Power and Energy in China, First Edition. Zhenya Liu.
© 2013 China Electric Power Press. All rights reserved. Published 2013 by John Wiley & Sons Singapore Pte. Ltd.

6.1.1   Overview of Energy Market Development

Due to the differences in industry features, development stages and reform progress, there is a varying degree of marketisation in the coal, electricity, oil and gas and other energy sectors in China. In general, the current building of China's energy market has the following basic characteristics.

6.1.1.1   A Diverse Competitive Market Situation Basically in Place

China's coal industry has a high degree of marketisation. Coal prices are basically determined by the market, giving rise to the steam coal, coking coal and other coal markets. As at the end of 2010, there were more than 10 000 coal enterprises of different forms of ownership across the country, resulting in tense market competition. China has clearly established its general ideas on tariff reform by initially creating a coal-electricity price linkage mechanism. In the competitive power development sector, a short-term pilot scheme was carried out, leading to an electricity market built on the basis of on-grid price bidding and a number of new market trading models such as cross-provincial electricity trading, 'replacement of small generating units with larger ones' for trading in generation rights, coal-fired/hydropower generation swap transactions, and direct trading between large consumers and power generation companies. Crude oil prices in China, independently set by enterprises with reference to international market prices, have been brought in line with the international market. Prices of domestic refined oil products are set according to government guided-prices or by the government, and have been indirectly brought in line with the crude oil prices in the international market in a controlled manner. Apart from the three major corporations, namely PetroChina, Sinopec and CNOOC, other central government-controlled corporations and many local state-owned enterprises, privately-run companies and foreign investment enterprises are also actively engaged in competition in such fields as oil exploration and exploitation, refining, refined product retailing and so on. Currently, the crude oil processing capacity of the local government-owned and privately-run refineries accounts for about 1/5 of the national total, and 40% of the gas filling stations are owned by enterprises other than the three major corporations. After the dual-track pricing policy for natural gas was abolished in 2010 with government-set prices and government guided-prices replaced by uniform government guided-prices, the pricing reform for natural gas has taken one step ahead towards marketisation.

6.1.1.2   The Scope of Market Trading and Resource Allocation is Widening

In order to adjust the geographical imbalance in the distribution of energy resources and energy demand in China, geographical restrictions on energy trading are gradually removed as a result of various driving forces like market competition and technological progress. Three regional coal markets have been established in the northeastern, northwestern and southwestern parts of China with the formation of a nationwide distribution network for coal transportation from north to south and from west to east to facilitate the shipment of coal from three key western coal-producing regions to the areas with coal shortage in eastern China. Large-scale, long-distance power transmission projects as well as oil and gas pipeline construction have laid a foundation for nationwide reallocation of electricity and oil and gas resources. Cross-provincial and cross-regional trading has become an important aspect of the electricity market, and electricity supply is gradually switching from locally balanced supply and self-balanced supply in certain areas to optimal allocation of supply on a greater scale. In 2011, the volume of electricity traded on the national electricity market amounted to 399.9 TWh, 2.4 times the figure for 2006.

6.1.1.3 Interactions between Domestic and International Energy Markets are Intensifying

China's economy is increasingly integrated with the global economy against the backdrop of intensified globalisation. The domestic energy market has gradually become an integral part of the global energy market, being subject to the impact of the international energy landscape as well as having a profound impact on the international energy scene. China has been active in the import and export of coal. The international market has become an important supplement to the domestic coal market, as evidenced by a net coal import of 145.8 million tonnes in 2010. Electricity import and export trade between China and its neighbouring countries has also been developing quite rapidly in recent years. The development prospects look very promising despite the fact that the transaction volume remains relatively low; in particular, the transmission of electricity from Russia, Mongolia and other countries to China has a great potential. The international markets for oil and gas have a more significant impact on China's energy market than the coal and electricity markets as China has already become the world's second largest crude oil importer. China's reliance on foreign oil and natural gas was approximately 55% and 15% respectively in 2010. A more common view is that China's fast-growing demand for oil and gas import is an important reason for the ever-rising international oil and gas prices. In general, increasingly close connections with the international energy market have provided China with more options to secure supply for the domestic energy market, but have also posed challenges to the country in terms of mitigating various risks stemming from the international energy marketplace.

6.1.1.4 Establishment of Market Mechanism Remains to be Further Improved

Due to a host of organisational and administrative issues, the development of the coal market is rather disorderly. The production, transportation and marketing processes have failed to converge with each other effectively; the rate of contract fulfilment remains low for thermal coal, and there is no assurance of quality. For thermal coal, there are currently two pricing mechanisms. The first mechanism is for the key-contract coal prices to be set after moderate interventions through the convergence of production, transportation and demand under the guidance of China's national macro-control initiatives. The second mechanism is for prices to be set through independent negotiations in the spot market. Due to a number of factors in recent years, such as demand growth and numerous intermediaries, the coal prices in China have been rising sharply, driving the costs of coal-fired power generation continuously and substantially higher. With regard to tariffs, both on-grid tariff and end-user retail tariff are regulated and approved by the government. Although a coal-electricity price linkage mechanism has been established, it remains difficult to use the system to alleviate the impact of rising power generation costs in a timely and effective manner. The conflict between 'coal prices set by market demand' and 'electricity prices under government planning' remains a prominent issue as the tariff has not been able to play its role fully and effectively in adjusting supply and demand. Due to the absence of a scientific tariff-setting mechanism, there is a lack of effective tariff protection for grid development and electricity trading across larger regions, with the result that a unified, deregulated, competitive and orderly electricity market system has not been established. Since the oil price-setting mechanism is still some way from true marketisation, the time lag in the adjustment of refined product prices can hardly reflect in a timely manner the relationship between supply and demand in the domestic market and the change in consumption structure. Moreover, China still does not have a say that befits a major oil consuming and importing nation over international crude oil prices. Given the irrational price relativity between natural gas and crude oil, among other energy alternatives, and the great price disparity between the domestic and import markets, an effective competitive market model has yet to be developed.

6.1.1.5 Market Regulation should be Further Improved

In the process of separating government functions from enterprise management and implementing energy market reforms, the regulation of the energy market in China is gradually drawing more attention and being reinforced. However, the current efforts to build a regulatory regime for the energy market has failed to catch up with the rapid development of the energy market. The outstanding problems with energy regulation in China are three-fold.

Firstly, regulatory functions are over-fragmented. At present, the regulatory functions governing the energy market in China are decentralised widely among many government departments. There are no designated regulators for the coal, oil and gas markets, except for the electricity market. This results in overlapping and fragmented regulatory functions and the principal responsible party cannot be clearly identified. The positioning and functions electricity regulators are not well-defined, with a large number of functions in relation to market regulation still distributed among relevant government departments.

Secondly, the building of a regulatory regime and relevant legislation are falling behind. Energy legislation is lagging behind, with energy regulation being mainly carried out based on administrative approval and policy documents. Such regulation lacks transparency; it is excessive and marked by instances of under and over-regulation and certain market distortions cannot be resolved effectively.

Thirdly, there is a lack of market regulatory measures and an effective implementation mechanism for unified energy planning. The establishment of an energy reserve system is still in the start-up stage. There is a lack of effective market control capacity in various segments such as coal exploitation and power source development.

6.1.2 Basic Thinking Behind Energy Marketisation

Energy marketisation is an important approach to effectively guide the behaviour of key market players and facilitate the optimal allocation of energy resources. It is also a specific requirement for China to improve its socialist market economic system and participate in competition and cooperation in the international energy market. In order to change its energy development mode and transform its energy strategy, China needs to place great emphasis on and play an active market regulatory role in vigorously carrying out the energy marketisation process.

The building of an energy market is more fundamental, strategic and complex than the building of a commodity market of a general nature. Moreover, as it is related to national security, both Western developed countries and the majority of developing countries take a positive, prudent attitude towards the pursuit of energy market reforms. China's economic development, energy endowments and basic economic system as well as the global development environment and the experiences and lessons of other nations in energy marketisation have indicated that the energy marketisation process in China must be based on its own national conditions while being realistic and forward-looking at the same time. China not only needs to unswervingly keep moving towards the direction of market reform, to be good at seizing the right timing for reform and to resolve reform-related problems in a timely manner, but also fully consider the impact of energy marketisation on the economy and people's livelihood. It must refrain from making any impetuous move that would bring unwarranted extra costs to the reform. China must show more determination and initiative towards reform, be mindful of the strategic importance of reform, and enhance the scientific approach to reform.

In the future, the furthering of the basic thinking behind energy marketisation in China should start with the promotion of China's energy strategy transformation and sustainable development. In the process, energy pricing reform should take centre stage, with a focus on the establishment of a market system and a regulatory regime. The key is to develop an effective system for energy operations based on organic integration of market allocations and macro-control measures. The objective is to provide sustainable energy support for China's modernisation programme by safeguarding national energy security, enhancing the efficiency of energy development, improving the quality of energy services and sharing the achievements of energy development. While advancing the energy marketisation process, China should continue to help meet the needs for energy supply to safeguard national energy security, help promote the development of the energy industry to improve energy efficiency, address the global energy situation, and improve competitiveness in the international energy market.

6.2 The Building of Coal Market

Among the various segments of China's energy sector, coal has the highest degree of marketisation. In the future, China needs to focus on measures to regulate the order of the coal market, promote the building of a market system for coal, and step up regulatory control to promote the healthy growth of the coal market.

6.2.1 Management of Coal Market Order

Good market order is a prerequisite for the full realisation of the coal market's role. In order to build good order in the coal market, China needs to shape qualified market players and improve the concentration of the coal industry by regulating market access and pursuing reorganisation and integration. It needs to rectify the coal distribution process for orderly integration of production, supply and marketing; it also needs to minimise inappropriate interventions in the coal market by governments at all levels to avoid various forms of market distortions.

6.2.1.1 Raise the Concentration of the Coal Industry

Since the 1980s, the development of the global coal industry has shown two trends of conglomeratisation and large-scale production. Coal companies in some countries have developed into large multinational coal conglomerates through mergers or the acquisition of coal mining interests. The market share of the top four coal enterprises in major coal-producing countries such as the USA, Russia and Australia has risen to more than 40%.

In recent years, Chinese coal enterprises have stepped up reorganisation and integration as industry concentration has continued to rise, driven by governments at different levels and market forces. Total coal production of the top four coal enterprises accounted for 22% of the national total in 2010 (see Table 6.1), but this industry concentration remains low as compared to that of other major coal-producing countries in the world. As at the end of 2010, there were still more than 10 000 coal-producing enterprises across the country with an average output capacity of less than 300 000 tonnes. This suggests that the coal industry remains very 'fragmented' and 'disorderly'.

China should further rationalise its coal industry by encouraging coal enterprises to start cross-sectoral, cross-regional and cross-ownership cooperation, with stronger efforts in the merger, syndication and restructuring of enterprises to accelerate the construction of large-scale modern coal mines and promptly build a batch of coal enterprises of a substantial size and with a capacity of 100 million tonnes as well as modern coal mines with a capacity of 10 million tonnes in accordance with the plans of relevant competent authorities.

Naturally, because of the regional differences in coal endowments, space for growth must be reserved for medium and small-sized coal mines in those regions with relatively

Table 6.1 Top 10 coal enterprises in China in 2010 by coal output. Unit: 1000 tonnes.

Ranking	Company name	Coal production
1	Shenhua Group Corporation Ltd	356 960
2	China National Coal Group Corp.	153 700
3	Shanxi Coking Coal Group Co., Ltd	102 140
4	Shanxi Datong Coal Mine Group Co., Ltd	101 180
5	Shaanxi Coal and Chemical Industry Group Co., Ltd	100 390
6	Henan Coal Chemical Industry Group Co., Ltd	74 010
7	Jizhong Energy Group Co., Ltd	73 320
8	Shanxi Lu'an Mining Group Co., Ltd	70 980
9	Huainan Mining (Group) Co., Ltd	66 190
10	Kailuan (Group) Co., Ltd	60 870
	Total	1 159 740

Source: The data in the table are extracted from the *Notice on Top 100 Coal Enterprises in China in 2011* published by the China National Coal Association in 2011.

scattered coal resources. Appropriately raising the level of industry concentration may provide an impetus to stimulate more efficient and intensive utilisation of coal resources, eliminate coal enterprises with obsolete capacity, facilitate the development of a modern corporate system for coal enterprises, and encourage coal enterprises to conscientiously undertake social responsibilities and consciously conduct business according to the law.

6.2.1.2 Straightening out Coal Distribution Process

In China, coal prices have been rising rapidly in recent years, producing a severe impact on the development of key coal-consuming industries, such as electricity, iron and steel, chemicals and building materials. The excessive rise in coal prices is both the result of the rapid growth in coal demand and the price increases in each part of the coal distribution process. Relevant studies have shown that the sum of noncoal costs for the coal distribution process such as transportation costs, taxes and levies as well as price mark-ups have accounted for 55% to 60% of coal prices at the end-user level.

To create good order in the coal market and avoid exorbitant prices and supply and demand imbalance arising from intermediate price mark-ups, China needs to further straighten out the coal distribution process in the future to enable dynamic integration of production, transportation and sales.

Firstly, the transportation and sales systems for coal should be reformed. Coal demand must be met with supply by encouraging coal mines to directly enter into contracts with end-use consumers in the corporate sector.

Secondly, unlawful practices in the coal distribution process must be curbed, such as hoarding and profiteering by coal dealers and intermediaries, resale of supply contracts, bidding up coal prices in collusion, as well as various rent-seeking activities in the coal transportation process.

Thirdly, a modern coal logistics system should be established. By relying on the large-scale regional coal trading markets or coal distribution centres, a modern coal logistics management platform should be established and collaborative operation of various modes of transportation should be promoted to better meet the demand for coal shipment.

Fourthly, additional efforts should be devoted to build more bases for coal and electricity in accordance with the principle of simultaneous coal transportation and power transmission while accelerating the development of power transmission to promote in-situ conversion of coal to electricity for large-scale and long-distance power transmission to external regions so as to ease the stressed transport capacity caused by an over-reliance on coal transportation.

6.2.1.3 Regulating Coal-related Charges and Administrative Intervention Measures

Even though coal pricing in China has been deregulated, various coal-related levies and charges are now still being collected without a uniform charging standard. These charges not only put an extra burden on coal enterprises, but have also become a major driving force behind soaring coal prices. In addition, strong administrative interventions in coal sales are seen in some coal-exporting regions, effectively controlling the quantity and price of exported coal with an impact on the healthy operation of the coal trading system.

To regulate market order, China needs, on the one hand, to weed out various coal-related charges, with efforts to rationalise charging standards and administrative interventions in strict compliance with the relevant requirements. On the other hand, it needs to check and verify the costs of coal in a scientific way and actively study various subsidy schemes at the national level for coal-exporting regions to enhance local capacity for sustainable development.

6.2.2   Coal Market Trading

At present, improving the development of China's coal market system has become the core task of establishing a modern coal exchange. This task should be accomplished in sync with the direction towards developing a large coal market with high turnover for building a scientific coal trading system in China.

6.2.2.1   Establishing a Unified Nationwide Coal Trading Platform

Coal trading in China is currently conducted mainly by means of convening coal trading conferences (for instance conferences on coordinating production, transportation and demand of coal, conferences on regional coal ordering, national general conferences on key coal contracts, etc.). This is a relatively elementary and unitary trading model rife with shortcomings such as a lack of transparency of market trading, inconsistencies in trading information and failure of prices to reflect the supply-demand situation in a timely manner. Although a number of regional coal trading markets have been established in various regions by leveraging on the coal production and distribution centres, they fail to play any significant role in balancing nationwide coal supply and demand due to various constraints such that a countrywide coal trading platform with a real impact has yet to be built.

To adapt to the trend towards increasing trading volume and the scope of the Chinese coal market, there is a need to establish a unified nationwide coal trading platform for carrying out various functions such as coal price discovery, trading information disclosure and regulation of the supply-demand relationship to provide an open, fair and equitable trading environment for major coal traders. By establishing a market access system and carrying out standardised management of trading contracts, China can enhance the standardisation and transparency of the coal market, and change the existing disorderly competition in the Chinese coal market. Building a unified coal trading platform is also beneficial to the logical allocation of coal transport resources and minimisation of the intermediate layers of the trading process, while providing a necessary channel and means for government agencies to strengthen macro-control initiatives in the coal market. On this basis, the establishment of a sound coal price indicator system in China would be helpful for coal prices to play a role in sending market signals and for enhancing the capacity for forecasting market supply and demand and price movements.

6.2.2.2   Strengthening Trading in Long-term Coal Contracts

Short-term trading accounts for a disproportionately large portion of Chinese coal trading, which is not conducive to maintaining coal price stability. In order to ensure stable development of the coal and related industries, it needs to actively promote trading in long-term coal contracts. Given the continued growth of coal consumption for power generation in the USA since the 1990s, coal supply and demand has basically remained stable because more than 60% of the total thermal coal supply is secured by long-term supply contracts of five years or longer.

Differences in energy endowments and consumption features have determined that China's coal supply and demand is inevitably tighter than that in the USA so that the trading volume of long-term coal supply contracts may not be as large as that of the USA. However, for large coal-consuming enterprises with stable coal requirements, a certain proportion of coal consumption should be subject to long-term contract trading led by the government. This is conducive to preparing for the convergence of production and transportation in advance, and maintaining stability in the coal market.

6.2.2.3 Developing Coal Futures Trading

Coal futures trading is an effective risk management tool for both coal enterprises and power companies. By selling a futures contract, a coal producer can lock in the sales price for a certain quantity of coal that the coal producer plans to produce in the future, while a coal-consuming enterprise may also buy futures contracts based on its forecasts of coal price changes to hedge against a price hike. Since the mid-1990s, some major coal-producing or consuming countries have selected certain coal types for conducting research in the risk management of futures and futures options from the perspective of price discovery and hedging. Examples are thermal coal futures trading in South Africa, steam coal futures trading in Australia and coking coal futures trading in Japan.

With the increased demand for hedging price risks in the domestic coal market, China needs to actively pursue the establishment of a domestic coal futures market. Considering that coal futures trading is subject to the impact of bottlenecks such as the nonuniformity of delivery quality and rail transport capacity, a specific coal type should be selected for a trial in futures trading with the setting of relevant trading rules in a coal-producing region with relatively abundant transport capacity. At the same time, work should start on the construction of large-scale warehouses for physical settlement and storage.

6.2.3 Regulation of the Coal Market

In view of the fundamental role of coal in China's energy development and the different pace of market reforms in the downstream electricity industry, China should place great emphasis on building capacity for market regulation while driving the establishment of the coal market.

6.2.3.1 Steadily Pushing Forward Reform of the Coal Pricing Mechanism

Since 2009, prices of various coal products in China, except for thermal coal, have been mainly determined by the market. Thermal coal prices are decided by a dual-track mechanism comprising key-contract prices and spot market prices. In recent years, given the continued sharp rise in the market prices of coal and China's stringent controls over tariffs, key-contract coal prices have played an important role in easing the pressure of rising fuel costs and maintaining the normal operation of power generation companies.

In the long run, thermal coal prices will eventually be decided by the market as well. But taking into account the price impact on downstream industries, the timing of thermal coal pricing reform should be coordinated with the tariff reform process. Before any material progress is made in the market reform of tariffs, key-contract coal pricing should be retained. Compared with coal, the marketisation of electricity tariffs faces more complex problems, and will require a longer process to achieve. At the present stage, a pricing mechanism based on autonomy in contact, negotiation and price determination between suppliers and purchasers should be allowed to coexist with key-contract coal prices subject to a certain extent of government intervention. In the future, with the programme of electricity tariff marketisation and continued improvement in the coal market, these two pricing mechanisms for thermal coal will be gradually unified to eventually form a market price-setting mechanism for coal that reflects the supply-demand relationship of coal, scarcity of coal resources, coal mine safety and environmental compensation costs for mining areas.

6.2.3.2 Improving Regulatory Measures for the Coal Market

Imposing Special Coal Gain Levy
A special coal gain levy is a form of regulatory control to restrain unreasonable coal price hikes through taxation. With regard to the specific circumstances of the coal industry and according to the different calorific values and quality levels of coal, China may determine the benchmark coal price as the base for imposing a special coal gain levy within a certain period of time, and for coal trading, a levy at progressive rates on the portion of price increase above the benchmark. A percentage of the collected special coal gain levy will be ploughed back to downstream coal-purchasing enterprises and the remainder channelled into the central government budget to support the sustainable development of the coal industry. This arrangement can minimise the impact of coal price fluctuations on downstream enterprises and help balance the interests between coal enterprises and downstream enterprises.

Establishing a Coal Reserve System for Emergency Response
By improving the government's coal reserve mechanism and the private-sector coal storage mechanism, a coal reserve of sufficient size should be established as a reserve pool to smoothen market changes between peak and off-peak seasons and to offset any irregular supply-demand changes and price fluctuations in the coal market. While securing the orderly operation of the domestic coal market and maintaining the stability of coal prices, the coal reserve can also help safeguard national interests in foreign trade by keeping the prices of imported coal at a reasonable level. Currently, the quantity of coal reserves in China appears to be lower than international standards. Taking thermal coal as an example, the standard amount of thermal coal reserves in China may be increased based on the risk level of the coal market and the needs to ensure power supply security with reference to the 40-day standard in the USA, a standard that sets the number of days of consumption sustainable by thermal coal reserves.

Encouraging Coal Import

Provided that it is economically viable, a number of target exporting countries should be selected to ensure China's coal import channels are smooth and a certain volume of import is maintained. Keeping the domestic coal prices at a reasonable level by converging domestic coal prices with international market prices will not only prevent exceptional price fluctuations in foreign coal markets caused by China's increased import, but also maintain the supply-demand balance and price stability in the domestic coal market. In order to alleviate the pressure from the supply of domestic coal resources and encourage coal-consuming enterprises to import coal, the import VAT rate for coal may be reduced from the current 17% to 13% or even lower with reference to the 13% VAT on coal products in the private sector. Favourable treatment applicable to transportation channels and capacity planning may be granted to imported coal under long-term agreements.

6.2.3.3 Actively Developing Joint Coal and Electricity Operations

The development of joint coal and electricity operations is an important choice for the sustainable development of coal and power enterprises. In the short run, the development of joint coal and power operations can help lower the trading costs for thermal coal, overcome the difficult coal and power situation, ease the business predicament of power generation enterprises and ensure a sustainable supply of electricity. In the long run, it can help coal and power enterprises complement each other and enhance their risk management capability to achieve a win-win situation. From a strategic perspective, the development of joint coal and electricity operations can help reduce energy production and supply costs, facilitate intensive development and integrated, efficient use of coal, and cultivate large energy companies with international competitiveness.

The development of joint coal and electricity operations can be achieved by mergers and acquisitions of upstream and downstream operations by power and coal companies or by direct merger of state-owned coal and power enterprises. China needs to strengthen policy guidance and give strong support through tax policy and other means for the development of joint coal and electricity operations, especially in the merger and restructuring of central government-controlled electric power enterprises and local coal enterprises, so as to balance the interests between the central government and the local authorities.

6.3 Establishment of an Electricity Market

The electricity reform in China began in the early 1980s. As the reform and opening-up policy continued, a series of reform initiatives such as a diversified approach to funding power operations and the separation of government functions from enterprise management were introduced successively, which effectively boosted the sustained and rapid development of the electricity industry. In the 1990s, the relaxation of government controls accompanied by market-oriented reforms on the privatisation of state-owned electricity assets grew into a trend in the global electricity industry. On the basis of the separation of government functions from enterprise management, the *Notice of the State Council on the Issuance of the Reform Plan for the Electricity System (Guo Fa* [2002] No. 5) was published in 2002 for the launch of a comprehensive market-oriented electricity reform. The internal and external environments for electricity reform and development have undergone profound change, compared with 10 years ago. Looking forward, China must remain committed to a policy on supporting development through reform. It must also conscientiously learn the lessons from China's past electricity reforms, grasp the international trend in electricity reform, refine reform strategies to keep in step with the times, and choose a model of electricity marketisation with Chinese characteristics in light of the actual conditions of China's electricity industry.

6.3.1   Reform of International Electricity Market

Against a complex political and economic background in the 1980s, some countries that believed in the neoliberal economic theory attempted to transform the electricity industry through the introduction of a market mechanism. Chile was the first country to reform its electricity market. As early as in 1982, the country established a wholesale electricity market through restructuring and privatisation of the electricity industry. The market reform of the UK's electricity industry started in 1990 and drew widespread global attention. Later on, a number of developed nations in Europe and America and some developing countries began to follow suit by conducting research on the electricity marketisation process, triggering a trend towards electricity market reform with implications for China's electricity market reform.

6.3.1.1   Basic Features of International Electricity Market Reform

No Uniform or Generally Accepted Standard Model for the Reform
The differences in national conditions and guiding ideologies of reform have given rise to a variety of electricity reform models. Different approaches have been taken to industry restructuring. In the UK, Argentina and Russia in the early years, the reform model adopted advocated a complete separation of the power generation, transmission, distribution and marketing processes. The model currently used by power companies in France, Japan, Brazil and the USA calls for consolidating power generation, transmission, distribution and marketing operations into a single conglomerate. In the UK, power generation assets were separated from power transmission assets in the early stage of reform, forming an independent market structure comprising power generation, transmission, distribution and marketing. Power generation was subsequently re-amalgamated with power distribution and marketing under the effect of market forces. As a result, six companies with integrated power generation, distribution and marketing operations accounted for 88% of the market. In terms of building a competitive power market, there is a full-capacity power pool model adopted in Britain during the early period and currently in Australia. There is also an electricity market model now generally accepted in EU countries dominated by bilateral trading and supplemented by centralised trading. There is also a power market model that features centralised trading, electric energy market and capacity market, as represented by PJM in the USA.

Large-Scale Privatisation of Electricity Assets is an Important Driver for Reform
Most of the countries pursuing electricity market reform, such as Chile, the UK, New Zealand and Argentina, have vigorously moved forward with the privatisation of electricity

enterprises while splitting up electricity assets. The process of reform in these countries is also a process of privatising electricity assets to some extent.

Committing to Reform Initially through Legislation and Driving the Process According to the Law

Electricity market reform involves the adjustment of various interest relationships and production relationships. Reform practices in different countries show that both developed and developing countries are committed to the belief that legislation should come first during the implementation of an electricity market reform. Reform programmes were first put forward by the government and sanctioned by legislation subject to full deliberation and discussion, before implementation in accordance with the provisions of the law. Relevant laws and regulations were introduced prior to the implementation of each step of the reform as a guide and standard to ensure that each important reform initiative has a legal basis. The responsibilities and authorities of electricity regulators are also defined by law. The electricity market reform in the UK initiated in 1990 was carried out pursuant to the *Electricity Act 1989*. The *Electricity Utilities Industry Law* in Japan has been amended three times since 1995 to ensure the implementation of electricity market reform in an orderly manner in accordance with the law.

Continuous Dynamic Change to the Reform Ideas and Focuses

Given the complexity of electricity market reform, countries around the world typically implement the reform in several stages, and dynamically modify the reform ideas and focuses according to the progress in different stages and the changes in the situation. In the UK, the New Electricity Trading Arrangement ('NETA'), a new market trading mechanism primarily based on bilateral trading, was introduced in 2001 to replace the 10-year-old full-capacity competitive bidding power pool trading model on the basis of past experiences in the reform. In 2005, it switched to the comprehensive implementation of the British Electricity Trading and Transmission Arrangements ('BETTA'), a unified electricity trading and transmission system, in England, Scotland and Wales. The EU also repeatedly adjusted its electricity and gas reform programmes. Act No. 2000 (1999) promulgated by the Federal Energy Regulatory Commission ('FERC') dealt with the problems arising from the deregulation of the electricity market after the introduction of Act No. 888 (1996). In 2010, power distribution companies in New Zealand were allowed once again to engage in the electricity retail business, and the assets restructuring of three major state-owned power generators was carried out.

Different Results of Reform

Thanks to the electricity market reforms, some countries achieved better results in the improvement of the electricity industry's operational efficiency and service levels, while other countries or regions encountered post-reform problems such as power supply tension, fast-rising tariffs and inadequate investment. The market reform carried out earlier in the State of California is a typical example of a failed electricity market reform. There was a severe power crisis in this richest state of the USA from 2000 to 2001 due to serious problems in reform design and market regulation. According to an analysis by the American Public Power Association ('APPA'), from 1997 to 2010, retail tariffs in the American states having implemented electricity reforms rose by an average of 4.0 cents/kWh, significantly higher than the average increase of 2.7 cents/kWh in the states not having undergone such reforms. Upon completion of the segregation and restructuring processes of the Russian electricity industry in 2008, residential tariffs rose by an average of 41.9% in the following three years, and are expected to rise sharply in the future. Inadequate investment in power infrastructure under a competitive market environment is a common problem faced by many developed countries following the implementation of electricity reforms. No large power plants or transmission projects have been built and commissioned in nearly a decade after the electricity reform in California. According to an analysis by the Office of Gas and Electricity Markets ('OFGEM'), the marginal capacity in the UK is expected to be less than 5% by 2020 under the existing market mechanism. In addition, power grid assets in some countries are controlled by foreign interests. For example, the two largest electricity distributors in the UK are respectively owned by Li Ka Shing's Hong Kong-based consortium and PPL Electric Utilities of the USA. Moreover, PPL also owns another electricity distributor in the UK. These overseas consortia and corporations jointly control about 50% of the electricity supply in the UK.

6.3.1.2   New Trends in International Electricity Market Reforms

Many Countries Become More Cautious about Electricity Deregulation
A number of global power crises in the 21st century have prompted many countries to rethink on their domestic electricity market reforms. Some countries, like the USA, have slowed down the pace of electricity market reform. By summing up the previous lessons and experiences in the reform, Brazil strengthened supervision of the medium- and long-term planning for power development and electricity supply security. In 2008, Japan decided to hold back the full deregulation of the retail market by deferring discussions of the liberalisation of supply options for consumers using less than 50 kW and for residential customers until 2013. *Energy Law Volume III* adopted by the European Union in 2009 on electricity and natural gas reforms no longer forces energy companies to split up, but instead allows power generation, transmission, distribution and marketing operations to be centrally managed within a single group. As the representative of many countries, the UK, having substantially relaxed controls on electricity in the early stages, has reinforced its intervention in the electricity market as a result of the problems emerging from the reform and the new situation in energy development.

Promoting Development of Clean Energy and Renewable Energy as an Important Goal of Reform
In recent years, as the issues of international energy security and global climate change are becoming increasingly prominent, more and more countries are regarding the development of clean energy and renewable energy and sustainable energy supply as an important goal of electricity reform. This has created a profound impact on the global electricity marketisation process. In 2011, the UK released the *Electricity Market Reform White Paper*, with plans to improve the electricity market mechanism and attract more low-carbon investments for the promotion of new energy and clean energy. Many countries have introduced relevant pricing mechanisms and subsidy policies, such as the fixed tariff system in Germany, the green certificate trading system in the UK and the

premium tariff system in Spain, in order to improve the market competitiveness of clean energy and renewable energy.

The Size and Scope of Electricity Market are Expanding

In the context of economic globalisation and with the changes in the international energy supply and demand situation, the optimisation of resource allocations over larger areas has created the internal demand for global energy utilisation. Cross-border and transregional power grids and long-distance power transmission projects have received great attention. The boundary and trading scope of the electricity market are continuously expanding. The EU plans to establish a unified EU energy market before 2014 to ensure an unrestricted flow and supply of electricity and natural gas within the EU. Electricity (energy) and other exchanges in France, Belgium, the Netherlands, Germany and the Nordic countries have realised joint trading operations and businesses amalgamation to form an integrated market covering Northern, Central and Western Europe as well as some of the regions in Southern Europe. The regional electricity markets in the USA are expanding as well. The Regional Transmission Organization ('RTO') has strengthened the coordination of trading mechanisms and rules between adjacent markets. In 2010, FERC approved the electricity market expansion proposal jointly submitted by five independent system operators, including PJM and its four counterparts in New York.

6.3.2   The Principles for China's Electricity Market Reform

Since the start of another round of electricity market reform in China in 2002, positive progress has been made but some problems have emerged as well. In the future, to push the electricity market reform further, China needs to take a solid approach based on its national conditions and realities and pursue an electricity reform path with Chinese characteristics, rather than simply copying any reform model from foreign countries.

Compared to the majority of countries having implemented electricity market reform, China's electricity market reform faces a number of different conditions. *The first is difference in the stage of development.* Driven by rapid socioeconomic expansion in China, the demand for electricity is growing fast and strongly. There is a great pressure to increase electricity investment, boost electricity development and secure electricity supply. *The second is difference in resource endowment.* Coal is the main energy resource of China. Energy demand and energy resources are distributed reversely. It is an arduous task to optimise the distribution of coal and electricity, develop clean energy, promote green and efficient exploitation and utilisation of energy resources, and optimise resource allocations over larger areas. *The third is difference in the basic economic system.* Since China applies a basic economic system principally featuring public ownership and builds socialism with Chinese characteristics, the electricity market reform is not aimed at privatisation.

Considering both the experience in international electricity market reform and the actual national conditions of China, the following principles must be upheld in the process of further driving the electricity market reform in China:

6.3.2.1   Promoting Scientific Development of Electricity Industry

Power consumption per capita in China is much lower than that in the developed countries. With continued industrialisation and urbanisation driving strong long-term growth in electricity demand, the development of the electricity industry presents a challenging task. Unlike the guiding ideology of reform in the electricity markets in most developed countries in Europe and the USA, priority must be given to promoting the development of the electricity industry and ensuring supply security in the electricity market reform in China.

China must, by deepening the reform, improve the institutional mechanism, attract investment in electricity, and facilitate integrated planning, rational distribution and coordination development for power grids and sources. It must also promote the large-scale development and utilisation of clean energy as well as the clean and efficient utilisation of fossil energy for power generation. The construction of Strong Smart Grids and the optimised allocation of electricity resources on a nationwide basis must be stepped up while the harmonious development of electricity in urban and rural areas must be promoted.

6.3.2.2 Safeguarding Security of Electricity System

Electricity security is important for economic development and social harmony. Safeguarding the security of the electricity system is both a prerequisite for the electricity market reform, and a basic principle that must be upheld. Electricity security in China is exposed to various risks due to a number of factors such as the uneven development of power grids and power sources, increasingly complex system operations, and external impacts. Electricity market reform must be carried out on the premise that the risks associated with electricity security are controllable, can be controlled and are under control. The reform must not be carried out at the expense of system security. Through the reform, the strengths of China's institutional mechanisms must be consolidated and developed, with the establishment of an institutional mechanism conducive to safeguarding electricity security.

Integrated power transmission and distribution and integrated grid operation and dispatch facilitate a free flow of information and efficient and coordinated operation between power grids at different levels and between power grids and dispatch centres. As a result, power grid dispatching operates in a more flexible way and troubleshooting is handled timely and effectively, thereby minimising the risks of power system break-down and widespread blackout. Over the past ten years, there have been frequent widespread outages overseas. In contrast, despite their structural weaknesses, sharply higher loads and increasing risk factors, the power grids in China are operating safely primarily because of the advantages of integrated power transmission and distribution as well as integrated power grid and dispatch operations.

6.3.2.3 Leveraging the Combined Effects of Market Mechanism and Macro-control Initiatives

Electricity is an important infrastructure industry. Promoting reform in the electricity market demands the employment of market-based means to improve the institutional mechanism for the electricity industry and improve the quality of electricity and the efficiency of resource allocation while strengthening and improving macro-control initiatives simultaneously to maintain government control over electricity development and to correct market failures. In terms of industry characteristics, the electricity industry consists of segments that can be opened to competition and segments that are natural monopolies.

Market regulation should therefore be strengthened when a market mechanism is being introduced to the industry. A scientific and efficient industry administration system with an operation mechanism can be developed by coordinating the invisible hand of market forces with the visible hand of macro-control.

6.3.2.4 Enhancing the International Competitiveness of Power Enterprises

In carrying out electricity market reform, the fostering of power enterprises with international competitiveness must be a key consideration. Out of consideration for maintaining the competitiveness of Electricité de France (EDF) in the international market, the French Government insisted that the power generation, transmission and distribution businesses be retained in a single enterprise.

China's basic economic system oriented principally towards public ownership and the challenging international situation in energy security argue against following the 'fragmented' reform path of de-amalgamation and privatisation pursued by some countries. Instead, China should preserve the dominant role of the state-owned economy in the electricity industry, cultivating power companies with economies of scale and international competitiveness, so as to participate more in international competition for energy.

6.3.2.5 Committing to Total Design Approach and Driving Steady Progress in Accordance with the Law

Electricity market reform is a systems engineering project requiring a total design approach and phased implementation. Great emphasis must be placed on strengthening the top-level design and scientifically determining the objective, framework, focus and implementation paths of the reform according to the national conditions and the realities of development in China. At the same time, the democratisation and transparency of reform decision-making should be enhanced, responsible parties for the reform clearly defined, and the scientific approach to decision-making ensured.

The market economy is an economy ruled by law. The electricity market reform demands a commitment to legislation as a priority and driving progress according to the law. The enactment of laws and regulations for electricity must be vigorously strengthened. Major reform ideas should be elucidated in legal form, and then promoted and implemented in accordance with the law. The situation where electricity legislation lags behind the implementation of reform should be changed gradually to push the electricity market reform on the road to legalisation, making the reform a more serious undertaking with little casualness.

6.3.3 Ideas on Building an Electricity Market System in China

In general, based on China's national conditions, a unified, open, competitive and orderly electricity market system needs to be developed under government supervision to break down the geographical restrictions on electricity trading, and to optimise resource allocations on a nationwide basis by establishing a unified and standard trading mechanism and fair market rules.

6.3.3.1 The Substance of a Unified, Open, Competitive and Orderly Electricity Market System

The core task of China's electricity market reform is to build up a unified, open, competitive and orderly electricity market system. This refers to a system with the attributes of unity, openness, competitiveness and orderliness. It is an effective operation system built for the purpose of providing a secured and efficient marketplace for the free movement and optimal allocation of power resources and various production factors of the electricity industry on a nationwide basis.

Unity means the employment of a set of unified market rules and operation mechanisms to facilitate the trading of electricity throughout the country and coordinate development between markets at all levels. Openness means to remove interprovincial and regional market barriers, break up all kinds of administrative and economic monopolies, provide market players with equitable access to the market and carry out open and transparent market operations. Competition means that market players can compete fairly and effectively on the marketplace and that the regulatory role of the pricing mechanism can work effectively on the market, not only to prevent acts of unfair competition such as the abuse of market positions, but also to avoid excessive competition. Orderliness means that the electricity market operates in an orderly manner under the laws and regulations and market rules; and that the market is able to make a reasonable response to any disturbance to maintain its normal and effective operation.

Among the four essential features of the electricity market system in China, unity is the common basis for the electricity market system to survive; openness is the prerequisite to ensure the vitality of the market; competition is the source of market efficiency, and orderliness is the assurance for market order.

6.3.3.2 The Thinking Behind Building a Unified, Open, Competitive and Orderly Electricity Market System

The basic idea for building a unified, open, competitive and orderly electricity market system is to 'Unleash Both Ends and Supervise the Middle', which means promoting development in power generation and sales to gradually introduce competition in the generation market and open up the power sales market in an orderly fashion. Through these efforts, the mechanism for scientific development and management of grids is perfected and grid regulation improved to ultimately form a market structure involving a multitude of buyers and sellers and build up an efficient electricity market system with Chinese characteristics.

Developing the Power Generation Market

The coal-electricity price linkage mechanism, the benchmark tariff and the renewable energy pricing mechanism should be perfected. The unified planning mechanism for power resource development should be defined and guidance to a rational distribution and scientific development of power resources provided. The development of the electricity trading platform should be strengthened to develop in an orderly manner a variety of trading products, promote cross-provincial and cross-regional electricity trading, accelerate the development of electricity trading based on standard contracts, codified procedures and market practices. The on-grid competitive bidding pilot schemes on the generation side should be promoted. Competitive power supply capacity should be gradually expanded with the implementation of an on-grid mechanism that combines generation dispatch for energy conservation and competitive bidding on the open market.

With the development of SSGs, the expanded scope of resource allocation and improvements to market rules, China should gradually push forward the integration and restructuring of electricity markets at all levels, accelerate the development of a national electricity market featuring the optimal allocation of power resources nationwide and an electricity market system consisting of balanced markets for security control areas (province-wide or cross-provincial). The transaction modes should mainly focus on medium to long-term trading cycles (12-month or longer), supported by short-term (monthly or daily) transactions and supplemented by real-time transactions.

Developing the Power Sales Market

Efforts should be maintained to steadily push forward a direct trading pilot scheme between large power consumers and power generators and incorporate direct trading conducted by large consumers into the provincial electricity trading platform for unified management. Key issues such as cross-subsidies, mixed transmission and distribution tariffs and ancillary services from direct trading by large consumers should be gradually resolved to lay a solid policy foundation for well-planned development of direct trading by large consumers. Reform in tariff mechanisms should be pursued in earnest and the mechanism of universal power supply services improved.

In the long term, provided that issues like cross-subsidies have been resolved and a systematic transmission and distribution tariff mechanism formulated, China should consider starting pilot schemes to remove restrictions on the options for large consumers, cultivating independent players for electricity sales, and gradually developing a competitive environment with a number of players engaged in electricity sales, including grid operators.

Perfecting Scientific Development of Power Grid and Standardising Management Mechanism

A unified planning mechanism should be established for power grids and power resources, the investment and management system of grid projects improved, and the project approval procedures streamlined. Efforts should be made to accelerate the improvement of an electricity transmission and distribution price formation mechanism well-attuned to SSG development and the optimal allocation of resources over large areas. Problems with rural electric power management mechanisms should be resolved in accordance with the principle of balanced development of urban and rural electric power to enhance grid capacity for sustainable development. China should continue to adhere to the integration of power transmission and distribution and the integration of grid dispatch operations, with the supervision of grid operations in accordance with the law.

In the longer term, China should gradually build up a mechanism for grid investment and development that is guided by unified planning, protected by a tariff mechanism and based on market-oriented principles. An operation mechanism for electric power should also be set up to contribute to electric power security, development efficiency, fair competition and balanced urban and rural development.

From the perspective of overseas experience, there is no standard model for grid management systems. In terms of the relationship between power transmission and distribution, there are instances of separate transmission and distribution operations, as well as instances of amalgamated operations. In terms of the relationship between grid operation and dispatch, there are instances of both amalgamated and segregated grid and dispatch operations. A particular system is selected based on both historical and practical factors. The practices in Japan, France and Canada show that in so far as an integrated enterprise is maintained, the reform target of introducing a market mechanism can definitely be achieved by establishing and refining the electricity regulatory mechanism and standardising the electricity market trading mechanism.

For China, adopting a management system that emphasises integrated power transmission and distribution as well as integrated grid operation and dispatch carries significance on several levels.

Firstly, it helps to safeguard power security. An important cause of the power outage in the USA on 14 August 2003 was poor internal communication as a result of decentralised grid management. Two weeks after the power outage in the USA, the East China Grid was close to suffering a major power outage, but ultimately succeeded in addressing the risk and crisis through timely and decisive measures thanks to grid management and unified dispatch. In Italy, during the early stage of reform, dispatch was once separated from grid operation, but these were reintegrated after the nationwide blackout on 28 September 2003.

Secondly, it is conducive to promoting disciplined grid planning and development. Integration of transmission and distribution as well as integration of grid operation and dispatch could bring about better overall grid planning at all levels, optimise the grid structures, promote development of transmission and distribution grids, balance development of urban and rural grids, and promote consumption of new energy and the optimal allocation of power resources over large areas.

Thirdly, it is conducive to improving efficiency. Separating power transmission and distribution sometimes may not improve the efficiency of the power industry. On the contrary, it may even bring efficiency loss. US scholar John Kwoka conducted research on the operating efficiency of 73 distribution companies in the USA from 1993 to 2004. It was found that the efficiency of the companies with segregated distribution operations went down by 4% after the reform, and the efficiency loss incurred increased year by year. Zhang Xinzhu of the Chinese Academy of Social Sciences conducted a quantitative analysis showing that separated transmission and distribution operations in China will increase costs by between RMB60 billion and RMB180 billion.

Fourthly, it is conducive to enhancing the international competitiveness of grid operators. As China does not aim at privatisation of distribution networks, there is little necessity of segregating transmission and distribution operations and the integration of transmission and distribution can maintain the economies of scale for grid operators to better participate in international energy competition and implement national energy strategy. For example, the EDF Group derives about 50% of its revenue from overseas business activities with a very strong competitive edge in the international electricity market, which is closely related to the fact that it is concurrently controlling generation, transmission, distribution and marketing assets. On the whole, the integration of transmission and distribution and the integration of grid operation and dispatch are logical choices in line with China's national conditions.

6.3.4 The Tariff System and Building of Tariff Pricing Mechanism

Tariffs are at the core of electricity market development. In order to build a unified, open, competitive and orderly electricity market system while promoting the sound and rapid development of the power industry, China should continue to deepen tariff reform, gradually build up a tariff system and a tariff-setting mechanism, and capitalise on the role of price levers in regulating the power industry.

6.3.4.1 China's Present Tariff Situation and Current Problems

China's current tariff system is comprised of several components, including on-grid tariffs, transmission and distribution tariffs, retail tariffs and tariffs for direct power purchase by large consumers. The on-grid tariff is set by the government, with the implementation of an operating-period tariff, a benchmark tariff, a coal-electricity price linkage system, etc. These are in addition to market pricing arrangements such as tender pricing, negotiated pricing for cross-provincial and cross-regional electricity transactions, etc. The transmission and distribution tariff comes in three different levels covering cross-regional grid, cross-provincial and provincial grids. The transregional and cross-provincial grids are subject to government pricing and internal negotiations at the corporate level. Pricing at the provincial grid level is set based mainly on the difference between the power purchase price and the retail price. The retail tariff is set by the government, based on the nature and use of power and the voltage class. Only a small number of tariff pilot programmes are currently underway for direct electricity purchase by large consumers, with the tariffs set through bilateral negotiations.

At present, China's electricity tariff system has the following problems.

Tariff Levels Generally Too Low to Encourage Power Conservation and Energy Efficiency Improvement

China's coal and oil prices are basically in line with the international market, but internationally, the electricity tariff is still at a low level. In 2010, electricity retail tariffs in China averaged RMB0.605/kWh (including funding levies and surcharges), or US$0.089/kWh and approximately US$0.008/kWh lower than that of the USA (see Figure 6.1). The low tariff directly contributes to low public awareness of power conservation and inefficient power utilisation. It has also boosted the demand for electricity and exacerbated the tight power supply situation.

In recent years, electricity tariffs in China have grown at a far lower rate compared with the prices of other energy and consumer goods. From 2003 to 2010, the Ex-Power Plant Price Index for Electricity rose 19%, a lower rate compared with many other industries such as coal (119%), oil (114%), metallurgy (47%), food (30%), and chemicals (26%). It was also lower than the Producer Price Index (PPI) (26%), the Consumer Price Index (22%), the Purchasing Price Index of Raw Material, Fuel and Power (56%), and the Price Index of Fixed Assets Investment (27%).

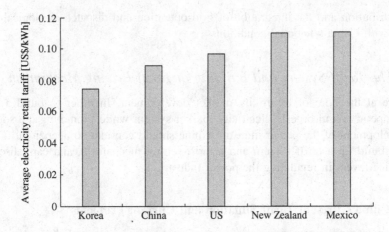

Figure 6.1   Average electricity retail prices of selected countries in 2010.

Irrational Tariff Structure Hampers Coordinated Development of Grid Power

The transmission and distribution price in China accounts for approximately 26.8% of the retail tariff, significantly lower than an average of 40% to 60% in overseas countries. In 2010, the transmission and distribution tariff in China was US$0.024/kWh (including tax and approximately the same as other countries), lower than the USA [US$0.0345/kWh, 2009], and also Brazil [US$0.0741/kWh, 2007], Mexico [US$0.0256/kWh, 2009] and other developing countries (see Figure 6.2).

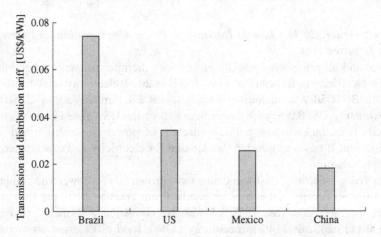

Figure 6.2   Transmission and distribution tariffs of selected countries in 2010.
Source: EIA of the US; ANEEL of Brazil; Report on the Monitoring of Power Tariff Implementation, State Electricity Regulatory Commission, The People's Republic of China; SENER of Mexico.

Unsound Tariff Mechanism Fails to Truly Reflect the Supply-demand Relationship and Production Costs

China has not yet established a systematic tariff-setting mechanism or a cost transmission mechanism. The on-grid tariff and the retail tariff are mainly determined by the government, while the transmission and distribution tariff is squeezed by both the on-grid tariff and the retail tariff. The established coal-electricity price linkage mechanism is imperfect and often fails to be implemented in a timely and adequate manner due to various factors. Affected by the rapid rise in coal prices, power companies suffer production and operational difficulties and many power generation companies are suffering serious losses.

No system for price adjustment in line with development costs has yet been developed for setting benchmark on-grid tariffs of wind power and other renewables. For example, with the decline in the prices of wind turbines, the total costs of wind farms went down by between RMB 1000 and RMB 2000/kW on average in 2011 from 2008, while the benchmark on-grid tariff for wind power was not adjusted accordingly.

Unsound Tariff System and Support Policies are not Conducive to the Development of Nonfossil Energy Resources

There is no tariff policy in China to coordinate peak-valley on-grid tariffs for large-scale wind power generation and encourage the development of pumped storage power plants. The rather low levels of tax surcharge for renewable energy can only meet 70% of the funding requirements of the subsidy scheme. For subsidies for renewable power generation projects, only the construction and operational costs of access networks for in-situ power consumption are taken into account, with no consideration given to such factors as long-distance power delivery from large-scale renewable power producers and grid expansion at the supply and receiving ends. Generally, peak-valley tariffs are not implemented on the power demand side. The tariff differential implemented is too small to have any meaningful impact on encouraging wind power consumption through peak load shifting.

6.3.4.2   The Goal and Thinking Behind Deepening Tariff Reform

The goal of China's tariff reform is to keep in step with the process of electricity market reform, formulate a well-structured tariff system with a tariff-setting mechanism, leverage the role of price signals in guiding electricity investment, consumption and resource allocation. It also aims to promote optimised planning, restructuring and scientific development of the power industry, in addition to the clean and efficient use of electricity resources.

The thinking behind deepening China's tariff reform is to gradually push forward price marketisation in competitive segments and price standardisation (systematic government pricing and regulations implemented) in monopoly segments based on efforts to improve established tariff policy and tariff mechanism. Specifically, the following initiatives should be taken.

In the area of power generation, China should improve the benchmark tariff and the linkage of primary energy prices with electricity tariffs. The aim is to establish the on-grid tariff for renewable energy and the integrated utilisation of power that is conducive to developing clean energy and reflecting cost changes. It is also designed to establish on-grid tariffs for coal-fired, nuclear power and gas-fired generation that can reflect changes in primary energy prices and promote the building of peak-regulation capacity as well as tariffs conducive to encouraging the construction of pumped storage power stations. In the longer term, along with the development of competitive power generation markets, market pricing should be gradually adopted not just for renewable energy and the integrated utilisation of power generation, but also for other on-grid tariffs.

In the area of grid operation, China should gradually establish cross-provincial and cross-regional transmission pricing systems and mechanisms that fully capitalise on the technical and economic advantages of UHV grids, promote the intensive development and power delivery of large coal-fired power, hydropower, nuclear, wind and other renewable power generation bases. It should also rationalise the relationship between cross-provincial/cross-regional grid pricing and provincial grid pricing, improve the price subsidy mechanism for access system projects for renewable energy, and incorporate the investment in EV charging and battery-swap facilities as part of the capital costs of grid development.

In the area of marketing, China should adopt a user tariff system and a tariff mechanism beneficial to large-scale utilisation of renewable energy, energy efficiency and emission reduction while establishing the linkage between retail tariffs on one side and on-grid tariffs and transmission and distribution tariffs on the other. The problems with tariff cross-subsidy should also be gradually resolved. In the long run, a sound market pricing mechanism should be established through bilateral trading, centralised auction and other arrangements as large users' supply options and the sales market are being deregulated.

6.3.4.3 Key Measures of Recent Tariff Reform

Improving the Coal-electricity Price Linkage Mechanism and Refining the Price Linkage Mechanism for Primary Energy and Electricity

In 2004, in order to rationalise the price relation between coal and electricity, the government introduced the coal-electricity price linkage mechanism. In principle, a period of at least six months is set as a price linkage cycle; if the average coal price within the cycle changes by 5% or more compared with the previous cycle, the on-grid tariff shall be adjusted in accordance with the changes in coal prices subject to 30% of the coal price increase being absorbed by the power enterprises. Following an adjustment to on-grid tariffs, the grid operator's retail tariff should be adjusted accordingly.

The establishment of the coal-electricity price linkage mechanism has helped to effectively ease the disparities between coal prices and electricity costs. However, problems like untimely and ineffective linkages have been exposed in the actual implementation process. There are three main reasons for these phenomena. Firstly, the rise in coal prices is often accompanied by a rise in consumer prices, so the government often postpones or avoids the implementation of the coal-electricity price linkage mechanism as part of its attempt to control consumer prices. Secondly, due to the short-term actions of companies and the bottleneck of railway capacity, the distorted market prices of coal stand in the way of properly implementing the coal-electricity price linkage mechanism. Thirdly, the tariff has long been the subject of many public policy objectives, and low public expectations of a price increase have made the implementation of the coal-electricity price linkage mechanism more difficult.

In the future, China should step up research on the following issues after summing up the experience of implementing the coal-electricity price linkage mechanism.

Refining the Coal-electricity Price Linkage Mechanism and Enhancing Implementation

Through the linkage mechanism, the impact of coal price changes can be passed onto users, which is conducive to removing disparities between coal and power prices, regulating the supply-demand relationship, and promoting the healthy development of the coal and electricity sectors. Therefore, China should continue to refine this mechanism. After an adjustment to on-grid tariffs, the electricity retail tariff should be adjusted accordingly.

Expanding the Scope of the Linkage Mechanism and Formulating a Primary Energy-electricity Price Linkage Mechanism

With the large-scale development and utilisation of natural gas resources, China's gas-fired generation will see rapid growth. In order to ease the impact of gas price changes, it is necessary to expand the coal-electricity price linkage mechanism into a primary energy-electricity price linkage mechanism to ensure that tariffs fully reflect the changing price of fuel for power generation.

Reforming the Electricity Billing System to Intuitively Reflect the Impact of Price Increases in Primary Energy

Currently, the electricity billing method is one of uniform charging with payment based on consumption levels and the tariff rates approved by the government. Implementation of the coal-electricity price linkage mechanism is likely to cause misunderstanding. China can learn from the experience of implementing the coal-electricity price linkage mechanism in the USA and the fuel surcharge for domestic airfares, to reform the way electricity charges are structured by dividing the tariff into two components, namely: a basic tariff and a fuel surcharge. When the coal-electricity price linkage mechanism or gas-electricity price linkage mechanism is implemented, it can reflect the impact of primary energy price changes on the end-user tariff through adjusting the fuel surcharge to make the tariff adjustment more comprehensible and acceptable to the users.

Establishing a Balancing Account for On-grid Tariffs to Avoid Extreme Fluctuations in Tariffs

As frequent adjustments to tariffs will result in excessive workload and additional costs, no country around the world where prices are controlled by government will adjust the on-grid tariffs in real time to reflect changes in fuel prices. When the fluctuations of upstream prices are drastic, end users would find it difficult to accept tariff readjustments of the same magnitude. Through establishing a tariff balancing account system, the direct impact of the changing upstream costs on the tariffs can be eased and the tariff adjustment cycle appropriately extended.

Adapting to the Development Trend of Renewable Energy, and Refining the Tariff Mechanism of Renewable Energy

The power generation pricing mechanisms of renewable energy implemented abroad can be classified into four types. The first is a fixed tariff mechanism where the government directly determines the power generation price of renewable energy. The second is a premium tariff mechanism where the power generation price of renewable energy is calculated based on the floating price in the competitive electricity market multiplied by a ratio or plus a fixed-price subsidy. The third is a tendering system where the government invites open bids for one or a group of renewable energy generation projects and determines the on-grid tariff with reference to the prices offered by the final successful bidder. The fourth is a green-electricity tariff mechanism where the government determines the power generation price of renewable energy based on the opportunity-cost method and comes up with specified prices at which electricity consumers can voluntarily subscribe for power supply.

At present, China's power-generation and on-grid tariffs of renewable energy are determined by the price control authority according to the power-generation characteristics of different types of renewable energy and the specific conditions of different regions according to the principle of benefiting the development and utilisation of renewable energy and economic viability. Electricity generation prices of wind, solar and biomass energy have been successively determined by tendering and benchmark pricing, and the latter is the main pricing mechanism currently in use. At the same time, China can levy renewable energy surcharges on retail tariffs to subsidise such renewable energy generation projects as wind power.

In the future, there are three different areas where the tariff policy for renewable energy can be refined:

Studying the Establishment of a Mechanism for Year-by-Year Reduction of On-grid Tariffs

China should learn from international experience and build up a mechanism as soon as possible where the on-grid tariff of wind power decreases with lower construction costs year after year, so as to encourage wind power companies to continue to reduce costs and adopt technological innovations for the benefit of society.

Studying the Establishment of a Dynamic Adjustment Mechanism of Surcharge (Funding Levy) for Renewable Energy Tariffs

According to China's plans for renewable energy development, it is expected that the growth of renewable energy demand in the future will be much faster than the growth in surchargeable capacity, and a subsidy shortfall will appear and grow. China should study and establish a dynamic adjustment mechanism of surcharge (funding levy) for renewable energy tariffs so as to ensure the healthy development of renewable energy.

Strengthening Support for Grid Access Projects for Renewable Power Generation

The current subsidy of RMB 0.01 to 0.03/kWh for grid access projects for renewable power generation is not sufficient to cover investment and debt service costs in these grid access projects. China may consider adopting different policies for recovery of investment and operating costs to support the power grid projects in the wind power bases. For the grid access projects of those smaller wind power bases, China can adopt the benchmark method for setting the subsidy level for grid access projects and recoup the subsidy according to the eventual on-grid wind power capacity. For those massive wind power bases designed to transmit electricity to other regions over long distances, the subsidy level may be individually assessed, and the construction and operational costs of those support grid projects that are higher compared with conventional energy grid projects may be apportioned through imposing a surcharge on renewable energy tariffs on a nationwide basis.

Gradually Establishing Independent Transmission and Distribution Tariff Formation Mechanism Based on Pilot Schemes
In order to ensure sustainable development of the power grid and promote the optimised allocation of energy resources over large areas, it is necessary to establish an independent transmission and distribution tariff formation mechanism. The establishment of an independent transmission and distribution tariff mechanism is not necessarily related to transmission and distribution grid management systems.

Moving with Caution towards Independent Transmission and Distribution Tariff Reform for Provincial Grids
China should select a number of typical provinces to start pilot schemes for independent transmission and distribution tariff reform on provincial grids, and should research and implement programmes for further reform based on the outcome of the pilot schemes.

Setting up a Tariff Information Dissemination Mechanism for Cross-regional/ Cross-provincial Grid Transmission
Uniform transmission tariff and security tariff should be formulated for cross-regional and cross-provincial grids. The transmission tariff can be recovered from the retail tariff as part of the power purchase cost of the power purchasing province, while the security tariff should be borne by the interconnected provincial grids concerned and incorporated into the provincial transmission and distribution tariffs and recovered from the retail tariffs.

Rationalising the Relationship between On-grid Tariff, Retail Tariff, and Transmission and Distribution Tariff
The independent transmission and distribution tariff mechanism should be incorporated for consideration as part of the development of a unified, open, orderly and competitive electricity market system, and be converged with reforms focusing on power generation market competition, direct power purchase by large consumers, and the liberalisation of the sales market. Linkage mechanisms for retail tariffs, on-grid tariffs and power transmission and distribution tariffs should be established. In the near term, it is particularly important to speed up the establishment of a regular transmission and distribution tariff adjustment mechanism that is able to ease tariff distortions in a timely manner.

Refining the Retail Tariff Mechanism and Appropriately Alleviating Tariff Distortions
In the longer term, electricity retail tariffs should eventually be determined by market competition. In the short term, the following measures can be taken for retail tariff reform.

Refining the Electricity Retail Tariff Formation Mechanism
The government-set electricity retail tariffs are the sum of on-grid tariffs, transmission and distribution tariffs, transmission and distribution power losses, government funding levies and surcharges. After the implementation of a linkage between the retail tariff and the on-grid/transmission and distribution tariffs, any adjustment to the overall level of retail tariff should reflect changes in primary energy prices.

Continuing to Push Forward the Reform in Retail Tariff Classification
The same tariff should be charged for the same voltage and load factor among commercial, industrial and other users. Peak-valley tariffs or seasonal peak-valley tariffs are implemented for power consumption in industrial, commercial and other areas, with increased peak-valley or seasonal price differentials. Meanwhile, EV charging and battery-swap is regarded as a new category of electricity utilisation and this tariff is included in the government's catalogue of tariff charges.

Promoting the Reform of Domestic Tariffs
A progressive tariff structure for residential customers should be adopted with a gradual transition to Peak-Valley TOU tariffs.

Rationalising the Tariff Relationship between Different Customer Segments
China should gradually resolve the historical debt problems and cross-subsidy issues by reforming the electricity retail classification system and domestic tariffs.

6.4   Development of Pricing Mechanism for Oil and Gas

China's crude oil prices have converged with the international market. The major problem of a market-oriented oil industry is how to improve the pricing mechanism for refined products, and to minimise the impact of drastic international oil price fluctuations on the domestic oil market. Compared with the petroleum market, the natural gas market in China is still at a stage of rapid growth. The future reform should focus initially on the enhancement of control of the natural gas industry rather than deregulation, and should aim for a market-oriented system in the long run. The focus and difficulty of such reform lies in the rationalisation of natural gas prices.

6.4.1   Reform of Pricing Mechanism for Refined Products

The refined product pricing mechanisms in countries around the world are unique in their own ways. Yet in general, they are set either by the market or the government. At present, the USA and other developed countries in Europe have adopted market-oriented pricing mechanisms. The reform of Asia's refined oil market started late, with an increasing number of countries and regions, such as Japan, Korea and Taiwan, having undergone a smooth transition to market pricing systems.

Table 6.2 Adjustments to China's refined product prices since 2010. Unit: RMB/tonne.

Date of adjustment	Adjustment amount for gasoline	Gasoline price after adjustment	Adjustment amount for diesel	Diesel price after adjustment
14 April 2010	320	7420	320	6680
01 June 2010	−230	7190	−220	6460
26 October 2010	230	7420	220	6680
22 December 2010	310	7730	300	6980
20 February 2011	350	8080	350	7330
07 April 2011	500	8580	400	7730
09 October 2011	−300	8280	−300	7430
08 February 2012	300	8580	300	7730
20 March 2012	600	9180	600	8330

Source: Data in the table are based on relevant documents issued by the NDRC.

6.4.1.1 Problems in China's Refined Product Pricing Mechanism

China has implemented a pricing mechanism where under government control, refined product prices are partially integrated with international crude oil prices. For example, if the moving average of crude oil prices for Brent, Dubai and Minas varies by more than 4% for 22 consecutive working days, the maximum retail prices for domestic refined products will be adjusted. The Chinese Government will take into consideration factors including the macro-economic situation and CPI when adjusting the prices of domestic refined products.

Under the domestic refined product pricing mechanism, China has effected 9 adjustments to domestic refined product prices since 2010, with 7 upward and 2 downward adjustments (see Table 6.2).

The current domestic refined product pricing mechanism has its inherent flaws. Firstly, the domestic refined product prices submissively follow international oil prices which cannot directly and timely reflect the supply/demand relationship or the changes in consumption patterns of the domestic market. Secondly, domestic price adjustments always lag behind the movement of international oil prices. In addition, the transparency of the pricing mechanism easily invites market speculation such as hoarding that is not conducive to maintaining stability of the refined product market.

6.4.1.2 Gradually Enhancing Marketisation of Refined Product Pricing Mechanism

China should learn from the experience of other countries to further enhance the refined products pricing mechanism towards a market–oriented system. The two options for making improvement are:

Promoting Flexible Adjustments to Domestic Refined Product Prices

The government can shorten the time lag between the movement of domestic oil prices against the international market by reducing the current 22-day reference period and lowering the 4% price variation requirement. In an attempt to move towards a more reasonable fully market rules-based system, the government approval system should be changed to a system for filing by the enterprises concerned of prices readjusted based on the government-approved Refined Oil Product Pricing Formula. Domestic refined products could gradually achieve full market pricing with the above transitional measures. However, the adjustment cycle should not be too short in order to shield the domestic market from the abnormal fluctuations of international oil prices. An appropriate adjustment cycle can ensure efficient integration of domestic refined product prices with the international market and also filter out the effect of abnormal international price fluctuations and ensure supply stability in the domestic market for refined products.

Strengthening Corporate Autonomy in Price Readjustment

The upper and lower limits for refined product prices shall be set based on the domestic supply/demand relationship. As long as the price fluctuations are within the prescribed range, the government only needs to perform its monitoring role. When the refined product prices go beyond the specified limits, the government can intervene with appropriate administrative measures. Private enterprises should be allowed to determine the timing and magnitude of price adjustment based on the approved price adjustment formula, enabling the market to gradually play a fundamental role in price adjustment. At the same time, the government's role will gradually shift from one of direct pricing control to one of market coordination and supervision. It can positively influence the oil prices by means of monetary and fiscal policies while retaining the power to directly implement price controls when oil prices have moved beyond the level sustainable by the macro-economy, so as to ensure healthy development of the domestic refined product market and associated industries.

A market-oriented system is the basic direction for China's energy pricing reform. In the long term, refined product prices will be determined by market demand. Taking into account the robustness of the market system, social affordability and national energy security, the pace of reform should be both constructive and prudent. The government should gradually lift the restrictions on the import and export of refined products for the wholesale and retail markets, hence increasing the flexibility of market supplies. With the formation of a competitive market structure and continuous improvement to the system, refined product prices will ultimately be determined by market conditions. In the marketisation process of refined product prices, the government should improve the applicable laws and regulations as well as tax and subsidy policies to strengthen the governance of the oil industry, and to step up macro-controls over the domestic oil market to safeguard market order.

6.4.2 Natural Gas Pricing Reform

In the next 20 years, natural gas will be the fossil energy seeing the fastest-growing demand in China, which, according to IEA estimates, will record average annual growth of approximately 6.4%. In order to meet the fast-growing demand for natural gas and to promote the rapid and healthy development of the natural gas industry, China should reform the natural gas pricing system to rationalise the price relativity between natural gas and its energy alternatives. China should also improve market supervision, gradually introduce a competitive market system, fully leverage the guiding role of prices, readjust the relationship between production and consumption, and promote a high-quality and low-carbon energy structure.

6.4.2.1 International Natural Gas Pricing Models

Currently, there is no uniform model for natural gas pricing in the world. There are 6 different international natural gas pricing models (see Table 6.3). Among these, Type 1 is applicable to both international and domestic trading, Types 2 to 4 to international trading, and Types 5 and 6 to domestic market pricing.

In international natural gas trading, the North American region and the UK employ competitive pricing for multiple gas sources and Europe adopts oil price index pricing. Japan, Korea and countries in Northeast Asia utilise JCC-linked pricing whereas Russia and the Central Asian region mainly operate a bilateral monopoly pricing system. In the domestic markets of countries around the world, the USA, Canada and the UK implement competitive pricing, and other countries and regions mostly follow a netback pricing system based on alternative energy prices.

6.4.2.2 China's Natural Gas Pricing Mechanism and Existing Major Deficiencies

At present, China's imported natural gas prices are determined by the market. Domestic natural gas prices are mainly determined by cost-plus pricing supplemented by

Table 6.3 International natural gas pricing models.

No.	Mode	Features
1	Competitive pricing	Price based on free competition in the open market
2	Oil price index pricing	Price based on a formula linked to oil fuel prices
3	Linked to average price of Japan's imported crude oil (Japan Crude Cocktail, JCC)	Price set according to the pricing formula linked to JCC with upper and lower price limits
4	Bilateral monopoly pricing	Buyers and sellers determine the price through negotiations
5	Netback market value pricing*	The retail price, gate price, ex-plant price or border price are determined based on the market value of natural gas
6	Cost-plus pricing	Price is determined by costs in each component of the production process plus approved profit margins

*Netback market value pricing is used to determine an affordable market price of natural gas based on the market value of energy alternatives, which is then used to derive the ex-plant price and border price after accounting for transportation, distribution, storage and other expenses.

government-guided prices. In particular, the ex-plant price and the pipeline transportation price (the sum being the city gate price) are based on government-guided prices, while the end-use consumer price (the sum of city gate price and local distribution fee) is approved by the local government.

China's natural gas pricing mechanism is characterised by the following major deficiencies:

Firstly, natural gas prices are set too low and out of tune with the domestic supply-demand correlation and the international market prices. In recent years, the demand for natural gas has been growing rapidly in China. But natural gas prices do not fully reflect the relationship between supply and demand where the retail prices are sometimes lower than the city gate prices in some areas, which has exacerbated the pressure on natural gas supply. Frequent gas shortages in recent years are closely related to this. In addition, low domestic natural gas prices have also given rise to a substantial price difference between domestic and imported gas prices. The average ex-plant price of the gas supplies from the West-East Pipeline Project is RMB $0.71/m^3$, and the gate price in Shanghai is RMB $1.41/m^3$. In China, the duty-paid price of piped natural gas imported from Central Asia is more than RMB $2.00/m^3$.

Secondly, the price relativity between natural gas and oil or its energy alternatives is unreasonable. The data provided by the relevant government departments indicate that domestic gas prices are equal to 1/4 of the price of LPG of equivalent calorific value, and 1/3 of fuel oil prices. The calorific value vs energy price analysis between natural gas and crude oil in 10 countries including the USA, the UK, Japan and China (see Tables 6.4 and 6.5) demonstrate that from 2005 to 2009, China's average price ratio between natural gas and crude oil for industrial use was 0.69 and that for residential use was 0.68, representing respectively the second lowest and the lowest level amongst the 10 countries in the study.

Thirdly, the pricing system is devoid of a solid scientific basis. The pricing of pipeline transportation is not supported by open and transparent pricing principles or procedures. The classification of end users is not scientifically-based and cross-subsidies exist. The retail price is only classified according to different usages and not backed by a differential pricing mechanism, such as peak and off-peak tariffs, seasonal tariffs, interruptible or noninterruptible user tariffs, etc.

In conclusion, China's current natural gas pricing mechanism and management model are lagging behind the development of the natural gas industry. Natural gas prices have failed to fully reflect the supply and demand balance and the scarcity of resources in the market. The fundamental role of the market mechanism in resource allocations in the natural gas sector cannot be fully fulfilled.

6.4.2.3 Steadily Pushing Forward China's Natural Gas Pricing Reform

The ultimate goal of China's natural gas pricing reform is to deregulate the ex-plant price of natural gas, which should be determined by market competition while the government should only manage the price of natural gas pipeline transportation in a natural monopolistic sector. Through deregulation and open competition, China can gradually transition

Table 6.4 Price ratio between natural gas and crude oil for industrial use in selected countries.

Country	2005	2006	2007	2008	2009	Average
Italy	1.05	1.11	1.07	1.02	1.41	1.13
Korea	1.20	1.21	1.20	0.78	1.20	1.12
Japan	1.19	1.04	0.99	–	–	1.07
Greece	0.95	–	0.97	1.05	1.12	1.02
France	0.96	0.99	0.88	0.95	1.09	0.97
Spain	0.77	0.89	0.85	0.79	1.11	0.88
UK	0.85	0.90	0.69	0.69	0.79	0.79
USA	1.02	0.78	0.67	0.60	0.53	0.72
China	0.72	0.60	0.63	0.53	0.95	0.69
Canada	0.86	0.65	0.56	0.54	–	0.65

Source: The data from 2005 to 2006 on China are the average prices of natural gas for industrial use in 36 major cities in the Price Yearbook of China. Data from 2007 to 2009 are provided by the Price Monitoring Centre of the National Development and Reform Commission. Data on other countries are obtained from the *Energy Prices & Taxes 2nd Quarter 2010*.
Note: All energy prices are inclusive of tax, and the unit used is USD/standard oil per ton.

Table 6.5 Price ratio between natural gas and crude oil for residential use in selected countries.

Country	2005	2006	2007	2008	2009	Average
Japan	3.70	2.96	2.70	–	–	3.12
Italy	2.46	2.29	2.22	–	2.67	2.41
Greece	1.83	–	2.19	1.99	2.65	2.16
Spain	2.07	1.90	1.93	1.66	2.37	1.99
France	1.71	1.70	1.67	1.44	2.11	1.73
UK	1.37	1.52	1.56	1.27	1.96	1.54
Korea	1.61	1.56	1.56	0.99	1.32	1.41
USA	1.54	1.37	1.14	0.83	1.20	1.22
Canada	1.26	1.15	1.06	–	–	1.16
China	0.75	0.64	0.64	0.49	0.86	0.68

Source: Data on China from 2005 to 2006 are the average prices of natural gas for industrial use in 36 major cities in the Price Yearbook of China. Data from 2007 to 2009 are provided by the Price Monitoring Centre of the National Development and Reform Commission. Data on other countries are obtained from the *Energy Prices & Taxes 2nd Quarter 2010*.
Note: All energy prices are inclusive of tax, and the unit used is USD/standard oil per ton.

to a market-oriented pricing mechanism in the natural gas sector. The natural gas prices will then not only accurately reflect the domestic supply-demand situation and a reasonable price relativity between natural gas and its alternative energies, but also effectively link with the natural gas prices on the international market to better integrate the domestic market with the international market.

Currently, China's natural gas market is still undergoing a transition stage into maturity. There are onerous tasks ahead in natural gas resources exploitation and pipeline network development. Faster growth and higher productivity are still the primary objectives of

China's natural gas industry and a fully deregulated market is still not feasible in the near future.

For some time to come, China's natural gas pricing reform should be focused on: 1. Natural gas price—China should appropriately raise natural gas prices, gradually rationalise the price relativity between natural gas and its alternative energies such as petroleum, and narrow the price difference between domestic and imported natural gas supplies. Energy conservation and emission reduction should also be promoted. 2. Pricing structure—China should determine the appropriate weightings of the ex-plant price, the pipeline transportation cost and the distribution charge in the retail price. Resource exploitation and pipeline network development should be promoted while cross-subsidies at the end-user level should be rationalised gradually and a differential pricing system implemented. 3. Pricing mechanism—China should gradually abandon the cost-plus pricing system for a more market-oriented approach with a netback market value pricing mechanism. It should establish a dynamic price adjustment mechanism that links with the prices of alternative energies and can fully reflect the supply and demand situation and the scarcity of resources in the market. These efforts are important for China as it seeks to move away from government-determined prices to a pricing model oriented mainly towards market determination and supported by government guidance. 4. Government policy—China should gradually ease control over the exploration and exploitation of natural gas resources, and develop diversified gas supply channels.

In general, China's natural gas pricing reform should be based on a realistic approach to seize opportunities and move ahead. Convergence with the pricing mechanism for alternative energies must be considered, so must the coordination with the overall progress of the natural gas market reform. It is not only necessary for China to facilitate natural gas exploration and exploitation, pipeline construction and gas imports, but also to take into account factors including social affordability and inflation. Although the international natural gas prices in recent years have been lingering at low levels due to the rapid development of unconventional natural gas in North America and the weakened global demand after the international financial crisis, China's natural gas prices will inevitably increase in the near future as a result of domestic price distortions, strong demand and the nonnegotiable long-term contracts for importing natural gas.

6.4.3 The Bargaining Power in International Oil and Gas Pricing

China is a major oil importer with continuous and rapid growth in natural gas imports. Fluctuations in international oil and natural gas prices have a significant impact on China's energy supply and socioeconomic development. Currently, China does not have the ability to influence international oil and natural gas pricing, with a lack of bargaining power in bilateral or multilateral trade. China often finds itself in a passive or submissive position in negotiations of oil and gas import prices. It needs to take measures to gradually reverse this situation.

6.4.3.1 Developing Oil Futures Markets and Increasing Bargaining Power in International Oil Market Pricing

Currently, around 13 billion tonnes of oil are traded globally, with up to 11 billion tonnes on the futures markets, which play a key role in determining international oil prices. In the international commodities markets, the oil futures prices are often used as benchmark prices. The two benchmarks that influence market oil prices more are the US West Texas Intermediate Crude Oil traded on the New York Mercantile Exchange and the UK's North Sea Brent Crude traded on the International Petroleum Exchange of London. Despite its short history, oil futures trading on the Tokyo Commodity Exchange is rapidly increasing with a growing influence in the Asian region.

Without its own oil futures market, China finds it difficult to provide feedback on its domestic oil market conditions to the international market. The successful operation of fuel oil futures on the Shanghai Commodity Futures Exchange has helped China gain valuable experience for development of a domestic oil futures market. In the future, by building on the experience in operating fuel oil futures, China should actively look into ways of promoting the development of a domestic oil futures market and establishing a futures trading market covering floor trading and over-the-counter trading for a variety of products, including crude oil, fuel oil and refined products. China should also strive to secure a greater say on international oil market trading and settlement rules, thereby boosting its influence on international oil pricing. China's huge oil production and consumption every year provides an important foundation for securing a greater say and a stronger influence.

To optimise market investor composition is an important part of building an effective oil futures market in China. In recent years, participants in the international oil futures market have not been limited to oil exploration, refining and oil trading companies. Many financial institutions have also been actively involved in the oil sector and been playing an extremely important role. China should learn from the experience of the international oil futures market, and gradually ease the restrictions on institutional investment in commodities futures so as to attract institutional investors, including investment funds, investment banks and insurance companies, thereby improving market liquidity and contributing to the scientific approach to price formation. At the same time, China should prepare domestic funds for participating in the international oil futures market with a stronger influence on international oil prices.

6.4.3.2 Expanding International Cooperation and Strengthening Bargaining Power in International Oil and Natural Gas Trade

China should expand international cooperation in the energy sector, aggressively launch 'energy diplomacy', actively participate in regional and international oil communities, maintain connections with international oil and natural gas organisations, establish communication and cooperation mechanisms with major oil and natural gas consumers and producers like the USA, Russia and China's Asian neighbours. Through the establishment of systems for bilateral, multilateral, regional or international energy cooperation, China should aim to develop an arbitration mechanism built on mutual protection and restraint and improve its bargaining power in international oil and natural gas trade.

Multilateral international energy supply systems should be established to achieve variety and diversity in supply sources and trading channels. China's oil imports come mainly from the Middle East, with little elasticity of demand and so putting itself at a disadvantage in international trade. As the proportion of imported natural gas has been increasing over the years, China must establish strategic oil partnerships with other oil and natural gas exporters by expanding oil and natural gas sources, signing long-term oil and gas supply contracts, increasing the proportion of oil imports from other countries, and enhancing its bargaining power over oil and natural gas prices. Apart from direct purchase, China can also consider loan-for-oil deals or joint development with investment to be paid back with oil. Through direct investment, equity participation and M&A, foreign oil and gas shares and mining rights can be acquired to increase the development and utilisation of overseas oil resources so as to own or access more oil and natural gas resources and strengthen the influence of the buyer's market.

6.4.3.3   Strengthening Energy Information Management and Establishing Authoritative Energy Information Release System

Energy information is an important factor that affects the operation of the international energy market. China is the largest energy producer and consumer in the world with continuous growth in energy imports. Information on energy exploration, production or consumer demand draws widespread attention, with an increasing impact on the international energy market. In order to create an external environment beneficial to China's energy security, particularly its oil industry, China should learn from the experience of other countries, with a strategic focus on the management of energy security information and the release of energy information in an organised and controlled manner. Prior to releasing information, in-depth analyses should be conducted of the impact on the international energy market with response preparedness plans in place.

6.5   Regulation of Energy Markets

A move towards energy market reform and innovation in energy market regulation are two aspects of a solution to the problems of energy development. Energy market reform can help rationalise the relationship between the government and the market with the objective requirements for improving the regulation of the energy market while innovation in market regulation provides a powerful assurance for resolving market failures and promoting healthy development of the energy market. In response to the existing problems in management of energy markets, it is necessary to adapt to the faster development of the energy industry and the continued progress towards a market-oriented system. In the future, China should be innovative in developing a regulatory system and a model of regulatory control, coupled with stronger efforts in formulating a unified planning mechanism and governing rules to strengthen regulatory control over price, safety, quality and other aspects.

6.5.1   Building a Big Energy Regulatory Framework

The so-called Big Energy Regulation means, subject to specific socioeconomic conditions, the regulatory efforts of particular government departments or professional regulatory bodies targeting the energy industry including coal, oil, natural gas and electricity, as well as the oversight and management based on relevant laws and regulations of all interested parties in the energy market. The term 'Big' here refers to the regulation of the entire energy industry, as opposed to the current regulatory system based on the types of energy. Secondly, it means that the industry-wide regulatory functions are consolidated into a certain department or a professional regulatory body assigned by the government, to avoid conflicting instructions or regulations issued by different energy regulators.

6.5.1.1   The Necessity of Building a Big Regulatory Framework

The Economic and Technological Attributes of Energy Resources Require Nationwide Overall Management
Energy production, transportation, supply and consumption are closely linked, and the coal, electricity, oil, gas, new energy and renewable energy industries are closely correlated, especially in respect of the intimate and complex relationships surrounding production, transmission and substitution of electricity. Therefore, only through overall management can one ensure reasonable convergence and integration among various sectors to enable full optimisation of the energy structure and also efficient and coordinated development. In addition, given the significant economies of scale of the energy industry, especially the economic advantages of size and network of power grids and oil/gas networks that make them natural monopolies, the government should exercise effective regulation.

The Strategic Position of Energy Resources Necessitates the State's Overall Management
Energy is a nation's infrastructure industry and economic lifeline, with an imperative strategic position. Any hitch in energy development may seriously affect China's modernisation and national security. In view of the increasingly complex national energy security situation, the evolving problems of global climate change and a new revolution in energy technology marked by the rapid growth of new energy and smart grids, the energy industry now plays a critical role in China's move to break through resource and environmental constraints, achieve a peaceful emergence and secure a commanding position in the global energy economy. Only through overall management at the state level can China achieve better planning and resource allocation to concentrate efforts, highlight priorities and accelerate development.

Multiple and Decentralised Regulatory System Cannot Meet the Needs of Energy Development
After several rounds of reform, the present regulatory system in China remains one of multiple and decentralised management. Due to the decentralised regulatory functions, the policies introduced by various government departments often lack coordination, which compromises administrative efficiency and leaves energy companies at a loss as to what to do. There is also a lack of coordination between different industries and between segments of an industry without an in-depth study of major issues of energy strategy. In the future, China needs to show determination in developing a big regulatory framework to support a strategic transition in, and the continued healthy growth of, the energy industry.

Table 6.6 Classification of international energy administration systems.

Types	Features	Administrative bodies	Representative countries
1	Administration Model with High Level of Centralisation	The National Energy Department or the Department of Fuels And Energy	US (Department of Energy), Russia (Ministry of Industry and Energy), Venezuela (Ministry of Energy and Mining), Mexico (Ministry of Energy), Thailand (Ministry of Energy), Saudi Arabia (Supreme Petroleum Council, under which are Ministry of Petroleum and Mineral Resources, Ministry of Industry and Electricity)
2	Administration Model with High Level of Decentralisation	The National Department for Coal, the National Department for Petroleum	India (Ministry of Petroleum & Natural Gas, Ministry of Coal and Ministry of Power)
3	Administration Model with Lower Level of Centralisation	Bureau of Energy and other departments under the Ministry of National Economy	Japan (Agency for Natural Resources and Energy, Ministry of Economy, Trade and Industry), Germany (Ninth General Bureau, Federal Ministry for Economics and Labour), UK (Bureau of Energy under the Department of Trade and Industry)

Source: Data compiled based on the *Research Report on the Chinese Government's Energy Administration Systems from the Perspective of Energy Security*, published by Shi Hong-yan

6.5.1.2   The Thinking Behind Development of a Big Energy Regulatory Framework

Learning from Experience in Energy Regulatory Systems Abroad

Despite the difference in energy administration systems (see Table 6.6), most countries, especially the majority of developed countries, have adopted a big regulatory system, with energy as the target of overall regulation. This provides a useful reference for China's efforts in readjusting its energy administration system and developing a big energy framework.

Establishing a Centralised and Authoritative Energy Administrative Body

In order to adjust its energy administration system and develop a big energy regulatory framework, China needs to fully rationalise the government's present energy administrative functions and, on this basis, consolidate the energy administrative functions of various departments into a single administrative body as far as possible. For those administrative functions that cannot be centralised, the authorisation limits and the scope of responsibilities have to be clearly defined, with the establishment of a communication and coordination mechanism. At the same time, China needs to step up efforts in developing its energy administrative body and optimising the balance of powers between the energy administration and other government departments and between the central and local energy administrations. China should explore and establish a comprehensive energy regulatory system covering coal, electricity, oil, natural gas and other energy resources. When defining the administrative functions, the relationship between the government and the energy market has to be clarified, and the functional roles of the government and the market in the development of energy resources well-positioned. The government should play a leading role in reforming the energy system, refining the relevant policies, developing and overseeing the implementation of applicable regulations, simplifying the administrative approval process, and minimising direct intervention in the operation of the energy markets in order to fully leverage the fundamental role of the market in allocating resources. At the same time, China should pay attention to strengthening the building of capacity for energy regulation and maintaining control over the energy market so as to effectively correct any market failures.

Refining the Legal System Governing Energy Regulation

Market regulation in accordance with the law is the fundamental principle in countries around the world for energy market development. Compared with the Western developed countries with a mature market economy, China's laws and regulations for energy development are lagging behind. A lack of sophisticated energy market rules, together with poorly-coordinated efforts in dealing with new situations and problems arising in the course of energy and market development, has created a situation where there is no legal basis to rely on, with inconsistencies and conflicts in established laws and regulations. The result is casualness in regulatory work with a lack of coordination in market development for the coal, power and other energy industries. There have also been instances of under- or over-regulation. In the future, China should accelerate the legislative process for the energy sector and coordinate the development of the energy market and energy legislation while establishing the terms of reference for energy regulators based on the law and building a legal basis for the discharge of regulatory duties to highlight the seriousness of regulatory work and the authoritativeness of the energy regulators.

6.5.2 The Thinking Behind Energy Market Regulation

6.5.2.1 Coordinated Development of Energy with Economy, Society and the Environment Needs to be Considered for Achieving Regulatory Goals

Safeguarding Sustainable Energy Supply as a Fundamental Target

The energy regulators should heed and scientifically analyse and assess changes in domestic energy supply and demand and energy investments. Energy regulatory policy should be adjusted in a timely manner, including import and export policies for energy resources and related products, with the focus, distribution and structure of energy resource development optimised. All this is designed to promote efficient development and utilisation of domestic energy resources, encourage dynamic integration of energy production, supply and sales and the development of an energy reserve system, while identifying more diversified energy sources for import to ensure the smooth operation of China's energy markets with supply security.

Promoting Development of Clean Low-carbon Energy as an Important
Regulatory Target

China should adapt to the transformation of energy structures and the trend towards low-carbon energy. Policy support for investment, market entry, pricing and taxation should be implemented to encourage development of new energy and clean fuels. Guidance and regulation should be provided in a timely manner on problems encountered throughout the development of new energy and clean fuels. Meanwhile, the requirements of energy efficiency and emission reduction should be fully met throughout the life-cycle of energy exploration, production, processing, transportation and consumption.

Actively Monitoring the Performance of Corporate Social Responsibility

Apart from optimised investment and efficiency, the corporate performance of social responsibility covering employee health, labour relations, production safety, environmental impact, energy efficiency and services to the public should also be monitored to achieve social harmony, energy conservation and environmental protection.

6.5.2.2   Overseeing the Industry Chain

Strengthening Demand Side Regulation

End users are the most important participant in the energy market with an important influence on the operation of the market. Without cooperation and support at the end-user level, the energy market cannot grow healthily. With the development of smart grids and automation of energy systems, stronger efforts to guide end-user behaviour and fully leverage demand side resources carry special significance for improving energy utilisation efficiency, energy conservation and emission reduction. To strengthen demand side regulation, it is necessary to influence user behaviour to maintain market order and the safety of energy systems. On the other hand, it is necessary to establish a demand response mechanism to maximise the development and utilisation of demand side resources to promote load shifting and the efficient and clean utilisation of energy, in addition to improving the operational standards of energy systems. At the same time, it is necessary to introduce a variety of incentive policies to guide energy demand at the end-user level and promote the utilisation of electric power in the field of transportation to reduce the dependence on oil.

Improving Supply Side Regulation

Based on the law and market rules, active regulation of energy exploitation, production, conversion, transportation, distribution and sales should be enhanced to maintain market order, ensure a fair market and facilitate efficient market operations. The regulation of energy prices should be strengthened, the reform of energy pricing systems promoted, the price relativity among various types of energy rationalised and price violations penalised. Service quality should be monitored, the energy service standards and the quality standards of coal and refined products strictly enforced, and inappropriate competition and behaviour harmful to public interests penalised. Monitoring of production safety should be strengthened. From an overall and strategic perspective, monitoring of nuclear safety should be enhanced and a legal system for managing nuclear safety equipment be set up as soon as possible to ensure safe utilisation of nuclear energy. A responsibility system for safe production should be enforced to ensure safety of coal mines and raise safety standards by eliminating small coal mines not meeting industry policy or safe production requirements or having an impact on the environment. Stringent safety supervision should be performed on oil and natural gas exploitation and transportation to mitigate the risks of major accidents of production safety and ecological damage. A long-term mechanism should also be established to identify, eliminate and manage hidden safety hazards in power operations. Emphasis should be placed on unifying dispatch operations and discharging safety responsibility for power generation, supply and consumption. Stringent standards should be imposed on grid connections while the generation and distribution sectors should be better coordinated to ensure normal operation of power plants and the safety and stability of power grids.

6.5.2.3 Equal Emphasis on Ex-ante Supervision and Ex-post Supervision

Establishing a Unified Planning Mechanism for Energy Development
Based on the characteristics of and correlation between different energy industries, a government-led and participative system for centralised planning should be created to coordinate the development of coal, electricity, oil, natural gas, nuclear energy, new energy and renewables. This will lead to the formation of an energy planning system featuring integrated and ad hoc planning to converge with planning for socioeconomic development, transportation and technology development. The system will promote coordination between energy and socioeconomic development, between primary energy and secondary energy, between energy production, transportation and utilisation, and between urban and rural energy development so as to lay a foundation for healthy and orderly development of the entire energy industry.

Refining Operating Rules and Technical Management Standards for Energy Market
China should refine the rules for gaining access to the energy market, formalise the eligibility requirements for participation in the energy market, and streamline the project approval process. Based on the principles of transparency, fairness and justice, China should improve the operating rules and technical management standards for the energy market, including rules and standards for investment, network access and transactions. The responsibilities and obligations of different market players should be specified, the relevance and operability of the rules and standards strengthened to promote fair trade, orderly competition, safe and efficient operation of the energy market, and to ensure the effectiveness of market regulation and healthy development of the energy industry.

6.5.3 Building Support System for Energy Market

6.5.3.1 Enhancing the Information Disclosure System for Energy Market

China should enhance the openness of government information on energy, improve the notification system for energy prices, and increase the transparency of the government's decision-making on nonconfidential energy issues. An information disclosure system combining voluntary and mandatory disclosure should be established to guide energy companies on disclosure of information on energy products and services to customers and the community in a timely manner, while regularly publishing CSR reports. An energy market information platform should also be established to release real-time data on the operation and trading activities in the energy market.

6.5.3.2   Refining the Energy Market Credit System

Based on the standards for evaluating the behaviour and creditworthiness of market players, a dynamic energy industry information system should be established to reflect in real time the credit position of energy companies and major users in the energy market. In addition, credit rating information should be used more effectively to link credit ratings with financing costs and customer relationships, so as to raise awareness of self-discipline among the energy market players and maintain market order.

6.5.3.3   Improving Early Warning and Emergency Response Mechanisms for Energy Market

China should step up monitoring and analyses of energy market operations and issue early warning against potential major change in the market situation so as to remind market players to stay away from risks in time and stabilise market order. At the same time, China should strengthen the development of an emergency response system/mechanism to timely neutralise or mitigate the impact of any contingency on the energy market and minimise any loss.

7

Energy Early Warning and Emergency Response

Moving into the 21st century, the energy issue in China has become increasingly complex and uncertain, and is exposed to a number of energy security risks. To pre-empt energy security risks, promptly eliminate the potential hazards to energy security, enhance the capacity to cope with emergencies and ensure a safe and reliable supply of energy, China must vigorously strengthen the building of capacity for early warning and emergency response.

7.1 Importance of Building Capacity for Energy Early Warning and Emergency Response

After the two oil crises in the 1970s, the world's major developed countries have placed great emphasis on energy early warning and emergency response management by incorporating them into their national energy strategic policies as an essential means to secure energy supply and maintain national security. Energy security in China is exposed to a number of risk factors due to the impact of the international situation as well as the internal and external environment. To learn from international experiences and strengthen the building of capacity for energy early warning and emergency response is of great significance to ensuring the security of energy supply and achieving sustainable socioeconomic development of China.

7.1.1 Risks Posed to Energy Security

The energy system in China is enormous and complex, covering a wide range of areas and subject to many risk factors that affect energy security. In terms of the sources of risk, the risk factors are primarily in the following three aspects.

Electric Power and Energy in China, First Edition. Zhenya Liu.

7.1.1.1   Energy Security Risks Triggered by Complex International Energy Political and Economic Situations

Energy development in China in the next 20 years is primarily characterised by increasing energy import volume and growing reliance on overseas supply. Understanding this trend is of positive significance to the coordination of the use of two resources in both the international and domestic markets for creating a situation to complement energy supply internally and externally. Increased reliance on imported energy means the international energy political and economic situations have an increased impact on energy security in China, causing potential risks that should not be ignored.

Firstly, the risk triggered by unrests in energy-exporting countries and regions. In the context of globalisation, the international energy issue has become increasingly prominent due to geopolitical factors. Major powers are competing with one another in some energy-rich countries and regions. The situation is complex and confused. Energy supply in China will be hit once the situation becomes unstable in some of the major energy-exporting countries and regions rife with prominent political and social problems. Since the 1960s, there have been more than 10 oil and gas supply disruptions in the world, most of which were caused by wars or conflicts as a result of geopolitical reasons, such as the Fourth Middle East War, the Iran–Iraq War, the Gulf War, the war in Iraq, etc.

Secondly, the risk possibly triggered by the over-concentration of energy import transport channels. Oil and gas import channels are highly concentrated in China. Taking oil as an example, most of the crude oil imports to China need to be shipped through the routes of the Indian Ocean and Malacca Straits. This relatively singular energy import transport pattern is very risky. Furthermore, assuming an important strategic position in the global competitive landscape, these areas have always been the focus of the struggle between great powers, where international relations are complex. Piracy is also very rampant in these waters. Transport channel is also an important risk factor that jeopardises China's energy security, in particular oil security.

Thirdly, the risk triggered by financial speculation in energy. There is an increasing energy financialisation trend following penetration and integration between the international energy market and the international financial market. Constant involvement of international speculative capital in the international energy market, especially in the international oil market, has been a major driving force behind the wild fluctuations in international oil prices in recent years. According to the statistical report of the Commodity Futures Trading Commission ('CFTC'), speculators held long positions much more than those held by ordinary traders during the oil price hikes in 2008. When the price reached a high level, speculators turned to short positions so that the substantial change in the positions led to dramatically increased market price fluctuations. Moreover, exchange rate fluctuations in the US dollar, a major currency of price in the international oil market, can also cause fluctuations in international oil prices. These have a major impact on energy security in China, thus increasing the risks and the costs of ensuring domestic energy supply.

7.1.1.2   Energy Security Risks Triggered by Unscientific Energy Development Methods

The rapid development of the energy industry in China in recent years has satisfied socioeconomic needs fairly well. However, problems arising from the unscientific and irrational approach to development have become increasingly apparent and contributed to increasingly acute security risk factors.

Firstly, the risk triggered by irrational allocation of energy. A robust energy transport system and scientific energy allocation determined by the reverse distribution of energy resources and energy needs in China are crucial to ensuring energy security in China. At present, China is lagging behind in the establishment of an energy transport system, and its energy allocation method is one of over-relying on coal handling. Thermal coal transport goes through a number of processes from the coal-producing areas in Western and Northern China and eventually reaches the coal-fired power plants in the coastal areas of Eastern China. This has both pushed up the retail prices and increased the risks of transport as well as the uncertainties over energy and power supply in Eastern China. The important reason why there have been recurring tensions between coal, electricity and transport in recent years in China is the irrational energy allocation pattern which is one of over-relying on coal handling.

Secondly, the risk triggered by the irrational structure of the energy system. A rational structure of the energy system is the foundation to ensure energy security, while an irrational one often contains potential risk factors that may affect energy security. Taking electric power as an example, the long-term emphasis on power generation, rather than power supply, has resulted in an imbalance in power sources and power grid development, a serious outstanding debt incurred by power grid development, an irrational network structure, a weak cross-regional backbone network and a weak capacity to withstand severe natural disasters and risks of accident. Given the rapid growth in demand, the operation of a large number of new generating units and the increasingly complex external environment, there is a real risk of widespread blackouts across power grids.

Thirdly, the risk triggered by the lack of standardised guidance for the development of new energy. To cope with the tremendous pressure from depleting fossil energy and a deteriorating environment, proactive development and utilisation of new energy sources has become an imperative. However, the initial development and utilisation of any new energy must be guided and regulated as necessary, otherwise there will be potential security risks. In particular, since wind power and solar power generation are intermittent, their access to power grids on a large scale is an important test to system security. In recent years, the development of new energy such as wind and solar power generation is rapid in China. But this development has also revealed some problems such as the setting of technical standards falling behind the development of the industry. In an off-grid incident on 24 February 2011 in Jiuquan, Gansu Province, 274 wind turbines in 10 wind farms went off-grid due to a lack of Low Voltage Ride Through capacity, thereby causing a series of reactions. Lost output amounted to 54.4% of the wind power output in Jiuquan prior to the accident, causing the main network frequency of the power grid in the northwest to reduce from 50.034 Hz prior to the accident to a low of 49.854 Hz.

Fourthly, the risk triggered by the weak foundation of energy security management. Various energy production safety-related accidents have occurred frequently in recent years in China. During the period of the 11th Five-Year Plan, there were a total of over 100 serious coal mine accidents, and over 22 major accidents each involving a death toll of more than 30. These accidents happened partly because of the poor coal mining geological conditions and lagging technical standards. But more importantly, it was because of safety supervision and management. Accidents like oil and gas leaks and explosions have shown an increasing trend due to ineffective production safety management. With urbanisation and development of the automobile industry, more petrol stations, gas-refilling stations and battery-charging stations will be located across urban and rural areas. As a result, security risk management will cover more locations and aspects, with greater difficulty. Nuclear accidents are very destructive to both personal and environmental safety. Since the inherent safety of the technology in the development and utilisation of nuclear energy has yet to be achieved, there should be higher requirements and more stringent standards for security risk management at nuclear power plants.

7.1.1.3 Energy Security Risks Triggered by Natural Disasters and Damage by External Forces

In recent years, external factors such as natural disasters and damage by external forces have an increasingly marked impact on energy development in China. In particular, some unexpected factors have caused significant risks to the safe supply of energy.

Firstly, the risk triggered by serious meteorological disasters. As a result of global warming, extreme climatic events such as high temperatures, low humidity, droughts, storms, floods, snowstorms and hurricanes have become more frequent in recent years. Extreme climatic events tend to be more frequent in China as well, and their extent of damage is increasing, posing a great challenge to the security of energy supply. The severe snowstorms in Southern China in 2008 hit transportation, electricity and communication so heavily that a large number of power facilities were damaged, and energy and electricity supplies in more than ten provinces and cities were affected seriously.

Secondly, the risk triggered by serious geological disasters. Geological disasters tend to be more frequent in recent years. Since 2008, geological disasters such as the earthquakes measuring 8.0 Richter scale in Wenchuan, Sichuan (12 May 2008), the earthquake measuring 7.1 Richter scale in Yushu, Qinghai Province (14 April 2010) and the mudslides in Zhouqu, Gansu Province (7 August 2010) have caused the loss of many lives and property of residents, and severely undermined energy generation and supply. The impact of secondary disasters caused by geological disasters on energy security should not be ignored as well. Nuclear leaks at the Fukushima nuclear power plant in Japan (March 2011) were caused by a tsunami created by an earthquake. Lessons from accidents abroad have sounded an alarm bell to us.

Thirdly, the risk triggered by damaging effects of various external forces. The extensive distribution of energy facilities throughout urban and rural areas has increased the difficulty in security protection. Various acts such as external illegal operations, enforced construction, theft and damage to energy facilities, whether knowingly or unwittingly, may cause harm to energy security. The number of energy security accidents caused by the damaging effects of external forces in China has remained high in recent years. This risk should not be ignored.

7.1.2   Significance of Strengthening the Building of Energy Early Warning and Emergency Response

To resolve the risks posed to energy security in China, the fundamental approach is to rely on scientific development, with equal emphasis placed on strengthening the building of capacity for energy early warning and emergency response. Energy early warning means to conduct real-time monitoring and analysis of potential risk factors that jeopardise energy security, analyse and judge the alerts promptly and give a forecast of the alerts when the system is running in an abnormal state by issuing an early warning signal to take response measures to eliminate and resolve security risks. Energy emergency response means to maintain basic energy supply and normal spending activities to ensure smooth economic performance by taking various initiatives to cope with emergencies proactively during the normal process of production and livelihood, such as acute shortage of energy supply, supply disruptions and wild fluctuations in energy prices. To strengthen the building of capacity for energy early warning and emergency response is of great significance.

7.1.2.1   Improving Initiative to Deal with Energy Security Risks

To strengthen the establishment of an energy early warning system and mechanism, enhance the capacity for energy early warning and improve the level of energy early warning is conducive to identifying any signs of risk factors or accidents in advance that may jeopardise energy security. To avoid devastating consequences, an analysis should be conducted to assess the scope of the possible impact and the degree of harm, and then measures should be taken promptly to nip a crisis in the bud. The most favourable outcome may be secured for some risk and accident factors, if not all resolved and eliminated, by coping with them proactively and effectively.

7.1.2.2   Enhancing Controllability of the Impact of Emergencies on Energy Security

The ability to promptly respond to emergencies to minimise impact and loss is a major component in measuring the level of energy security in China. The serious natural disasters in recent years have deepened our understanding of the importance of capacity building for energy emergency response. To establish a sound energy emergency response system, strengthen the setup of energy early warning and various emergency response systems and improve the scientific process and effectiveness in dealing with energy emergency response for ensuring the order in energy generation and supply is conducive to keeping the impact of emergencies and the loss to a minimum. Relevant statistical analysis indicates that the effective operation of an emergency response system can reduce the incident rate to 6% of the level recorded when an emergency response system is absent.

7.1.2.3   Maintaining Social Harmony and Stability

Energy security events are closely related to the general public and society. When an energy emergency is handled improperly, it is easy to trigger a chain reaction that affects every aspect of society, economy and livelihood, and disrupts social and public order. The strengthening of capacity building for energy early warning and emergency response is not only related to energy supply, but also to the safety of the lives and property of residents, the order in production and livelihood, the natural and ecological environment as well as the overall creation of a harmonious society.

From an international point of view, the USA, European Union and Japan have all placed great emphasis on issues of energy early warning and emergency response. The USA has established a mechanism to cope with energy crisis within the framework of national crisis management and energy security strategy by organically combining energy crisis response with overall national crisis management and energy security strategic management to carry out the planning and coordination of the three at the macro and micro levels. The USA has also formulated an array of initiatives for energy-related events through legislation and other forms to improve the early warning preparedness and emergency response procedures. For example, the USA has developed a set of emergency response procedures specifically for the safety of nuclear power plants in accordance with the requirements of the International Atomic Energy Agency. To cope with unexpected problems regarding electricity supply, the rules regarding electricity safety and reliability were expressly introduced under the US Energy Policy Act of 2005. The European Union has established a sound regulatory regime governing energy emergency response and set sustainability strategic goals of assuring the security of energy supply, improving the competitiveness of the energy industry and maintaining energy supplies. Japan has treated energy early warning and emergency response management from a strategic height by vigorously strengthening the building of capacity for energy emergency response since the first oil crisis, thus gradually establishing a comprehensive energy emergency response management policy that comprises a petroleum reserve policy.

7.2 Energy Early Warning Mechanism

China places increasing emphasis on energy early warning. One of the main duties of the Office of the National Energy Leading Group formed in May 2005 is to track and understand the status of energy security as well as forecast and issue an early warning about macro and major energy problems. In January 2008, the State Council published the *Opinions on Strengthening Energy Forecasting and Early Warning* with a clearly-worded proposal for the development of a sound energy statistical system, steadily pushed forward the establishment of an energy forecast and early warning information system, and focused efforts on raising the capacity and level of energy forecast and early warning. One of the major duties of the National Energy Administration formed in August 2008 is to carry out energy forecasting and early warning. In accordance with the requirements set out in the State Council's documents, governments at all levels have been actively exploring ways in strengthening energy forecasting and early warning. Domestic research institutions, industry associations and energy companies have strengthened energy forecast and early warning research at different levels. Generally speaking, China has made a good start on energy early warning. However, given that this work was started not long ago, many aspects need to be strengthened and improved upon continuously before a scientific and sound energy early warning mechanism can be established.

7.2.1 Focus of Energy Early Warning

The energy early warning procedures generally involve analysing the possible occurrence of any abnormal state (i.e. a specific alert) in energy security and energy operation, identifying the root causes that lead to such abnormal states (i.e. identifying the source of the alert), then monitoring and analysing the risk signs (i.e. analysing the signs of the alert) and finally issuing a warning promptly according to the severity of the alert (i.e. forecasting the degree of the alert).

Energy early warning is generally divided, by time, into short-term early warning and long-term early warning. Short-term warning is primarily to issue an early warning about the risk factors that may have a significant impact on energy security for the current or recent period, generally within five years. Long-term early warning is to issue an early warning about the risk factors that may have a significant impact on energy security for a long period of time in the future, generally within 10 years, 20 years or beyond.

7.2.1.1 Short-term Energy Early Warning

Short-term energy warning primarily means issuing a risk alert on any abnormal states such as potential energy supply disruptions during energy operation, acute imbalance in energy supply and demand, wild fluctuations in energy prices and disastrous energy-related accidents. It should be noted that there is an inherent correlation between the above-described four states, in which a risk factor (e.g. a severe natural disaster) is likely to lead to the occurrence of several or all of the four states at the same time.

Strengthening Risk Early Warning of Energy Supply Disruptions

From exploitation to end-users, energy has to go through a number of processes such as processing and conversion, transport and distribution as well as sales and supply. The occurrence of problems in any of these processes may cause energy supply disruptions. In particular, the weak links, important parts and key segments of energy generation are the potential sources of risk associated with energy supply disruptions.

To properly handle risk early warning about energy supply disruptions, there is a need to focus on the tracking, research, monitoring and analysis in the following aspects: (1) the political, economic and social situations, the strategies of great powers and the geopolitical trends in energy-exporting countries and regions; (2) the security status of main energy import channels (including sea and land channels) and the operating conditions of energy import carriers; (3) the security status of the key energy generation and processing bases in China; and (4) the security status of major energy transport channels (including energy transport routes, backbone power grids and backbone oil and gas pipelines, etc.) and storage hubs for energy transhipment in China.

Strengthening Risk Early Warning of Acute Imbalance between Supply and Demand

The factors that lead to an acute imbalance between supply and demand may be caused by the energy supply side or the energy demand side, or both the supply and the demand sides. In principle, the factors that lead to energy supply disruptions could cause an acute imbalance between supply and demand. Moreover, policy adjustments, changes in the natural environment and so forth could lead to an acute imbalance between energy supply and demand in the short term. In particular, since China has not yet developed a scientific and efficient energy operating mechanism that organically integrates market adjustments with macro controls, the risks associated with an acute imbalance between energy supply and demand due to inappropriate policies and price distortions should not be ignored. Changes in the natural environment will affect energy supply as well as energy demand. For example, the water level during flood or dry seasons will affect hydropower output, and the cold or hot climate will affect heating and air conditioning load. Air conditioning load during summer currently accounts for 1/3 of the urban maximum power load. The continuously intense heat of summer will cause the demand for electricity to rise sharply in summer, exerting tremendous pressure on electricity supply.

Based on the analysis of the causes of risk leading to an acute imbalance between energy supply and demand, to properly handle risk early warning about an acute imbalance between energy supply and demand, besides paying close attention to the risk factors that lead to energy supply disruptions, there is a need to focus on the tracking, research, monitoring and analysis in the following aspects: (1) the national macro-control policies, especially the policies related to the restructuring of energy-consuming industries (including industrial policy, fiscal policy, investment policy, import and export policy, etc.); (2) the national and local energy policies (including the energy-control policies of resourceful provinces, such as the policies of major coal-producing provinces to limit the export of coal to other provinces, etc.); (3) the changes in energy investment (including the changes in the total investment as well as the changes in the investment portfolio); (4) the utilisation of energy equipment (the number of annual utilisation hours of power generation equipment, etc.); and (5) the changes in the natural environment (mainly the weather factor).

Changes in energy investment and utilisation of energy equipment are important indicators for the analysis and forecast of energy supply and demand. Increasing energy investment indicates rapid growth of energy supply capacity in the future while a reduced proportion of the transport segment in the energy investment portfolio reflects the future possibility of energy supply being constrained by transport bottlenecks. Structural differences among the regions where energy investment is made mean different conflicts between energy supply and demand among the regions. A sharp rise in the utilisation rates of energy equipment indicate growing tension between energy supply and demand, and a continued and significant decline in the utilisation rates of energy equipment forebode an oversupply on the horizon.

Strengthening Risk Early Warning of Substantial Fluctuations in Energy Prices

For general commodities, prices are largely determined by supply and demand. Changes in energy prices are subject to supply and demand on the market as well as various nonmarket factors. For China, the domestic electricity and natural gas prices are subject to stringent government control. The risk of substantial price fluctuations is primarily due to the changes in the pricing mechanism, i.e. the marketisation of energy prices. The risk of energy price volatility in the near term is mainly due to coal imports and various imported energy resources (mainly crude oil, natural gas, etc.) Changes in the policy of the Organisation of the Petroleum Exporting Countries ('OPEC'), geopolitical trends in major energy-generating areas, energy policy adjustments by major energy-importing countries, financial speculation in the international energy market, changes in the exchange rates of the world's major currencies, local wars and other factors all have a significant impact on international energy prices, and are a major factor causing the risk of price fluctuations in China's imported energy.

Based on the analysis of the causes of risk leading to substantial fluctuations in energy prices, to properly handle risk early warning about substantial fluctuations in energy prices, besides paying close attention to the risk factors that lead to an acute imbalance between supply and demand in China, there is a need to focus on the tracking, research, monitoring and analysis in the following aspects: (1) the reform initiatives for the marketisation of domestic energy prices; (2) the changes in OPEC's policy as well as the political and economic situations and the energy export policies of major energy exporting countries; (3) the geopolitical movements (especially the political and economic situations of the major countries of origin for China's overseas energy) and the local war risks; (4) the movements of speculative capital in the international energy market and the trends in the exchange rates of the world's major currencies (particularly the US dollar); and (5) the energy policies of major energy importing countries (traditionally the great powers with energy demand, such as the USA, Japan, Germany and France as well as emerging economies like India).

Strengthening Risk Early Warning of Disastrous Energy Incidents

Disastrous energy incidents such as incidents at coal mines, nuclear leaks, dam breaks at hydropower plants and power grid failures as well as incidents involving damage to oil and gas pipeline networks and major oil refining facilities will cause great hazards to the security of energy supply, and have an enormous impact on the socioeconomic development as well as the production and livelihood of residents. They are a major factor that triggers an energy crisis and even a social crisis. The risks associated with disastrous energy incidents mainly originate from meteorological disasters, geological disasters, damaging effects of external forces, illegal production, etc.

Based on the analysis of the causes of risk leading to disastrous accidents, to properly handle risk early warning about disastrous incidents, there is a need to focus on the tracking, research, monitoring and analysis in the following aspects: (1) the changes in meteorological conditions (including the date, place and probability of the occurrence of various disastrous weather conditions and their degree of hazards, such as high temperature and droughts, rainstorms, hurricanes, thunder strikes, low temperature and frosts); (2) the changes in geological conditions (including the date, place and probability of the occurrence and the hazard level of various geological disasters such as earthquakes and mudslides; (3) the factors of damage by external forces (including unruly work behaviour and sabotage); and (4) the weaknesses of production safety management (including the potential hazards of major and extraordinary serious incidents, and symptomatic factors).

In summary, the proper handling of risk early warning about disastrous energy incidents requires efforts in two areas. Firstly, the communication of information must be strengthened between the energy department and the meteorology and geology departments to carry out coordination and interconnection between risk early warnings of natural and geological disasters and risk early warnings of energy security. Once the obvious signs of a serious natural or geological disaster are identified, a warning must be issued to the energy department promptly so that it can make preparations to cope with it. Secondly, the early detection and prevention of any symptoms of damage by external forces or a major accident must be strengthened. There have been obvious early warning signs in various production safety-related major incidents in China in recent years.

7.2.1.2 Long-term Energy Early Warning

Long-term energy early warning primarily involves the forecast and assessment of the risks, at the strategic level, associated with unsustainable energy development, and the issue of an alert about such risks. Its main focus covers the support capacity for domestic resources, environmental carrying capacity, energy system security, the capacity for access to foreign resources, etc.

Support Capacity for Domestic Resources
The prospective energy resource reserves, geological reserves, exploitable reserves and the reserve-production ratio are major indicators that reflect the support capacity for energy resources. Of these indicators, the first three are absolute quantity indicators while the last one is a relative quantity indicator. In terms of absolute quantity indicators, coal, petroleum, natural gas and other conventional fossil energy reserves in China are all among the top in the world. But in terms of relative quantity indicators, the support capacity for domestic resources is not promising because China exploits and consumes a large amount of energy annually. According to the calculations based on BP statistics, China's coal reserve-production ratio was 35 years, petroleum reserve-production ratio was 10 years and natural gas reserve-production ratio was 29 years in 2010, accounting for only 29.7%, 21.4% and 49.5% of the world average respectively. With increased efforts on resources exploration, coal, petroleum and natural gas reserves are expected to increase further. In particular, the space for increasing natural gas reserves is relatively large, but because of the continuous growth in the demand for energy consumption, the possibility of substantially increasing the reserve-production ratios of various conventional energy reserves including natural gas reserves is minimal. The lack of support capacity for conventional energy resources presents a great challenge to future energy development in China.

In addition to the aggregate indicators, structural indicators must also be taken into account when measuring support capacity for domestic resources, such as the proportion of oil and gas and other quality energy to energy reserves. China's coal reserves are relatively abundant while quality energy reserves are obviously insufficient. This structural problem is also a major factor that undermines support capacity for resources.

Environmental Carrying Capacity
Since mankind began using fossil energy on a large scale, the environmental development issue has become associated with energy development. On one hand, the environmental burden has become heavier due to the excessive use of energy resources. On the other hand, a large amount of pollutant emissions has caused a substantial negative impact on the environmental quality, resulting in increasingly prominent problems—a deteriorating ecological environment and a weakened environmental carrying capacity due to energy development and consumption. For example, the damage by the exploitation of energy resources to the ecological environment in mining areas, the massive emissions of greenhouse gases and other waste gases after energy combustion, the damage by oil

spills to the marine environment, the substantial impact of accidents at nuclear power plants on the environment, etc.

Moreover, the carrying capacity of the environment for energy development in China must also be considered in the context of combating global climate change. Control of greenhouse gas emissions and mitigation of global warming are major issues facing all countries around the world, and have also become major constraints on energy development in China.

To achieve sustainable energy development, great emphasis must be placed on the monitoring, analysis and evaluation of environmental carrying capacity, and on the timely issue of risk early warning. The main scope of monitoring covers the impact on and damage to the ecological environment of the development and utilisation of different energy resources; the overall situation of the nationwide ecological environment; the environmental capacity and ecological restoration capacity of different regions, especially the ecological conditions in major energy exporting regions and major energy consuming regions. The differences of economic development in different regions must be considered, a comprehensive assessment of energy development and utilisation as well as environmental factors must be conducted, risks promptly identified, regulatory control, rational distribution and moderate development strengthened, and the environmental costs of energy development reduced.

Energy System Security

Energy system security should be examined and risk early warning issued from a long-term perspective, focusing on the monitoring, analysis and evaluation of major structural risks accumulated during the long-term development process of the energy system. These major structural risks may originate from: (1) the irrational planning for and uncoordinated development of energy generation and transport (such as the disproportion of coal transport and power transmission); (2) the irrational physical structure of the energy transmission network; (3) the excessively high proportion of imported energy to total energy consumption (i.e. reliance on energy resources from abroad); and (4) the excessively low proportion of renewable energy to the primary energy structure.

To achieve sustainable energy development, emphasis must be placed on the monitoring, analysis and evaluation of energy system security, with a focus on tracking and analysing various structural risk factors that impact energy system security and forecasting an alert promptly in order to cope with the risk factors in advance and enhance the capacity of the energy system to prevent major security risks.

Capacity for Access to Foreign Resources

A country's capacity for access to foreign energy resources is a comprehensive reflection of the strength of that country. In the context of globalisation, higher capacity for access to foreign energy resources often means a higher level of assurance of national energy security. With China's increasing reliance on energy from abroad, the scientific analysis and assessment of China's capacity for access to foreign energy resources, and the timely issue of early warning while such capacity is decreasing have significant implications for ensuring China's energy security and sustainable development.

To assess China's capacity for access to foreign energy resources, consideration should be given to the following aspects: (1) the cooperative relationships with resource-exporting countries, including the cooperative relationships with the private sector; (2) the influence on the countries through which imported energy is transported and the military support capacity for major energy import channels; (3) the collaborative relationships with major energy-importing countries; (4) the influence among international energy agencies; (5) the global competitiveness of China's energy companies; and (6) the foreign exchange reserves, etc.

7.2.2 Organisational Structure and Management System of Energy Early Warning

7.2.2.1 A Sound Organisational Structure of Energy Early Warning

Unified organisation and management must be carried out at the national level for energy early warning due to the many factors involved and its wide-ranging impact and great significance. To this end, upon completion of the smoothening of the existing energy early warning system, the establishment of a sound, structured, unified and efficient energy warning organisational system with clearly defined duties must be accelerated. The organisational system may be led by the national energy competent department, with involvement of relevant Chinese ministries and commissions, local governments, energy industry associations, research institutions and large energy companies. It should be accountable to the National Energy Commission and commence work under the leadership of the Office of the National Energy Commission. The system should be able to cover all aspects of energy early warning, and the participants should have clearly defined responsibilities and tasks to ensure the whole organisational system operates in a coordinated and efficient manner.

The provincial governments should also establish a sound, corresponding energy early warning organisational system to organise and manage energy early warning within their provinces (autonomous regions and municipalities directly under the central government). The provincial early warning organisational system is an extension and an integral part of the national energy early warning organisational system, and carries out its work under the unified guidance of China.

7.2.2.2 Establishing a System for Analysing and Judging Energy Situations

The energy issue is a very complex issue and involves various long-term and short-term goals and balance of interests, while the closely related international energy competition and the climate change issue also involve strategic games between great powers. Therefore, to judge the situation of energy development in a scientific manner and enhance the scientific level and accuracy of energy early warning and decision-making, a high-level system for analysing and judging energy situations must be established by regularly or irregularly deploying relevant leaders and senior experts to conduct an in-depth analysis of the energy development trends as well as some hot, sensitive and deep-rooted issues, and a comprehensive assessment of national energy security, forecast development movements and formulate forward-looking response initiatives.

The system for analysing and judging energy situations can be established in two ways. The first way is to set up a ministerial joint committee on cross-department national energy security early warning to attract energy industry associations and large energy companies to the committee, and promptly discuss, analyse and coordinate major energy security and early warning issues. Another way is to set up a national energy security policy committee as a consultative and deliberative body of the Office of the National Energy Commission or a leading national energy department to collect opinions for the accurate analysis and judgement of the energy security situations and the prevention of major energy risks in advance with scientific decision-making and prompt early warning.

7.2.2.3 Establishing a Unified System for the Issue of Energy Early Warning

The release of energy early warning information to the public promptly and accurately is a key part in preventing and countering energy disasters, and a major move to effectively reduce casualties and property losses. As early warning information is serious, sensitive and forward-looking, the timing and scope of the release of such information must be assessed in a scientific manner, the possible impact must be comprehensively considered and the normal order ensured.

Energy early warning information must be published on a unified basis and exclusively by leading energy departments in accordance with the level of energy warning and an authorisation hierarchy to ensure energy early warning is issued in a serious and authoritative manner. A unified platform for the release of energy early warning information should be set up, on which any information in relation to national energy security, major energy emergencies and their processed information will be released to the public on a unified basis on government websites and in announcements, conferences, news media and other channels. For the early warning of emergency energy disastrous events, the approval process must be streamlined and a fast-release 'green channel' set up for making immediate releases to the public free of charge.

Energy companies should take the initiative in strengthening communication with various departments such as the telecommunications, transport, railway, water conservancy, meteorology, earthquake, public security, safety supervision, land resources and environmental protection departments as well as the army, armed police and local governments at all levels. Any early warning information on the possible impact of energy emergencies in different areas must be given in a timely manner, with active cooperation with government departments in the release of information on energy early warning.

7.2.2.4 Improving Energy Statistics System

Research on energy early warning must be based on quality statistical data. At present, due to the weakness of statistical work on energy in China, energy statistics can hardly reflect the full picture of energy performance completely and accurately, thus causing great difficulties to research on energy early warning. To meet the requirements of energy early warning, a sound statistical system should, under the unified leadership of China, be established as soon as possible to cover the entire energy generation, supply, transport and consumption processes, and clearly define the responsibilities of government departments, industry associations and companies in energy statistics. A unified energy data collection and statistical platform at the national level should be developed to enhance the timeliness, comprehensiveness and authoritativeness of energy statistics.

7.3 Energy Emergency Response System

The main purposes of energy emergency response are to mitigate or eliminate the impact of emergencies on energy supply and consumption, minimise loss, ensure energy supply and maintain economic and social order. China has been continuously strengthening energy emergency response and significantly improving energy emergency response capacity by drawing lessons from major security incidents at home and abroad, and summing up the work experiences in coping with the tensions in coal, electricity, oil, gas and transport and in carrying out operations against ice disasters and other emergencies to ensure uninterrupted power supply. Looking to the future, the energy security situation in China is complex and challenging, accompanied by numerous uncertainties. China must continue to strengthen the building of capacity for energy emergency, continuously improve the energy emergency response system and conscientiously act as a gatekeeper to ensure national energy supply.

7.3.1 Organisational and Management Structure of Energy Emergency Response

China has effectively coped with one crisis after another in recent years by building on the strengths of its emergency response management practices and fully utilising a strong capacity for mobilisation. Energy emergency response management has been improved during this process as well. Generally speaking, there is still a relatively large gap in the level of energy emergency response management between China and developed countries due to a lack of an institutionally sound and efficiently-run system for organising energy emergency response.

In the future, China should fully draw the mature experiences from developed countries in energy emergency response management by setting up an energy emergency command and coordination office at each level based on a top-down approach, improving the organisational and management structure of energy emergency response and exercising unified command over major energy emergencies and energy emergency response. Nationwide energy emergencies should be under China's centralised command and coordination, with division of responsibilities among relevant departments, local governments and energy companies. Regional energy emergencies should be specifically handled by the governments of provinces (and autonomous regions and municipalities directly under the central government), with instructions given by relevant national authorities which will help coordinate cross-provincial matters. On one hand, energy companies should establish internally a sound emergency response system, and effectively connect it with the energy emergency response systems of the governments at all levels. On the other hand, they should actively participate in the government's energy emergency response project, play a key role in energy emergency response and effectively perform their corporate social responsibilities.

7.3.2 Emergency Response Programmes for Energy Emergencies

From 2003 to 2010, relevant departments of China had successively completed 25 emergency response programmes for national projects and 80 emergency response programmes for departments under the unified arrangement and guidance of the State Council's working group on emergency response programmes. Large domestic energy groups had also set up emergency response programmes for emergencies. In this way, an emergency response programme system has been established, with the characteristics of the respective companies and a focus on the prevention of major security-related incidents.

At present, most of the emergency response programmes, whether at the national or corporate level, are just aimed at handling emergencies per se and best able to cope with a single type of emergencies only. A multidimensional and well-integrated emergency response management system has not yet been formed, resulting in poor convergence between some adhoc programmes and departmental programmes.

Given the new situation and new requirements for energy emergency response, China needs to improve and perfect the relevant energy emergency response programmes as soon as possible to form a 'totally horizontal and vertical' system for responding to energy emergencies. This system should primarily comprise: (1) A general programme at the national level for energy emergency response to define the responsibilities and obligations of governments at all levels, companies and the general public in energy emergency response, to determine the order in basic energy supply, to maintain the operation of major national agencies, national defence facilities, emergency response command office, transport and communication hubs, medical emergency and other vital departments, and to assure the availability of energy needed for daily life and production. (2) Emergency response programmes for energy projects, including programmes to handle disruptions of energy imports into China; emergency response programmes for integrated coordination of coal, electricity, oil and transport; emergency response programmes for widespread blackouts on power grids, and emergency response programmes for mining disasters. (3) The improvement and perfection of the energy emergency programmes of relevant departments and local people's governments at all levels. The emergency response programmes of all departments and localities must be aligned with the general emergency response programmes for national energy emergency response and the emergency response programmes for projects. (4) An emergency response programme for energy companies. (5) An emergency response programme for dealing with energy disruptions for key users. Key users include major energy-consuming companies and work units involving public safety such as hospitals, schools, large communities, transport and communications hubs, large-scale infrastructure facilities and busy public places. Refer to Figure 7.1 for the system framework of national energy emergency response programmes.

To ensure the energy emergency response programmes are relevant and effective, an analysis and assessment of the programmes should be conducted on a regular basis, drawbacks should be rectified promptly and the programme system should be enriched on an ongoing basis to ensure relevance with the actual situation and a high level of operability to lay a foundation for ensuring effective energy emergency response.

7.3.3 Supplies Reserves for Energy Emergency Response

Emergency supplies are essential guarantees to cope with energy emergencies. When handling emergencies, developed countries have plenty of reserves of supplies and goods as backing. For example, an emergency response supplies reserve and a regular rotation system have been established in Japan, and a reserve of emergency response supplies and medical supplies has been established in the USA.

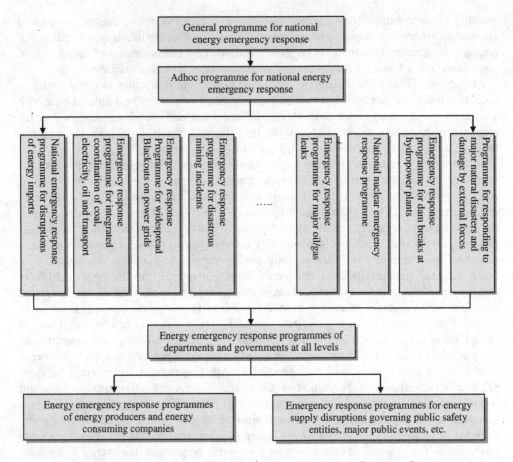

Figure 7.1 System framework of National Energy Emergency Response Programmes.

Since the Zhangbei earthquake in 1998, disaster relief supplies reserve systems have been established across the country. A total of 10 central livelihood relief supplies reserve warehouses have been established so far, which have played a key role in a number of relief operations for serious natural disasters. In light of the present situation of relief supplies reserves and with reference to the experiences and practices of developed countries in emergency response supplies reserves, China needs to start building an energy emergency response supplies reserve based on the principle of unified management, scientific distribution, information sharing and unified allocation.

7.3.3.1 Subject Entities of Supplies Reserves for Energy Emergency Response

Targeting corporates, a multilevel, diversified system for securing energy emergency supplies reserves should be built. In particular, the system should be designed to strengthen the local governments' energy emergency response reserves, improve the mechanism for assuring and coordinating energy emergency response supplies between local governments and companies, and enhance the support capacity for energy emergency response supplies.

7.3.3.2 Planning for Energy Emergency Response Supplies Reserves

Focus should be placed on strengthening the energy emergency response supplies reserves and increasing the number of emergency response supplies reserve depots in China's major economic circles (such as the Bohai Economic Rim, the Yangtze River Delta Economic Rim and the Pearl River Delta Economic Rim), key cities (such as Beijing, Shanghai, etc.) and major natural disaster-prone areas to ensure the supply of energy emergency response supplies.

7.3.3.3 Types of Energy Emergency Response Reserve

Priority should be given to reserving basic supplies such as supplies for livelihood, emergency rescue for engineering projects, fire control, communications, lighting and flood control. For large-scale mechanical engineering equipment and means of transport for energy emergency response such as emergency mobile generators and other specialised products, equipment and materials, a tracking system and an underlying database for preparing energy emergency response supplies should be set up so that these supplies can be leased and transferred on the spot or to the nearest places where they are urgently needed.

7.3.3.4 Management of Energy Emergency Response Supplies

China should actively explore other means of energy emergency response reserves such as physical reserves, reserves by agreement and dynamic reserves. The amount of reserves should be determined reasonably and the purchase, storage and rotation of emergency response supplies handled properly. Under special circumstances, social resources should be promptly mobilised and expropriated in accordance with relevant laws or regulations to replenish energy emergency response supplies.

7.3.4 Energy Emergency Response Publicity Campaign and Emergency Drills

7.3.4.1 Carrying out Energy Emergency Response Education and Publicity Properly

Compared with other social emergencies, energy emergencies are highly risky, highly specialised and very difficult to handle. They easily cause serious psychological panic among the general public. To enhance the energy emergency response capacity of residents through regular energy emergency response, publicity and education is the best precaution against contingencies. Education on energy emergency response awareness must be strengthened by using various media and means of communication to widely publicise the knowledge of handling energy crises and raise public awareness of energy crises. Basic education on energy emergency response and training in emergency response capacity must be strengthened, the public helped and guided to know how to identify energy risk factors and major sources of energy risk, the structure of an energy emergency response programme and its implementation procedures so as to cultivate and strengthen safety awareness. General training that focuses on the basic skills of crisis prevention, self-help and mutual medical help must be conducted. Energy emergency response professional knowledge and skills education must be strengthened, and education on safety knowledge specifically about dangerous chemicals, coal mines, nuclear and power must be carried out, as should specialised, intensive training in basic expertise for the general public and professionals involved in crisis response and management. For example, for the residents near nuclear power plants, general education on nuclear safety, radiation protection and emergency response knowledge must be strengthened, and public understanding of nuclear radiation in a scientific and rational approach must be reinforced to seek public understanding and cooperation.

7.3.4.2   Conducting Regular Energy Emergency Drills

Emergency drill is an indispensable, integral part of an emergency response programme, and is one of the most important activities under an emergency response management system. A good emergency response programme alone does not guarantee effective response by the government, companies or individuals to an energy emergency. As energy emergencies tend to develop rapidly, emergency rescue allows for no delay, making it impossible to carry out operation by referring to every rule under an emergency programme. We can continuously raise the level of emergency rescue and successfully achieve emergency rescue targets only by regularly conducting energy emergency drills and familiarising ourselves with the rescue skills and gaining experience in the process.

Regular energy emergency drills are generally conducted in western developed countries. Drills are often conducted at the federal and local levels of the USA based on crisis response plans and programmes, and these plans and programmes are modified and improved in line with the situation to strengthen collaboration between departments and disaster relief skills of professionals. Among the specialised emergency drills conducted specifically for different types of power facilities, the relatively large-scale and more frequent emergency drills are those emergency drills for unexpected security-related incidents at nuclear power facilities. Approximately eight nationwide emergency exercises for nuclear incidents have to be organised in France every year, while emergency simulation drills are often conducted specifically for oil emergencies in Japan.

In recent years, China has been increasing the intensity of energy emergency drills by conducting comprehensive emergency combat exercises specifically for major energy incidents such as nuclear emergency response, widespread blackouts on power grids and marine oil spills successively. To cope with the development needs of the nuclear industry in China, test the effectiveness of the nuclear emergency programmes and their execution procedures, and maintain and enhance the capacity for nuclear emergency response, China conducted its first national nuclear emergency exercise code-named 'Aegis 2009' in November 2009. More than 2000 people, including those from the national and provincial authorities and nuclear operators together with soldiers and members and expert advisory group of the National Nuclear Emergency Coordination Committee, the military, the nuclear emergency response office in Jiangsu Province, Tianwan nuclear power plant and residents participated in the successful combat exercise. State Grid Corporation of China places great emphasis on emergency drills by regularly organising and conducting emergency drills. Upon completion of an emergency command centre in August 2009, State Grid conducted a large-scale joint exercise on flood and typhoon prevention. In August 2011, State Grid successfully conducted a joint emergency drill and an emergency combat rescue exercise in response to an earthquake in Northern China. The power companies in various provinces (and autonomous regions and municipalities directly under the central government) under State Grid Corporation of China and some of its direct affiliates organised and conducted their respective targeted drills. The drills were primarily combat drills, supplemented by tabletop exercises, such as the joint emergency drill of Shanghai Municipal Electric Power Company on a rail power failure, and the comprehensive emergency drill of Fujian Electric Power Co., Ltd. on flood and typhoon prevention for the power grids.

But generally speaking, China has not committed adequate resources to energy emergency drills. In particular, the resources for combat drills are not enough. In the current period and the future, China stills needs to further strengthen the establishment of an energy emergency drill mechanism and policy. Firstly, it needs to develop proper plans for energy emergency drill programmes to enhance the effectiveness of drills. Secondly, it needs to actively secure enough funding for energy emergency drills to ensure strong manpower, material and financial sources are available for energy emergency drills. Thirdly, it needs to assess, analyse and sum up experiences from energy emergency drills conscientiously and properly, while modifying drill plans and energy emergency programmes promptly to ensure the emergency programme system is effective.

7.3.5 Scientific Management of Energy Emergency Response

Energy emergencies are highly probable and destructive in a modern society. For example, nuclear leaks, explosions of dangerous chemicals and widespread blackouts are seriously hazardous, making energy emergency response more difficult to handle. To align with the energy emergency response management practices in major developed countries, China needs to enhance the scientific approach to handling energy emergency response.

7.3.5.1 Committing to Unified Command and Scientific Deployment

After 9/11, the USA focused on the use of emergency response capacity at the national and community levels by forming the Department of Homeland Security to place agencies such as the Federal Emergency Management Agency under the administration of the Department of Homeland Security for carrying out unified command and coordinating response. The USA concentrates on a disaster relief system that spans from the central to the local administration by establishing an integrated command and dispatch system that comprises military, police, fire control, medical and civilian rescue organisations for immediate and rapid mobilisation and rescue in the event of a disaster to minimise losses. The British government has proposed integrating the police, fire control and medical emergency services into a single emergency system. The member states of the European Union have established a robust mutual support energy network and a collaborative energy emergency response management system across the regions under unified leadership.

From an overall perspective, there is still a wide gap between energy emergency response management and the reality of current energy security in China due to poor coordination in some aspects. Energy emergency rescue forces are scattered, emergency command functions overlap, and local and community-wide emergency rescue forces are not fully mobilised. If a cross-regional, cross-industry or cross-national major energy incident or one involving a variety of disasters happens, it is difficult to share resources and join forces together to make a coordinated response. China should actively learn from the mature experiences of developed countries in energy emergency response management by sticking to unified leadership and command during the handling of energy emergencies. Moreover, China should scientifically deploy and fully mobilise the resources and strengths of local governments and communities to create an energy emergency response handling mechanism with involvement of the government, companies and the public by guiding the relevant parties to develop a response through concerted efforts in an orderly manner.

7.3.5.2 Conducting Rescue with a Scientific Approach and Enhancing Emergency Response Handling Capacity

China has been laying a good foundation for the prevention, investigation and handling of accidents for a long time, but has started rather late in emergency response management and rescue operations. This has been a weakness. During the entire management process of energy emergency response, it is a very essential task to carry out rescue in a scientific manner to minimise and reduce disaster losses and avoid the occurrence of secondary disasters.

The focus should be on the use of scientific and technological means to enhance the capacity for handling energy emergency response. In the USA, for example, an advanced and reliable emergency communications system plays a key role in dealing with an emergency response. During emergency dispatch and command, advanced internetworking connection devices enable communications between the police or fire departments to make interconnection possible, especially to ensure the efficiency of emergency command and dispatch. China should vigorously develop emergency response using scientific and technological means. In particular, support capacity for the communication system should be enhanced and advanced communications technologies, including computer systems, database systems, geographic information systems, satellite positioning systems, remote sensing systems and video systems, should be developed and used. The standard of monitoring, early warning, emergency response handling and other technical equipment should be raised. Energy emergency response equipment should be improved to effectively enhance the capacity for handling emergency response.

Strong emphasis should be placed on rescue, using scientific means to prevent the deterioration of the situation and the occurrence of secondary accidents or disasters. After the occurrence of an accident or a hazard, the handling of emergency response using scientific means will effectively avoid the deterioration of the incident and mitigate losses and casualties. On 23 December 2003, the Luojia 16H natural gas wells in Kaixian, Chongqing, exploded during a drilling operation. A large amount of natural gas containing highly concentrated hydrogen sulphide was emitted and dispersed. But for the failure to ignite the gas promptly, massive poisoning and deaths would have been reduced or avoided. Moreover, how to avoid radiation from radioactive substances in a nuclear power plant incident, and how to avoid electric shocks during rescue in a power incident rescue operation entail the strengthening of response and rescue using scientific means to minimise disaster losses.

7.3.5.3 Committing to Information Disclosure and Guiding Public Opinions in a Correct Way

In successfully dealing with all kinds of unexpected and public events in recent years, China has acquired the important experience of disseminating information to the public promptly and accurately, and guiding, organising and mobilising the masses proactively. In handling energy emergency response, great emphasis should be placed on disclosing information and guiding public opinion on energy emergencies to protect the public's rights to information access and to facilitate the creation of an environment conducive to handling emergency response. On one hand, information should be made available through conventional and contemporary media such as radio, television and the Internet to promptly notify the public of the progress, and take the initiative in opinion-making to gain public understanding and support for the way energy emergency response is handled. On the other hand, the procedures for disclosing information on energy emergency response should be regulated. Unified management and release of such information should be strengthened to ensure such information is accurate, comprehensive, timely and authoritative. The credibility of the information disclosed should be enhanced. Public interests and social stability should be safeguarded. International impact should be taken into account when handling information on emergency response to foreign-related energy emergencies. A good international image should be established to better safeguard national interests.

7.3.5.4 Normal State Should Be Restored Timely after Completion of Emergency Response Activity

Emergency response activity indicates a temporary abnormal state. All work should be restored to normal in a timely manner upon completion of an emergency response activity. Otherwise, some of the emergency response measures are likely to become the source of potential hazards in the future. After a disaster, recovery and reconstruction work should be organised promptly to mitigate the damage and impact caused by the emergency, properly resolve any conflicts and disputes arising during the process of handling the emergency, and cooperate with all parties in restoring production, livelihood, work and social order to normal as soon as possible.

As the state of energy emergency response is being restored to normal, attention should be paid to the interface between various measures to avoid triggering new problems. Moreover, it is important to sum up the experience from the incident, improve emergency response initiatives and the emergency response programme system, promptly replenish emergency supplies that have been consumed, and adopt relevant improvement measures.

7.4   Energy Reserves

Energy reserves building refers to the activity of stockpiling a certain amount of energy products and energy resources at ordinary times and in a planned manner for the purpose of guarding against energy supply risks, including energy emergencies such as acute shortage of energy supply, supply disruptions and drastic price fluctuations, to prevent heavy socioeconomic losses and safeguard the normal operation of the national economy. In a sense, the building of energy reserves is a 'hematopoietic project' as well as an insurance policy. It is an important measure to ensure the security of energy supply. At present, energy reserves in China are still in the early stage of building. The reserves are small in size, the reserve capacity low and the reserve system not yet perfected. To assure the security of energy supply in China and meet the needs of the country's rapid economic growth, it is imperative to further strengthen the building of energy reserves.

7.4.1   Present Situation of Energy Reserves in China

Compared with Western developed countries, China is relatively late in emphasising the building of energy reserves. Moving into the 21st century, as the Chinese economy is heavily hit by the rising crude oil prices due to the increasingly tense international energy situation, and the oil security issue has become increasingly prominent, the Chinese government has embarked on a drive towards building a reserve system for petroleum and other energy options. The 11th Five-Year Plan and the 12th Five-Year Plan both outline plans for building petroleum and other energy reserves. It has now become a major strategic move to strengthen the building of energy reserves for safeguarding China's energy security.

7.4.1.1   Petroleum Reserves

In 2004, China started to work on building national petroleum reserves in full swing with plans to complete the construction of hardware facilities for petroleum reserves such as oil depots in three phases within 15 years. As at the end of 2010, the four national petroleum reserve bases in Zhenhai and Zhoushan in Zhejiang, Liaoning in Dalian and Huangdao in Shandong were completed under Phase 1 of the petroleum reserve construction project in China. Strategic petroleum reserves and commercial reserves amounted to 178 million barrels and 168 million barrels respectively, creating a reserve capacity to sustain 36 days of consumption. Construction of eight national petroleum reserve bases under Phase 2 is in progress and scheduled for full completion by 2012. Site selection for Phase 3 is underway.

The National Petroleum Reserve Centre, formally established on 18 December 2007, is responsible for building and managing the national petroleum reserve bases, undertaking the tasks of purchasing, storing, rotating and drawing on strategic petroleum reserves, and monitoring changes in supply and demand in the domestic and international petroleum markets.

In overall terms, China's current petroleum reserve capacity is still very limited. The petroleum reserves are small in size and the petroleum reserve system is still in the initial stage of building.

7.4.1.2 Natural Gas Reserves

China has just started to build natural gas reserves, which are limited in size. Nationwide gas storage capacity represents less than 3% of the total natural gas consumption.

Gas storage, gas fields and LNG are methods widely adopted internationally for building natural gas reserves. Underground storage is the major means of building natural gas inventory in China. The underground gas storages have created a peak-shaving capacity of 1.87 billion m^3, with the Bohai Region accounting for 1.8 billion m^3 and the Yangtze River Delta for 50 million m^3. During the 12th Five-Year Plan period, the underground gas storage programme in China will enter a stage of accelerated growth, with a cluster of gas storages planned to be built in Henan, Jiangsu, Sichuan and other provinces. As a peak-shaving facility, LNG reserves feature short cycles of consumption and can respond quickly to a shortage of natural gas supply, making them suitable for areas where there are no ideal geological conditions for the construction of gas storages or where gas reserves are inadequate. For example, the LNG peak-shaving plants built in Shanghai, Shenzhen, Shenyang and other places play a key role in peak-shaving with local natural gas storages. They can be used to supplement underground gas storages for proper development.

7.4.1.3 Coal Reserves

In June 2005, the State Council published a report *Several Opinions on Promoting the Healthy Development of the Coal Industry*, with a clearly-worded proposal to establish a policy governing the strategic reserves of coal resources, carry out protective exploitation of special or rare coal types, and launch a natural coal reserve programme.

In May 2011, relevant Chinese ministries and commissions published the *Interim Measures Governing Emergency Coal Reserves of China* to enhance coal supply capacity to deal with emergencies. Based on the principle of 'corporate ownership, government subsidy, strengthened supervision and emergency response', the Interim Measures specify a blueprint for building national emergency coal reserves. The first block of planned reserves is 5 million tonnes, with aggressive plans to bring the second block to an emergency reserve capacity of 10 million tonnes by the end of 2011 and approximately 20 million tonnes by 2013, in order to sustain gas consumption for 7 to 10 days.

7.4.1.4 Natural Uranium Reserves

As a clean energy source, nuclear power plays a key role in achieving China's target for energy conservation and emissions reduction. According to the national nuclear power programme, natural uranium demand in China will rise dramatically in the future. The building of uranium resource reserves is a major move to secure a stable supply of uranium resources.

Currently, the uranium resource reserves in China serve basically as the production reserves for companies. China National Nuclear Corporation and China Guangdong Nuclear Power Holding Co., Ltd. are two major players in China engaged in the exploration, development, operation and management of nuclear resources. The exploration, development and international trading in overseas countries of natural uranium, which can be used for both military and civilian purposes, are subject to stringent control for peaceful purposes only, and to the rigorous supervision by international organisations such as the International Atomic Energy Agency.

7.4.2   Experience in International Energy Reserves

OECD began building a petroleum reserve system with the formation of IEA in 1974. Since then, the system has undergone several major crises and tests over 30 years of development, and the IEA member states have developed increasingly robust energy reserve systems. Despite differences in the operating rules for building energy reserves, these member states have acquired rich experiences and practices as an important source of inspiration for China to commence the building of energy reserves.

7.4.2.1   Petroleum

Petroleum reserve is the focus of many countries in the building of an energy reserve system. Following a comprehensive survey and analysis of the petroleum reserve building in different countries, the major experiences and practices are summarised as follows.

Firstly, these countries have built a government-led, multilevel petroleum reserve system combining the strengths of government, companies and the private sector. The petroleum reserve systems around the globe generally consist of three parts: the first is the strategic petroleum reserves of governments; the second is the commercial reserves of oil companies, and the third is the reserves of associations or alliances lying between the first and second parts. Each country's petroleum reserve system features a different composition. In the USA, petroleum reserves are primarily composed of the strategic reserves of the government and the commercial reserves of companies. The government assumes full responsibility for the strategic petroleum reserves and authorises the Department of Energy to take charge of them. The strategic petroleum reserves principally consist of crude oil used mainly for the needs of national security and emergency response. The building of commercial reserves by companies is purely market-driven, for the purpose of regulating supply and demand as well as price fluctuations in the petroleum market. The strategic petroleum reserve system in Japan is an integrated system combining the government and the private sector. It is divided into three levels—government reserves, statutory company reserves and commercial reserves. Statutory company reserves are the reserves undertaken by companies as required by law. Under Japanese law, companies are required to stockpile petroleum of a quantity enough for use up to 70 days in the country. The German government does not stock up petroleum directly. Instead, it regulates the country's petroleum reserves by means of legislation and policy setting. Its national statutory petroleum reserves are undertaken mainly by petroleum reserve alliances. The alliance members include all the importers of crude oil and refined oil as well as oil refineries in Germany. The amount of reserves accounts for approximately 3/4 of the total petroleum reserves in Germany. Moreover, German companies can build production and commercial petroleum reserves based on their own needs and capabilities.

Secondly, these countries have established a legal system governing strategic petroleum reserves. To ensure smooth progress in building petroleum reserves, major Western developed countries have enacted relevant laws, such as the Energy Policy and Conservation Act of the USA, the Petroleum Reserve Law and the Japan National Oil Corporation Law of Japan, and the Energy Security Law and the Petroleum and Petroleum Product Reserves Law of Germany. These laws regulate the building of national petroleum reserve systems, the establishment of operational mechanisms, the determination of the amount of reserves and the requirements for drawing on reserves. They also contain specific provisions governing the duties and obligations of government authorities, localities, corporate entities, public institutions and citizens for energy reserves and supply.

Thirdly, these countries have selected reserve sites and storage methods that can facilitate transport and processing, featuring low storage costs and high security performance. For example, the Gulf of Mexico coast of the USA has been selected for its unique geological conditions and geographic location. Oil pipelines and refineries in the USA are concentrated along the Gulf of Mexico coast where oil tanks, pipelines, barges and piers are in place to facilitate transportation of petroleum products. The close proximity to the manufacturers of petroleum products not only significantly enhances the rapid response capacity of strategic petroleum reserves, but also reduces the costs of transportation for the release of reserves. Moreover, there are numerous large salt caves along the coast that make suitable sites for the storage of petroleum.

Fourthly, these countries have set out stringent conditions and procedures for drawing on petroleum reserves. In the USA, the power to draw on petroleum reserves is in the hands of the president. Once the president decides to draw on the reserves, the Department of Energy will announce the amount of petroleum to be released to the market by means of tender invitation to determine the winning petroleum company that will be allowed to purchase the petroleum. The Federal Government pumps strategic reserves into the market in three ways: firstly, to draw on reserves on a full scale (mainly to cope with acute supply disruptions); secondly, to draw on reserves on a limited scale (mainly to cope with supply disruptions on a large scale and for an extended period); and thirdly, to draw on reserves for test (mainly to test whether the reserve facilities and systems are operating normally). In Japan, petroleum reserves are drawn from private reserves and national reserves. When domestic petroleum supply falls short of demand or is disrupted, restrictions on demand will first be considered, followed by drawing on private reserves and then national reserves. The power to draw on reserves is vested with the Minister of Economy, Trade and Industry.

7.4.2.2 Natural Gas

It is common practice for major natural gas importing countries around the world to build their own natural gas reserves. Given the differences in the level of reliance on imported natural gas, geographical conditions and economic systems, different methods (see Table 7.1) and institutional mechanisms for building national gas reserves have been adopted. Each country is unique in terms of gas-reserve building experience.

Firstly, as a mode of storage, production reserves are the norm. Few countries have maintained strategic reserves. For example, the natural gas reserves in the USA, the UK, France and Spain are for the purpose of coping with seasonal peak shaving requirements and temporary supply disruptions. Only those countries with heavy reliance on natural gas imports have set up or are ready to set up strategic natural gas reserves. For example, Japan places equal emphasis on both natural gas production reserves and strategic reserves, and Italy is considering strategic natural gas reserves to eliminate as much as possible the adverse impact of war, major emergencies and other incidents on natural gas supply.

Secondly, as a method of storage, gas storages in most of these countries are underground gas storages. They have LNG reserves and gas field reserves as well. Different countries have selected different reserve methods in light of their geographic and resource conditions. On the whole, underground gas storage is the major method of reserve for its low costs, large capacity and low technical requirements. But the common problem is location. LNG reserves are more suitable for densely-populated and economically developed countries and regions where there is a lack of piped gas sources. Japan has to rely mainly on LNG reserves.

Thirdly, in terms of reserves management, the natural gas production reserves in these countries are primarily managed and operated by energy companies. Safety, environmental protection and technical operation standards are the focus of government attention. Strategic reserves are geared towards meeting national requirements.

7.4.2.3 Natural Uranium

There are three major operation systems for management of natural uranium reserves. These include direct undertaking by the government, a combination of government reserves and corporate reserves, and a combination of government reserves and institutional reserves. With the development of nuclear power, the USA, France, Japan and other countries have long raised their respective natural uranium reserves to a national level. In particular, Japan has been conducting surveys on overseas uranium ore resources since the 1960s. The country had largely secured a stable supply of overseas uranium resources by 2000, with the establishment of a comprehensive reserve system in the name of the government.

7.4.3 The Thinking Behind Building Energy Reserves in China

China is accelerating the pace of building energy reserves, and much of the work still needs to be further strengthened and improved. It needs to conduct in-depth research and analysis on the legislation and the mode of energy reserves, the method and planning for energy reserves, the size of energy reserves and so forth. On this basis, the building of energy reserves in China must be strengthened comprehensively.

7.4.3.1 Building a Sound Energy Reserve System

China should develop its energy reserves in light of its national conditions, systems and financial strength. Based on the principles of ensuring supply security and secure, economically efficient and diversified reserves, China should implement unified planning and active measures to take the timely opportunity of improving the multilevel energy reserve system in stages.

Table 7.1 Natural gas reserves in selected countries.

Country	Method of reserve	Size of reserve
US	Mainly underground gas storages, supplemented by LNG reserves	400 underground gas storages and 5 LNG receiving terminals. The volume of working gas at gas storages is 17.7% of annual consumption (2007)
Russia	Mainly underground gas storages	24 underground gas storages. The volume of working gas is approximately 25.0% of annual consumption (2005)
Canada	Mainly underground gas storages	52 underground gas storages and 1 LNG receiving terminal. The volume of working gas is approximately 17.8% of the annual consumption (2007)
UK	Mainly underground gas storages, supplemented by LNG reserves	5 underground gas storages and 4 LNG receiving terminals. The total volume of working gas is 4.6% of annual consumption (2008)
France	Mainly underground gas storages, supplemented by LNG reserves	12 underground gas storages and 2 LNG receiving terminals. The volume of working gas at gas storages is 27.7% of the annual consumption (2008)
Spain	Mainly underground gas storages, supplemented by LNG reserves	3 underground gas storages, 5 peak-shaving stations and 1 reliquefaction plant. The total volume of working gas is 10.6% of annual consumption (2008)
Italy	Mainly underground gas storages	9 underground gas storages. The volume of working gas is 18.4% of annual consumption (2008)
Germany	Mainly underground gas storages, supplemented by LNG reserves	45 underground gas storages and 1 LNG peak-shaving station. The total volume of working gas is 23.9% of annual consumption (2008)
Japan	Mainly LNG reserves	26 LNG receiving terminals and 5 LNG strategic reserve bases. The Japanese Government assumes responsibility for sustaining nationwide consumption for up to 30 days. The corporate sector is expected to maintain natural gas reserves sufficient for up to 50 days of consumption (2008)

Source: This table is compiled based on "An Analysis of the Situation and Experience in Respect of Overseas Natural Gas Reserves" by Shengli Ma and Fei Han. It was published in Issue 8 of *"The Natural Gas Industry"* in 2010.

The country needs to further improve its petroleum reserve system, strengthen the building of natural gas reserves, develop a moderate level of emergency coal reserves, and start building national natural uranium reserves as soon as possible. It should choose a mode that combines government reserves, obligatory corporate reserves and commercial reserves as the mode of petroleum reserves in China. The government reserves and obligatory corporate reserves are national strategic reserves which should not be drawn on unless at a very critical moment. Commercial reserves represent the corporate reserves less the obligatory reserves. In the reserve building process, commercial reserves are allowed to circulate but the amount of obligatory reserves stipulated by China should remain unchanged. China is currently building strategic petroleum reserves which are government reserves, while commercial reserves at the company level have been ignored, and private reserves have not yet been officially set up. In the future, China needs to build multilevel petroleum reserves. Given the fast-growing demand for natural gas in China, the need for building natural gas reserves is increasingly urgent in order to achieve a secure supply of natural gas. The building of emergency coal reserves in key coal transport hubs and coal consuming areas can enhance the capacity to secure coal supply during major natural disasters and emergencies. This, to some extent, can act as a buffer to regulate supply and demand in the coal consuming regions. China has a great demand for natural uranium due to its large-scale and speedy nuclear power development. To maintain stable and secure nuclear energy development, higher levels of strategic reserves are required, for which planning should be accelerated at the national level.

In the process of building energy reserves in China, a sound three-tier energy reserve system at the central government level, the local government level and the corporate level, should be set up under the centralised supervision of the central government. This system organically integrates the rights and obligations of the central and local administrations and companies towards reserves to form a collaborative structure comprising national strategic reserves, corporate and obligatory reserves, and company-based commercial reserves. The national strategic energy reserves and the corporate and obligatory reserves are primarily for ensuring energy security and coping with major emergencies, while the commercial reserves are mainly for maintaining day-to-day regulation of demand in the energy market.

7.4.3.2 Improving System for Energy Reserve Management

Since energy reserves involve a multitude of subject entities and interests, even the slightest mistake will cause heavy losses. For this reason, many countries maintain stringent management and policies for the building, purchase, storage, rotation and utilisation of energy reserves. China should make a conscientious effort to learn from the experiences of Western developed countries in energy reserve management by accelerating the establishment of a sound energy reserve management system, and reinforcing the supervision and management of the purchase, storage, rotation and utilisation of energy reserves. China should continuously optimise reserve planning, purchase and storage while designing the rotation process in a scientific manner, improving the operational efficiency and effectiveness of energy reserves and setting stringent requirements for powers and procedures to partially or fully draw on the reserves under government control in different circumstances for specific purposes.

China should further improve its management system for energy reserves by expressly specifying the responsibilities and obligations of the decision-makers, the execution level and the operational level for reserves. The enactment of laws and regulations in relation to petroleum and other energy reserves should also be expedited and energy reserves developed in accordance with the law.

7.4.3.3 Enriching Energy Reserve Varieties

The energy reserves of countries around the world are characterised by variety. In the USA, petroleum reserves comprise crude oil, refined oil and heavy fuel oil, with crude oil taking the lion's share. In Japan, petroleum reserves comprise crude oil and refined oil, with government reserves being all crude oil, while refined oil constitutes 56% and crude oil 44% of private reserves. In Germany, petroleum reserves are crude oil, gasoline and middle distillate oil. Approximately 50% of the existing reserves are crude oil and 50% are gasoline and middle distillate oil.

China should build energy reserves that comprise a rich variety of energy products, such as petroleum, natural gas and natural uranium. For reserves held in petroleum products, government reserves should principally comprise crude oil supplemented by refined oil. Based on their specific requirements, companies may develop a level of corporate reserves in crude oil or refined oil in proportion to company size, and may build more fuel depots, covering heavy oil, gasoline and diesel. For reserves in natural gas and natural uranium products, particular consideration should be given to strengthening energy consumption by the petrochemical industry, residents and nuclear power companies to meet China's economic development and the growing demand of urban and rural residents for natural gas and natural uranium products.

The reserves of energy resources must also be strengthened by conserving areas which may contain energy resources such as oil, gas, uranium and rare types of coal as back-up resources to be used in case of emergency, or by sealing up orefields (such as oil wells and oil fields with proven reserves, etc.) where production has or has not commenced, to be used in extraordinary times. Some gas fields containing high reserves and exploitation potential but involving great exploitation difficulties and long transport distances may be used as national reserves of natural gas resources. Moreover, in the areas where gas is supplied principally through LNG receiving terminals, consideration should be given to the building of a certain size of LNG reserves. The reserves of special and rare types of coal resources such as coking coal and fat coal should be strengthened. The reserves of uranium resources should be increased, the exploration and development of domestic uranium mines actively pursued, and the acquisition of overseas uranium resources accelerated to meet the needs of the sustained and stable development of nuclear power.

7.4.3.4 Determining the Reasonable Capacity of Energy Reserves

The capacity of energy reserves is determined on the basis of country-specific conditions with no standard criteria. Under IEA requirements, the strategic petroleum reserves of its member states should not be less than 90 days' net imports. In practice, however, the current capacity of the petroleum reserves of many countries is higher than the IEA requirements. The Japanese government and private petroleum reserves combined can meet consumer demand for more than 160 days of consumption. In terms of the level of gas reserves, the EU has proposed the requirement that the reserve of a country should be set at 10% of its annual imports. An LNG reserve system has been set up in Japan to ensure that the reserves of private companies are available for consumption for 50 days, and government reserves for 30 days.

In China, opinions differ on the appropriate level of domestic strategic petroleum reserves. According to China's plans for building petroleum reserves, the national petroleum reserve capacity will be upgraded to approximately 85 million tonnes, equivalent to 90 days' net petroleum imports, upon full completion of Phase 3 by 2020. Considering the level of petroleum reserves, imports and economic support capacity, this level of petroleum reserves is more in accord with the realities of China.

China needs to further increase the level of gas reserves in the future and improve the reserve system. With reference to the practices of different countries around the world, and based on the future demand for natural gas consumption in China, relevant Chinese departments have initially confirmed the suitability of a standard that sets the level of natural gas reserves at 20–25% of demand. Moreover, companies in charge of daily and monthly balances (subject entities) are required to share their responsibility towards maintaining a level of reserves corresponding to their respective positions.

As far as coal is concerned, China has abundant coal reserves. The coal supply tension in recent years is the result of a structural imbalance and not a problem of total volume. In May 2011, China issued the *Interim Measures Governing National Emergency Coal Reserves*, introducing the launch of the first block of emergency coal reserves with a planned capacity of 5 million tonnes. This move will be able to deal with disrupted supply or acute shortage caused by major natural disasters or emergencies. But in the long run, it cannot solve the structural imbalance due to the coal supply tension. The cost and capacity of coal reserves are the two relatively big issues. Coal is characterised by its vulnerability to weathering, deterioration or even spontaneous combustion. Moreover, coal stockpiling in open areas takes up large space and is easy to cause environmental pollution. The longer time the coal is stored in open areas, the higher the cost of storage is and the lower the value of the coal itself. Emergency coal reserve sites are extensively located. But coal is different from petroleum or natural gas in that petroleum and natural gas can be transported by pipelines but coal relies more on the more costly means of transport by rail and road. In extreme weather conditions or emergencies, transport costs will be enormous and will substantially reduce the cost effectiveness of emergency coal reserves. The fundamental way to solve this problem is to further increase the level of emergency coal reserves. Considering the relatively large installed capacity of coal-fired generation, the frequent occurrence of natural disasters and other situations, a capacity of 50 million tonnes for an emergency coal reserve is appropriate. On the other hand, the capacity of thermal coal storage at thermal power plants should be increased. Coal stockpiled at thermal power plants in the USA is generally good for around 40 days of consumption, while thermal power plants in Japan have a running inventory for around 10 days of consumption. In China, the current coal inventory at coal-fired power plants is generally available for 7–15 days of consumption. This volume can hardly cope in the event of emergencies. In order to ensure a safe and stable supply of electricity, the capacity of thermal coal storage at coal-fired power plants should be increased appropriately. In particular, in eastern and central China remote from coal supply bases, the days of consumption that the thermal coal stocks at coal-fired power plants can sustain should be increased to approximately 30 days.

7.4.3.5 Optimising Methods and Layout for Energy Reserves

In light of specific conditions such as energy import and export patterns, energy processing facilities, petroleum pipeline development, construction of transport facilities and distribution of consumer groups in China, energy reserve sites are to be further improved and optimised.

Petroleum reserves can, in the short run, be held in existing crude oil terminals and facilities in the coastal areas. In the long run, consideration can be given to the construction of reserve bases in the areas close to Russia and Kazakhstan and in the major petroleum producing areas in China. Refined oil reserves should be consumer market-oriented. They should be located mainly in big cities and petroleum product distribution centres. As to the storage methods, experiences can be drawn from the USA, Germany, France and other nations in crude oil storage. In the future, priority should be given to the use of underground salt caverns for storage.

China should build underground gas storages, mainly peak-shaving gas storages, for natural gas reserves in the large- and medium-sized cities in northeastern China, northern China, the Yangtze River Delta and the Pearl River Delta where the consumption of natural gas is high as well as in the natural gas distribution centres. Underground gas storages should be built along backbone lines, such as the West-East Pipeline Project for connection to the key natural gas backbone networks. Sites should be selected for building underground gas storages in major provinces and cities accessible by transnational natural gas pipelines. Reserve facilities should also be developed in big coastal provinces where massive volumes of LNG are imported.

On the geographical distribution of emergency coal reserves, key consideration should be given to locations where coal is produced, transported and consumed, while the level of accessibility and the distance from major economic centres should also be taken into account. Since coal production is concentrated in northwestern China while coal is consumed mainly in the southeastern region, key consideration should be given to the building of emergency reserves in coal transshipment ports, land terminals and economically developed areas in southeastern China. A study should be conducted on the specific locations in light of specific requirements. An appropriate number of coal reserve bases should be built in the coal producing areas.

Natural uranium reserves should be maintained in both the international and domestic markets. More reserves should be built with the capacity for producing economically exploitable natural uranium resources. Domestically, plans should be developed using a scientific approach to stepping up the exploration and exploitation of uranium resources. Extensive research should also be conducted on uranium production in overseas countries and the production conditions of uranium producers. On this basis, programmes for the development of overseas uranium resources in a scientific and rational manner can be set up to timely push forward the building of overseas uranium resource development bases and to gain control over more economically exploitable uranium resources through different channels such as equity participation and the acquisition of controlling interests and orefields.

8

Innovation in Energy Technology

Technology innovation is key to solving the energy challenges and energy crises faced by human development. Innovation capability in energy technology is one of the decisive factors in building competitiveness in the international energy market. To promote energy transition, China must thoroughly implement a national reinvigoration strategy based on science and education, pursue plans for building an innovation-oriented country, increase investment in energy technology innovation, and establish the strategic direction and focus of energy technology research. It must also improve the mechanism of energy technology innovation, master key technologies and core competences in the field of energy, narrow the gap with developed countries in the field of energy technology R&D as soon as possible, and to strive for building competitive advantages in technological innovation in the world of energy.

8.1 The Situation of Energy Technology Innovation

To better protect national energy security, address the climate change issue and seize first-mover opportunities in global competition, many countries in the world attach great importance to energy technology innovation. Under the *Action Plan on Science and Technology and Sustainable Development* approved by the G8 Summit in 2003, R&D and extensive application of energy technology were incorporated as one of the three key areas to achieve sustainable development. Guided by the national reinvigoration strategy based on science and education and plans for building an innovation-oriented country, work on energy technology in China goes from strength to strength with satisfying results. But compared to the situation of major developed countries in energy technology innovation, China still has a formidable task ahead.

8.1.1 Technology Innovations in International Energy Sector

In recent years, many countries in the world have regarded energy technology innovation as an important pillar of energy strategy and an integral part of technology strategy, with stronger policy support to build a significant competitive position and propel global energy

Electric Power and Energy in China, First Edition. Zhenya Liu.
© 2013 China Electric Power Press. All rights reserved. Published 2013 by John Wiley & Sons Singapore Pte. Ltd.

technology innovation into a new period of dynamic growth. This situation has imposed new requirements on China for strengthening energy technology innovation.

8.1.1.1 The USA Trying to Maintain Its Position as Global Technology Leader

The USA has a tradition of energy technology innovation. In the 1960s, the country began looking into unconventional technology for natural gas exploration and development and subsequently achieved continued breakthroughs to promote the development of technology for utilising coal bed methane and shale gas. R&D efforts in clean coal technology have continued uninterrupted since the 1980s. Moving into the 21st century, the USA has further increased support for energy technology innovation, targeting nuclear energy, hydrogen energy, renewable energy, superconducting power transmission and fuel cell technology. According to the *Energy Independence and Security Act* passed in 2007, the country's investment in clean energy and energy efficiency technologies is expected to increase to US$190 billion in 2025. After becoming US President, Barack Obama proposed a new *Energy Plan* to promote, at the national strategic level, R&D on renewable energy, smart grids and electric vehicles, among other technologies. From 1974 to 2009, the energy R&D expenditure of the US government amounted to US$159.86 billion, making the USA the largest country in terms of investment in energy technology innovation during the period.

The USA holds the implementation of energy technology policy in special regard. As the architect of energy technology innovation, the US Department of Energy has set up Energy Frontier Research Centers (EFRCs), Advanced Research Projects Agency—Energy (ARPA-E) and Energy Innovation Programme Office (EIPO), with responsibility for promoting basic research, applied research and commercialisation of energy technology.

8.1.1.2 Japan Seeks Technical Advantage in Dealing with Resource Disadvantage

Lacking in energy endowments, Japan pays special attention to energy technology innovation, often with foresight to develop forward-looking plans in this area. In the 1970s, Japan launched a new-energy technology development programme. In the 1980s, Japan carried out overall planning for R&D on domestic wind energy technology and industry promotion. In 2001, Japan announced an 18-year plan for the exploration and development of combustible ice. In February 2012, Japan took the lead in the global launch of combustible ice drilling lab projects. In recent years, Japan has further stepped up research and development efforts in new energy, low-carbon energy, electric vehicles and related technologies. Launched in 2008, the *Cool Earth—Innovative Energy Technology Programme* laid out the roadmap for the years up to 2050, with the development of innovative energy technology as a priority.

Japan hopes to achieve the diversification of energy structure and the efficient use of fossil energy by building an edge in energy technology R&D and the promotion of energy conservation. Japan also hopes to search for new domestic energy resources, build the world's most advanced structure for energy supply and demand, and alter the over-reliance on imports for its energy requirements. By increasing the export of energy technology, Japan wants to maintain a say over the global energy market and control the upstream energy industries to expand the market share of its energy products. Currently, Japan leads the world in R&D and industrialisation in the fields of energy conservation, clean coal, nuclear energy, hydrogen energy and renewable energy. From 1988 to 2007, the number of patents registered in Japan in relation to the use of clean energy is almost twice that of the USA during the same period.

8.1.1.3 Germany Leads R&D in Renewable Energy

Being the traditional powerhouse of industry, science and technology, Germany attaches great importance to energy technology innovation and accords priority to technology advancement for energy development. When the German government decided to develop large-scale offshore wind power in 2002, it channelled the majority of R&D funding for wind energy into the study of crucial issues of offshore wind turbine technology, ecological impact and grid access. The country leads the world in energy innovation in these areas. Before launching offshore wind power projects on a large scale, Germany has already accounted for more than a third of the world's market for offshore wind turbines.

To reduce the heavy dependence on imports for its energy requirements, Germany passed *Network Access Law for Renewable Energy* in the 1990s, stipulating renewable energy R&D as the fundamental of the nation's technology development. Today, Germany has built a strong competitive edge in new-energy technology, targeting to have 80% of its domestic electricity supply coming from renewable energy by 2050.

Germany's achievement in energy technology innovation is mainly due to strong policy and financial support as well as a sophisticated collaboration-based research mechanism. A platform dedicated to the coordination of energy technology research policy was set up for major energy technology research programmes. Special research funds and large-scale government procurement are instrumental in promoting energy technology research and commercialisation of the research findings.

8.1.1.4 Energy Technology Development Universally Held in High Regard

In recent years, new-energy technology has been accorded top priority in energy technology development in the UK and France. Renewable power generation, nuclear energy and greenhouse gas emission control are the focus areas of energy technology innovation in the UK. Britain has also decided to establish a network of technology innovation centres to promote cooperation between enterprises and research institutes in low carbon technology research. France hopes to repeat its success in nuclear technology development in the fields of wind power, solar energy and new-energy automobile sector. It has increased related investments in these areas.

Being resource-rich countries, Canada, Russia and Brazil also actively promote energy technology innovation. Currently, Canada has a distinct advantage in the R&D on hydrogen and fuel cells with more than 50% of the world's hydrogen fuel cell vehicles using Canadian technology. Russia has made a substantial investment to maintain its dominance in nuclear energy and combustible ice technology. In recent years, it has also taken energy conservation and improvement of energy efficiency technology as the focus of energy technology development. Brazil has achieved a leading position in bio-ethanol technology.

Overall, global energy technology innovation has entered a new period of activity where new-energy technology, clean energy technology, smart grid technology and basic research in energy has become the main focus of the world's R&D. The key to building a proactive role in this competition lies in developing government-led strategic plans for energy technology innovation, and meticulous arrangements in support policy and capital investment. Only when close links are developed between basic research, technology development, commercialisation of research findings and the industrialisation process can a country maximise the combined strengths of different parties to fully realise the supportive and leading roles of energy technology in the development of the energy industry.

8.1.2   The Situation of Energy Technology Innovation in China

8.1.2.1   China Has Always Valued Energy Technology Development

Energy technology has been an important part of China's work on technology development. After the founding of new China, the country has made a series of major achievements in oil and gas exploration and development, in the use of nuclear energy, and in large and medium-sized hydropower construction. Since the economic reform and opening up in the 80s, the status and role of science and technology as the primary productive forces have been highly valued and energy technology innovation has entered a new stage of development. With continued improvement in energy technology, China has completed its independent energy technology R&D system and equipment manufacturing mechanism. Building on the experiences from a large number of key projects, such as Qinshan Nuclear Power Plant, the Three Gorges Power Station and gas transportation from the west to the east, a host of core technical problems in energy development have been overcome. China has also become one of the few countries in the world with considerable strengths in R&D in almost all areas of energy technology.

Moving into the 21st century, with increasingly significant constraints on energy development by resources, the environment and development modes, there is tremendous pressure to develop sustainable energy amid ever-higher expectations for energy technology innovation. In 2006, the Chinese Communist Party's Central Committee and the State Council came up with a major strategic plan for building an innovative country, with energy as the key area of independent innovation. *The National Outline for Mid- & Long-term Scientific and Technical Development (2006~2020)* has further stated clearly the area and direction of energy technology innovation, giving energy, science and technology innovation unprecedented priority.

8.1.2.2   China Has Made Significant Achievements in Energy Technology Innovation

With the support of national policy and the concerted efforts among enterprises, universities and research institutes, China has achieved significant results in energy technology since 2006 by building on the solid foundation of the past few decades.

Firstly, a system for energy technology innovation is coming into being, with the successful development of a series of energy technology innovation platforms including research institutions, corporate R&D bases, the Science and Technology Park,[1] various laboratories and research centres.[2] A budding and capable energy technology innovation team is beginning to take shape. This has created a situation where industrial companies, academic organisations and research institutes work closely together on crucial and core technologies in energy to overcome key challenges.

Secondly, China has mastered a body of advanced energy technologies. It has made a series of major breakthroughs in energy exploration and development technology, clean energy technology, and energy transport and conversion technology. UHV transmission technology and equipment manufacturing has reached world-leading standards while 1GW-level ultra supercritical coal-fired power stations, fast reactors, the 10 million-tonne-level coal mining equipment, deepwater semi-submersible drilling platform technology and equipment manufacturing have also reached the world's advanced levels. It has basically mastered the manufacturing technology of wind turbines of less than 3 MW and has formed a relatively well-developed PV industry chain. Since 2006, the number of energy technology projects conferred National Science and Technology Progress Awards (First Class) has increased significantly (see Table 8.1).

Thirdly, there is an increasing level of domestic content in some of the important energy equipment. Refineries with a capacity of 10 million tonnes and ethylene plants with a capacity of 1 million tonnes can now be designed and manufactured locally. The level of domestic content in the crucial equipment for the improved second-generation nuclear power plants has exceeded 80%, compared with over 90% for the equipment of the Southeast Shanxi-Nanyang-Jingmen 1000 kV UHV AC transmission pilot project, and nearly 70% for the equipment of the Xiangjiaba-Shanghai ±800 kV UHV DC transmission demonstration project. All major components of the 10 MW photovoltaic power generation system are locally manufactured.

Fourthly, a lot of energy technology achievements have been promoted and adopted. Advanced technologies in power grid construction, such as large-scale power grid control and UHV transmission, have been applied. In the coal industry, advanced technologies such as underground coal gasification, heavy medium coal preparation and other advanced technologies are being promoted. China boasts the world's largest capacity of coal-to-liquids, coal-to-olefins and coal-to-natural gas equipment. A number of energy-saving technologies and equipment have gained popularity while a lot of scientific and technological achievements in the energy field have also achieved industrialisation with promising market prospects.

8.1.2.3 Shortcomings of China's Energy Technology Innovation

Overall, there are still gaps between China and major developed countries in the technological development of energy. China's energy industry has also failed to meet the need for sustainable development.

---

[1] Energy is the key focus of research efforts in four national independent innovation demonstration zones (Zhongguancun Science Park, Wuhan East Lake High-tech Development Zone, Hewubeng Independent Innovation Test Area, Shanghai Zhangjiang High-tech Industrial Development Zone).

[2] Including 18 State-level Key Laboratories, 8 National Engineering Research Centres, 6 Enterprise-level State Key Laboratories, 38 National Energy R&D (Experiment) Centres.

Table 8.1 Energy technology projects conferred National Science and Technology Progress Awards (First Class) during 2001–2011.

Year obtained	Number of awards	Winning projects
2001	2	Discovery of Kela-2 gas field and mountain UHP gas reservoir exploration technology
		Daqing Vacuum Residue Fluid Catalytic Cracking Technology Development and industrial applications
2002	2	Discovery of Sulige large-scale gas field and integrated exploration technology
		Development of complete set of 2 million tonnes/year residue hydrotreating (S-RHT) technology
2003	2	High-efficiency exploration technology and practices in Block 1/2/4 of the Sudanese basin of Muglad
		Modern mine construction and technology practices in Shendong
2004	2	Forming mechanism and exploration of subtle oil pools in terrigenous fault basins
		Design and construction of 600 MW nuclear power plant in Qinshan
2005	1	Research and application of development technology for high-pressure condensate gas field in Tarim basin
2006	2	Accumulation mechanism and exploration technology of deep marine carbonates and discovery of Puguang gas field
		10 MW high-temperature gas cooled test reactor
2007	3	Large area accumulation theory, exploration technology and major discovery of low-medium abundance lithostratigraphic reservoirs
		Research and application of ultra supercritical coal-fired power generation technology
		750 kV AC power transmission and transformation key technology, equipment development and engineering applications
2008	2	Equipment development and engineering applications of complete set of 600 MW supercritical thermal power equipment
		Key technology of Flexible AC Transmission Systems (FACTS) (Thyristor Controlled Series Compensator, TCSC) and promotion of application
2009	3	Development and application of Naphtha catalytic reforming packaged technology
		Development and industrialisation of major sets of HVDC transmission equipment
		Research, installation development and application of full digital real time simulation technology for power systems
2010	5	Exploration and development technology for efficient and stable production for capacity of over 40 million tonnes in Daqing oilfield in later high water-cut stage
		Three Gorges power transmission project
		West-East gas project engineering technology and its application
		Exploration and development of giant ordovician carbonate oil-gas fields
		Development of innovation system for offshore oil and gas exploration and development in China

Table 8.1   (*continued*)

Year obtained	Number of awards	Winning projects
2011	3	Innovation in geological theory for Qinghai-Tibet plateau and major breakthroughs in locating mines*
		Key technologies for efficient exploration of complex oil and gas fields under special circumstances
		Design technology development and engineering practice of HVDC projects

Source: The data in the table are sourced from the website of the National Office for Science & Technology Awards.
*The project has won a National Science Progress Award (Special Class).

Firstly, as far as basic research and frontier technology are concerned, China has not yet developed its own source of energy innovation, with the key areas of the coal, oil and electricity industries still heavily relying on foreign technology. Frontier research in the fields of clean coal technology, oil exploration, mining, refining, alternative technology, energy storage technology, combustible ice technology and superconducting technology, has yet to reach internationally-advanced levels.

Secondly, as far as equipment manufacturing is concerned, the design and production of highly-technological energy equipment still largely rely on introducing or replicating foreign technology. As China is still unable to develop its own innovative ability and manufacturing capability, this reliance remains strong particularly for high-efficiency power generation equipment, silicon-based solar cells and the key components of some thin-film solar cells, the core components of high-power heat pumps for using surface geothermal energy, and the key nuclear energy generation equipment.

Thirdly, China does not have a robust system or a strong forward-looking vision for energy technology innovation. While China's individual energy technology projects share a certain degree of continuity, there is not enough interaction between different energy innovation projects because of a low starting point and the lack of management experience. Due to a lack of overall planning for clean energy R&D with limited fundamental and ground-breaking achievements, there is still room for improving the level of energy technology innovation in China.

8.2   Principles and Focuses of Energy Technology Innovation

For energy technology innovation, China needs to thoroughly implement the strategy of national reinvigoration through science and education, with plans to build an innovative country focusing on providing guidance and support for energy sustainability. Based on a commitment to the principle of pursuing independent innovation, key breakthroughs, forward planning and an industry-driven approach, China should coordinate and promote fundamental research, key technology research and frontier technology research to resolve major technological challenges and achieve world-class research work.

8.2.1 The Fundamental Principle of Energy Technology Innovation

The next 20 years will be an important transition period for China's energy development strategy, with sustainable development as the major direction. The whole process of energy development, conversion, transport and usage will experience profound changes. The key to China's future energy development lies in enhancing the quality of energy development, improving energy efficiency, optimising structural adjustments and promoting the optimal allocation of energy sources. For this, the promotion of energy technology innovation must serve as the starting point for strategic transformation in energy, which will be driven by major breakthroughs in energy technology.

Energy technology innovation is a massive systems engineering project. In order to improve input and output efficiency, China must coordinate planning and rationalise organisational efforts to establish and uphold the principle of pursuing independent innovation, key breakthroughs, forward planning and an industry-driven approach.

Firstly, China must be committed to independent innovation. This approach is the soul of technological development and the key to enhancing the core competitiveness of the energy industry. In practice, it has repeatedly been shown that real core technologies cannot be bought in the competitive international energy market. As some technologies are nowhere to be bought, much will have to depend on one's own efforts to seek breakthroughs through independent innovation. Without its own intellectual property rights and with the continued dependence on imitating or importing foreign technology, China will never be able to change the situation where its energy industry is sizeable but not strong and its core technologies are subject to foreign control. Only by improving its independent innovation capacity can China realise leapfrog development in energy technology and complete its transformation from a big energy nation to a strong energy nation.

Secondly, China must be committed to making key breakthroughs. As a developing country, China's energy technology R&D is built on a weak foundation, with a scarcity of technology resources and a shortage of inputs in energy technology compared with R&D demand. In order to raise China's energy technology to world-leading standards as soon as possible, the organisation of R&D must be well-focused, with the selection of a host of core and key technologies as the major areas of pursuit where China has already built solid ground and an edge in these technologies and these technologies will have a significant impact on China's energy development. Efforts should be concentrated to resolve prominent issues in China's energy development and to surpass the technology standards of major developed nations.

Thirdly, China must be committed to forward planning. The characteristics of energy technology R&D dictate a period of less than or over 10 years or even longer between the start-up of research and the final application and transformation for production purposes. This requires us to focus on the current situation when organising work on energy technology innovation and to make active efforts to overcome the technical challenges of energy development. We are also required to take a long-term view by planning a number of basic research and frontier technology research projects so as to enhance energy technology capability and the momentum of innovation. These efforts will help us take the initiative in the future technology competition and fully leverage the guiding role of technology.

Fourthly, China must be committed to an industry-driven approach. From project initiation, full consideration must be given to the prospects of utilisation and market demand, focusing on the market relevance of technology innovation. The results of technology innovation must be promoted and applied in a timely manner and should be put into production as soon as possible. There is also a need to speed up the industrialisation process and actively cultivate the energy industry. China should also adhere to a mechanism for technology innovation based on tripartite collaboration involving the industrial, education and research sectors, with a commitment to a government-led, company-driven and project-based research system. China should also focus on realising the innovative role of enterprises and place enterprises at the centre of energy technology investment, innovation and application.

8.2.2 Focus Areas of Energy Technology Innovation[3]

As determined by the need for changing the way of energy development and also the guiding principles of China's energy technology innovation, the key areas of pursuit for energy technology innovation in the future are energy-related basic research, technologies relating to the exploration, development and conversion of traditional fossil energy, new-energy technologies, high-efficiency transmission and storage technologies, energy-efficiency technologies, control technologies for pollutants and greenhouse gas emissions, and energy-related frontier technology.

8.2.2.1 Basic Research

To surpass developed countries in energy technology innovation, China must put great emphasis on basic research related to energy technology and achieve more fruitful results from original research. It is because only through the important discoveries of basic science could new technology transformation and significant technology upgrading be realised. Only when continued breakthroughs in energy-related basic research are made can energy technology development attain lasting momentum and the strategic deployment of some key and frontier energy technologies be supported.

Important energy-related basic research covers: (1) theoretical basis of the efficient and clean use of fossil energy and its transformation; (2) the key scientific issues of high-performance thermal power conversion and storage; (3) theory and new ways for large-scale utilisation of renewable energy; (4) theory for the secure, stable and economic operation of grid systems; (5) the basic technology of large-scale nuclear energy and the scientific foundation of hydrogen energy technology; (6) new materials; (7) earth system and resources, environmental and disaster effects.

Basic research plays a critical role in the improvement and advancement of energy exploration, development, production and utilisation. It can also enhance public awareness of energy, prevent and reduce the adverse effects of energy consumption, directly promote new ways of energy usage, and help adopt better exploration, development, production and utilisation technologies. This is a prerequisite for China to achieve an overall enhancement of energy technology innovation. Research in the secure, stable and economic operation of

---

[3] According to *The National Outline for Mid- & Long-term Scientific and Technical Development* (Department of Science & Technology, 2006), *Energy Science & Technology in China: A Roadmap to 2050* (Science Press, 2009).

power grids can provide fundamental theoretical support for the development of efficient and intelligent transmission and distribution technologies and the safe and effective operation of power grids. The study of earth system, resources and environmental and disaster effects bears theoretical significance for projecting the impact of energy usage and development on atmospheric circulation, geological activities and the natural environment. This can also help effectively predict and prevent natural disasters induced by the development and utilisation of energy, while promoting more rational and scientific development and utilisation of energy. The geological study of marine oil formation will provide important theoretical support for improving the technology for exploration of low-permeability gas reserves and exploitation of unconventional oil and gas resources, the study of exploration and development technology for complex geological conditions and deep-sea oil and gas, and the technology for raising old oil field recovery.

8.2.2.2 Exploration and Development of Traditional Fossil Energy and Production Conversion Technology

The adjustment and transformation of energy structure entails a long process where the dominant position of traditional fossil fuels in the energy mix will not change over a long period of time. However, the exploration for traditional fossil energy, research on development technology and equipment manufacture will always be the key areas of China's energy technology innovation.

The key research areas covering the exploration and development of traditional fossil energy and production conversion technology include: (1) Technology for exploration and development of coal, oil and gas resources. It mainly covers efficient prospecting and rapid fine detection technology, the technology for exploration and development of traditional fossil energy (such as the core technology for deep-sea oil and gas exploration and development), integrated coal-mining equipment that is suitable for complicated and deep geological conditions, exploration and mining equipment for large-scale oil and gas exploration, drilling and production as well as equipment for large-scale offshore oil platforms suitable for deep-sea operations. (2) Clean production and efficient conversion technology of coal, oil and gas, mainly covering independent research on coal washing and screening technology, directional transfer control technology for the coal conversion process, 700 °C ultra-supercritical power generation technology, circulating fluidised bed technology, IGCC technology, large-scale air-cooled units, and natural gas power generation technology. (3) Accident prevention technology for coal, oil and gas exploration and development. It includes the study of accident prevention technology for traditional fossil energy resource development such as the prevention of major safety hazards in coalmines, handling techniques for offshore oil and gas incidents and the technology for preventing marine pollution in addition to independent research on relevant rescue equipment.

These technological advancements will boost domestic energy production and supply capacity and also reduce or eliminate the environmental impact of the exploration and utilisation of traditional fossil fuels.

8.2.2.3 New-energy Technologies

Increased development and utilisation of new energy is the global trend of energy development. China must align its energy technology innovation with the pace of world development by promoting R&D on the technology for development and utilisation of new energy and its industrialisation.

The key research areas of new-energy technology include: (1) Technologies for development and utilisation of renewable energy, including notably the design and manufacturing technology of the core components for large-scale wind turbines, collaborative control and grid technology of large-scale wind farms, offshore wind power generation technology, ocean energy generation technology, high-silicon solar cell technology, selection and cultivation techniques for energy-generating plants, biomass fuel technology, and high-temperature heat transfer core technology. Also included are independent research on high-efficiency heat absorbers resistant to high temperature and corrosion, high-power wind turbines and key equipment of the offshore wind power, geothermal energy, tidal and wave power generation equipment, advanced PV and solar thermal power generation equipment. (2) Technology for development and utilisation of unconventional fossil energy, including mainly the technology for exploration and development of unconventional oil and gas resources, including coal bed methane, shale gas, sand gas, water-soluble gas, combustible ice, shale oil and oil sands, in addition to independent research on related equipment. (3) Technology for development and utilisation of new nuclear energy, covering third-generation nuclear power, fourth-generation fast reactors, nuclear waste disposal and controlled nuclear fusion, and independent research of related equipment.

8.2.2.4 High-efficiency Transport and Storage Technology

Energy transport and storage are two critical issues affecting the development, utilisation and stable supply of energy in China. Unless these two issues are resolved, no fundamental change in the under-capacity situation of coal, electricity, oil, gas and transport can be expected and the development and utilisation of new energy and renewable energy will also be severely constrained.

The key research areas of energy transport and storage technology include: (1) Safe and efficient transmission technology, mainly covering large-scale AC/DC hybrid grid operation control technology, smart grid technology, real-time measurement technology, fully digital real time simulation technology, $\pm 1100\,kV$ UHV DC key technology, Flexible Alternating Current Transmission System (FACTS) and Superconducting Power Technology, and large-power grid security technology. Also involved are the research and manufacture of UHV power transmission equipment, power grid safety equipment, group control equipment for grid access for renewable power projects. (2) Oil and gas transmission technology, mainly including large-diameter, high-pressure pipeline technology, automation of dispatch management for oil and gas pipeline operations, and communication technology. Also included are the research and manufacture of long-distance oil and natural gas pipelines, large-scale crude oil tankers and LNG carriers. (3) Energy storage technology, mainly including research and manufacture of energy storage equipment, such as high-capacity hydrogen storage technology, compressed air energy storage technology, high-energy density storage batteries, super capacitors, etc.

8.2.2.5   Energy-saving Technologies

Energy conservation is of utmost importance for China's energy development strategy, with energy conservation technology being the key area of energy technology innovation. Given its low overall energy efficiency, China sees huge potential for energy conservation, giving energy conservation technology promising prospects for extensive application.

Industry is the most important energy-consuming sector in China. The industrial application of energy conservation technology should aim at promoting the use of energy efficient equipment. Examples include the use of efficient industrial boilers and motors, the adoption of advanced technologies to improve the energy efficiency of the production process, and a focus on the recycling and reuse of residual heat and pressure for industrial purposes. Secondly, China needs to focus on the application of energy conservation technologies in the production process. To achieve this, China should have the courage to change the traditional process flows, so that energy-saving equipment and technology can be better combined with industrial production processes to achieve the best possible outcome. Thirdly, China needs to focus on the application of energy-saving technologies in the production system and on the overall energy efficiency of the whole industrial production process. To this end, emphasis must be placed on optimising the production process design and changing the processing mode currently on a single-product basis. Finally, the processing and integrated utilisation of resources should be improved to ensure more rational use of energy.

In the transport industry, innovation of energy conservation technology comes in the form of improved efficiency of oil and gas consumption for conventional vehicles and development of new-energy vehicle technology, such as electric vehicles, and electric rail technology. The focus should be placed on the study of battery technology, motor drive technology, and electric vehicle integration control technology.

For the construction industry, energy conservation technology mainly refers to the development of new insulation materials and state-of-the-art energy management systems, R&D on advanced lighting technology and energy-efficiency air-conditioning technology, and the reduction of energy consumption for construction work.

8.2.2.6   Pollutants and Greenhouse Gas Emission Control Technology

The massive energy consumption in China has produced a severe impact on the environment. To ensure coordinated development between energy consumption and the ecology, efforts must be made to optimise China's energy consumption structure, with a focus on the development of pollution control technology, in particular desulphurisation and denitrification technology, precipitator technology and heavy metal pollutants control technology for coal-fired units.

Greenhouse gas emission is another hot issue facing the international community. Due to a high proportion of coal consumption, China's greenhouse gas emission is among the highest in the world. Carbon capture and storage (CCS) technology is an important technical means to control greenhouse gas emissions in China. The focus of R&D work on CCS technology is the technology for separation and capture of carbon dioxide from the flue gas of coal-fired boilers as well as the geological and chemical sequestration technologies of carbon dioxide.

8.2.2.7 Frontier Technologies

Areas of energy-related frontier technology where more monitoring and research efforts are required cover mainly: (1) fuel cell technology, new high-temperature superconducting materials and preparation techniques; (2) smart technologies such as intelligent sensory, self-organising network information technology, smart materials and structural technology; (3) exploration and detection technology, such as seabed multiparameter rapid detection technology, airborne geophysical survey technology; (4) deep-sea fishing technology, etc.

These frontier technologies are in accord with the future direction of China's energy development and utilisation, which merits close attention by the energy industry. For example, the new high-temperature superconducting materials and preparation techniques can be applied in the manufacture of motors, high-energy particle accelerators, power cables, energy storage equipment and controlled thermonuclear reactors.

8.2.3   The Goal of Energy Technology Innovation

The goal of China's energy technology development is planned to be fulfilled in two stages.

For the first stage, China intends to develop into an innovation-oriented country and one of the world's leading advanced energy technology countries by 2030.

The milestone of this stage is much higher capability of independent innovation in the field of energy technology, with both independent innovation and production capacity for major energy equipment reaching world-class standards. Building on the foundation of popularising and improving existing mature technologies, China should strive to enable a number of advanced energy technologies to reach the stage of industrial application and promotion. Meanwhile, China should also aim at making breakthroughs in the critical parts of some advanced energy equipment, and significantly enhancing the domestic content in China's major energy equipment.

In concrete terms, China aims at encouraging widespread utilisation of the technology for developing and utilising wind, solar and other renewable energies, the technology for efficient and clean utilisation of coal, UHV and smart grid technology, energy-efficiency technology for the industrial and construction sectors, large-diameter high-pressure pipeline technology, automation of dispatch management of oil and gas pipeline operations, and communication and information technology. At this stage, breakthroughs and extensive applications should be achieved for 700 °C ultra-supercritical power generation technology, circulating fluidised bed technology, IGCC technology, deep-sea oil and gas exploration technology, large-scale electricity storage technology and new-energy vehicle technology, with the domestic content in the related equipment raised to over 80%. China is a global leader in the efficient and clean use of traditional fossil energy, having established a cutting edge in some technological areas of new energy development and related equipment production. At the same time, China shall continue with research on superconducting power, fourth-generation nuclear power technology, coal gasification technology with adjustable hydrogen/carbon ratio, hydrogen energy, applied research on natural gas hydrates and practical technology.

From 2031 to 2050, Stage 2 aims at making China the world's leading energy and technology power and the world's major base for manufacturing and supplying advanced energy equipment, with overall R&D strengths in most frontiers of independent innovation in energy technology. China should also strive to achieve breakthroughs in all technology areas, covering nuclear fusion, natural gas hydrate development and utilisation, space solar power generation and large-scale ocean energy generation. On this basis, China can address the new issues and challenges in the energy sector and to embark on a new round of aggressive energy research and innovation.

8.3   Development of System for Energy Technology Innovation

To strengthen its innovation capability in energy technology, China needs to integrate innovation resources, improve the mechanism for energy technology innovation, strengthen the R&D team, adopt a strategy for energy technology innovation and develop a healthy and robust system for energy technology innovation, with a focus on breaking all the institutional constraints that restrict the vitality of technology innovation.

8.3.1   Integration of Resources of Energy Technology Innovation

8.3.1.1   Strengthening Infrastructure and Platform Development for Energy Technology Innovation

The basis of energy technology innovation is an energy infrastructure platform comprised of an energy experimental research base, major scientific facilities and instruments, scientific data and information, and energy technology resources. The building of an energy technology infrastructure platform is focused on the frontier research in the field of energy technology with the mission to develop an experimental research base that is well-staffed, technologically competent and academically superior in line with the requirements of China's national energy strategy and anchored by the State Key Laboratory Scheme, the National Engineering Research Centre, Key Laboratories of State-owned Enterprises and the National Energy Research (Experiment) Centre, based on a programme implemented by the State. Efforts should be made to launch a series of energy engineering projects and infrastructure facilities and promote the sharing and development of scientific instruments, facilities and key technologies. The latest information technology should also be fully utilised to develop an energy digital technology platform, along with efforts to develop standards for different energy technology resources and an efficient resource-sharing mechanism in energy technology. Based on the characteristics of the different energy technology resources, a flexible sharing model should be adopted to break away from the current situation marked by fragmentation, mutual exclusion and overlapping among different energy types.

8.3.1.2 Optimising Allocation of Technology Innovation Resources

Under government guidance, efforts should be devoted to fully leverage the role of enterprises in attracting the active participation in technology innovation of universities and research institutes with an exceptional professional edge, further strengthen an energy technology innovation system built on collaboration among the industrial, education and research sectors, and effectively allocate energy technology resources within the entire energy sector, which can provide sustainable momentum for the energy industry by sparking the innovative vigour of all parties involved. While greatly improving the innovation capability at the corporate level, efforts should also be focused on building a new mechanism that puts together research institutes and tertiary institutions to meet the corporate demand for innovative services and brings about collaboration among the industrial, education and research sectors in different forms. Within the framework of the Comprehensive National Science and Technology Programme, such as the *863 Project, 973 Project, National Science and Technology Support Programme* and *Torch Programme*, independent energy technology plans or topics should be developed. Technology development plans should be coordinated among the different sectors of the energy industry to prevent repeated wastage of technology resources. Targeting research projects or new inventions that require a higher level of funding and are more risky, a risk-sharing mechanism among the government, research institutes and enterprises shall be established. With government support and through a partnership between manufacturers, tertiary institutes and research organisations, an alliance between users and manufacturers and a power industry coalition, a number of consortia of common interests for independent technology innovation should be established.

8.3.2 Development of Mechanism for Energy Technology Innovation

8.3.2.1 Creating Innovative Atmosphere for Energy Technology

Innovation culture provides fertile soil for promoting innovation capability. In order to rapidly enhance the capability of independent innovation in China's energy sector, it is necessary to generate a creative ambience that reflects the requirements of the times and the national conditions of China. Through building a public and social environment in favour of creativity and leveraging the role of the government in strategic planning, promoting and service in energy technology innovation, innovation communities and chains should be founded for promoting the creation of a tripartite partnership involving the industrial, education and research sectors and also integrated research, application and promotion capabilities. This will help consolidate and expand the results of innovation gained by major corporations. At the same time, emphasis should be placed on building a social atmosphere that encourages success and forgives failure so that any aspiration for innovation that is conducive to energy industry development and technological progress is respected, creativity is allowed to flourish and the momentum of innovation is recognised. Only in this way can an inquisitive spirit, team work and cooperative spirit be fostered, with a consensus on innovation as the focus of the energy sector to truly raise the level of innovation in China's energy industry.

8.3.2.2   Improving Mechanism for Energy Technology Investment

Technology investment is the material foundation of technology innovation as well as an important precondition and basic assurance for technology sustainability. Today's investment in science and technology is an investment in a nation's future competitiveness. Nowadays, the developed countries and the emerging industrialised countries regard higher technology investments as a strategic initiative to enhance national competitiveness. In contrast, the level of technology investments in China's energy sector appears insufficient, with an irrational investment structure and a weak technology foundation. In order to secure a leadership position in energy technology innovation in the future, investment in energy has to be increased. On one hand, technology investment should be stepped up to increase the proportion of energy technology funding in the country's total investment in science and technology and to provide stable and sustainable financial support for basic energy research, frontier technology research and research on key technologies of common interests. On the other hand, enterprises should be guided to invest more in energy technology innovation. Through a sound appraisal mechanism and other measures, state-owned energy companies, particularly the major state enterprises, should be encouraged to promote energy technology innovation. National energy technology plans must reflect the major demand for technology in the corporate sector to attract greater participation from corporations. Tax, financial, land and other incentives should be provided to strongly support and encourage a drive towards energy technology innovation in different industries.

8.3.2.3   Improving System for Transforming Achievements in Energy Technology

Successful transformation of energy technology achievements requires the integrated utilisation of market-based initiatives and support policy to help create a system conducive to transforming technology findings. Different types of technology intermediaries and service agents should be actively developed and an information platform for energy technology built, together with the development of a trading market to effectively match supply and demand for technology products and services. Policy support for transformation of achievements in energy technology and the development of small and medium-sized technology enterprises should be strengthened, with tax reduction, funding support, subsidised interest and other measures. At the same time, funding channels in support of transforming technology findings should be expanded and venture capital, industry capital or financial capital should be allowed to participate in the transformation process to resolve funding bottlenecks. Science parks and industry bases should be set up with the characteristics of industry clusters to leverage their role in enabling transformation. The award system for appraising technology achievements should be improved, with a focus on the social and economic benefits of transformation. Alternatively, a reward system for transformation of technology achievements may be established to enable the awards to play a guiding role in the energy industry.

8.3.2.4   Protecting Intellectual Property Rights of Energy Technology

By improving the legal protection for intellectual property rights, and with reference to the characteristics of the intellectual property of energy, an energy industry certification system and a social credit system conducive to the protection of intellectual property rights should be developed to offer protection to the intellectual property of major energy technology under law. At the same time as the technology intermediaries and service agencies are being developed, knowledge sharing and technology transfer between industry peers themselves and between universities and research institutes should be encouraged. Also, efforts to organise better protection of intellectual property in the energy field should be improved to arouse greater awareness of intellectual property protection among the relevant entities and staff. A special review and protection system for intellectual property in energy technology should be established to avoid the loss of propriety intellectual property through corporate restructuring, M&A, and technology exchange and cooperation. In addition, awareness of confidentiality must be solidly strengthened to prevent the loss or theft of technology research findings in the course of cooperation with foreign parties in energy technology.

8.3.3 Building Talent Team in Energy Technology Innovation

8.3.3.1 Strengthening Training and Development of Talent Team in Energy Technology

Human capital is the most important of all strategic resources. Building a strong team is essential to the successful launch of energy technology innovation. The *National Programme for Mid- and Long-term Talent Development (2010–2020)* must be implemented and, based on major energy technology projects and energy engineering projects as well as high-level training programmes, like the *Chang-Jiang Scholars Programme*, training and development of champions in the energy field should be strengthened. More support should also be given to high-calibre experts in terms of job title and honour assessment, special government subsidy, postdoctoral working platform, research funding, etc. Attractive remuneration packages should be offered to senior experts who have made outstanding contributions in key areas of energy technology. The talent selection system in the energy field should be improved and refined to change the current rank-based approach and create an environment favourable for the growth of young talents. An energy technology team across all sectors, professions and age brackets should be set up to build an echelon that covers the full spectrum of professions with numerical strength and a rational structure. On this basis, an energy innovation team of combined and complementary strengths can be built to focus on new technology frontiers and major research projects.

8.3.3.2 Emphasising Recruitment of Overseas Talents

Focusing on strategic priorities and missions, proprietary research should be strengthened, along with efforts to implement programmes for introducing overseas talents to step up the development of an international team by making the best of foreign expertise. Through programmes like the *Thousand Talents Programme* and the *Thousand Young Talents Programme*, high-calibre technology talents in short supply should be employed. Overseas renowned scholars active in the international energy sector should be invited to come to China for joint research with funding support and for exchange and work activities. This will help enhance the global vision of China's energy innovation team and to improve their ability to understand issues of frontier technology and develop a forward-looking vision. The focus should be on launching international cooperation in energy technology covering strategic issues of energy technology, major energy equipment and important frontier areas and direction of technology. Channels for cooperation should be slowly expanded with innovative partnership models to improve the level of cooperation.

8.3.4 Innovation Strategy for Energy Technology

8.3.4.1 Coordinating Different Forms of Independent Innovation

Independent innovation can be achieved through original innovation, integrated innovation, and re-innovation by processing imported technology. A particular form of innovation can be selected on a case-by-case basis with reference to specific issues. Stronger original innovation in basic energy science and frontier technology can lay a solid foundation for future development. Using a variety of technologies, equipment and tools in the field of energy applications, and through the selection, optimisation and system integration of innovative elements and contents, greater economies of scale can be achieved. Increasing efforts in the adsorption, processing and re-innovation of imported technology is an important way to quickly enhance China's technology innovation capability, while other fruitful results of original innovation can also be availed of to improve innovation capacity. As one of China's weak links, re-innovation by processing imported technology is the most popular method in other countries, particularly developing nations. In 2010, the ratio of technology spending by large and medium-sized enterprises to the investment in imported technology was 0.43:1, compared with 3:1 or above for developed countries. Emphasis should therefore be placed on the absorption of imported technology and, on this basis, technology upgrading and innovation can be pursued and efforts better coordinated to achieve major innovation in all or parts of the value chain of the energy industry.

8.3.4.2 Expanding Established R&D Advantages by Leveraging Strengths and Rectifying Shortcomings

China is one of the few countries in the world with a relatively complete energy industry system and a relatively solid foundation in energy technology. It has built a competitive edge in some areas of the energy sector. While organising work on energy technology innovation and allocating the resources required for this purpose, the focus should be on rectifying shortcomings by committing more resources to research on the weak links and to areas where China has already established a leading edge so as to consolidate and expand its competitive position. The alternative to this is a quick loss of China's established leadership role. However, it will not be realistic to try to achieve a breakthrough in all areas of energy technology in a short time. China has already developed a strong position in the areas of Ultra High Voltage (UHV) and smart grid technology. By picking up momentum, China will become the international leader in this field. Doing otherwise will result in the loss of China's hard-earned leadership position, causing serious damage to the energy industry.

8.3.4.3 Actively Participating in International Exchange and Cooperation in Energy Technology

Energy security and climate change are common challenges faced by human society that call for the collective wisdom of mankind to address and resolve. China has been carrying out extensive exchange with the USA, the European Union, Japan, Russia and other countries and regions in energy technology. On this basis, it is necessary to further strengthen exchange and cooperation among different countries in the world in the field of energy technology, participate more actively in the major international energy technology research projects, and engage in bilateral or multilateral cooperation in energy technology. Energy enterprises should be supported in their endeavours to 'go global' and expand the export of their hi-tech energy technology and products. Enterprises should also be encouraged to set up R&D facilities and industrialisation bases overseas. By doing so, they can learn from the advanced experience of other nations and boost their own research capability. In the process, they can also share China's research findings with other nations and make a contribution to the resolution of the energy and environmental problems faced by human society.

9

Ensuring Energy Sustainability

For the Chinese energy industry to achieve sound and rapid development and gain the initiative in international competition, it is necessary to improve relevant laws, regulations and policies, build a unified and comprehensive standard system and develop a group of energy conglomerates with rather strong international competitiveness.

9.1 Energy Laws, Regulations and Policies

It is not possible to pursue energy sustainability in China without the assurance of a scientific and effective system. The continuous improvement and perfection of Chinese laws, regulations and policies on energy and the full fulfilment of the active regulatory and guiding role of the system are of great significance to the sustainable and healthy development of China's energy industry.

9.1.1 Establishment of a Legal Regime for Energy

9.1.1.1 Committing to Regulating and Guiding Energy Development in Accordance with the Law

Before the implementation of the reform and opening up programme, the development of China's energy industry was basically planned and guided by the government without the benefit of a legal regime. Since the implementation of the reform and opening up programme, the pace of the establishment of a legal regime for energy has accelerated gradually in China. Energy legislative activities were frequent in China in the 1990s, compounded with the separation of government functions from enterprise management and also the energy market reform, as witnessed by the successive promulgation of a large number of energy-related laws and regulations, such as the *Electricity Law*, the *Coal Law* and the *Energy Conservation Law*. Since the beginning of the 21st century, the establishment of a legal regime for energy in China has been enhanced on the whole. On one hand, new phenomena and issues such as climate change, sustainable development and energy security have been governed by laws, following the successive introduction of the

Electric Power and Energy in China, First Edition. Zhenya Liu.

Renewable Energy Law, the *Recyclable Economy Promotion Law* and the *Oil and Natural Gas Pipeline Protection Law* and the making of amendments to some promulgated laws to a varying degree. On the other hand, the enforcement of energy laws and regulations has been stepped up. The People's Congresses at all levels have reinforced the supervision and inspection of the enforcement of energy laws and regulations.

Because of a late start, there is still a gap in terms of energy legislation between China and the USA/European countries with a highly developed market economy, as manifested especially in three aspects. Firstly, China's energy legal system is not comprehensive, as fully demonstrated by the absence of a basic energy law that can fully reflect energy strategy and policy orientation, a basic legal system for some energy industries, and supporting laws and regulations for individual industries. Secondly, the quality of energy legislation in China remains to be improved. Some laws and regulations are not sufficiently coordinated and are even in conflict and discord with each other, while some others fail to meet realistic requirements due to the changes over time and in the situation. Thirdly, law enforcement remains to be further strengthened because there are still instances of noncompliance with the laws and law enforcement is weak.

As a result of the profound changes now taking place in the internal and external environment of energy development in China, it is an urgent task and situation to further strengthen the establishment of a legal regime for energy. Firstly, as various reform measures have been carried out in recent years, China's energy industry is becoming increasingly market-oriented, with more diversified market players and complicated interest relationships. Therefore, it is difficult to build effective and fair market order and form a proper interest relationship solely by administrative measures, industry codes, and self-discipline by market players. Instead, it needs to vigorously strengthen the establishment of a legal regime for energy and regulate the energy market in accordance with the law. Secondly, the energy industry is an important basic industry which is related to national economic security and the building of a harmonious society. Strengthening the establishment of a legal regime for energy and boosting energy reform and development in accordance with the law is conducive to clarifying and defining the objectives of energy development for effectively guiding and driving various forces towards the scientific development of energy. Finally, it is a fundamental national strategy of China to rule the country by law and build a socialist nation under the rule of law. The socialist market economy itself is an economy ruled by law. Since the establishment of legal regimes in various fields has been accelerated as a whole in recent years, the establishment of a legal regime for energy also needs to be coordinated and taken forward.

9.1.1.2 Improving the System of Energy Laws and Regulations

Introducing **Energy Law** *in a Timely Manner*
Energy Law signifies prominently a nation's sound energy legal system. It establishes the logical framework, basic context and overall legislative ideas of an energy legal system as a whole, and serves as a major basis for the formulation and revision of other laws and regulations in the energy sector. In the absence of specialised laws and regulations, the *Energy Law* can fulfil the role of a principled and rational legal norm and provide legal protection for energy development under specific circumstances. It is a top priority

to strengthen the establishment of a legal regime for energy, improve the energy legal system and push the introduction of the *Energy Law*.

The continued delay in introducing *Energy Law* over the years is primarily due to a number of reasons. The first reason is related to the subject of law. As the energy administrative system in China has been evolving, it is difficult to determine the party responsible for performing regulatory duties and providing the relevant public services at the same time. The second reason is about the legislative basis. The lack of highly forward-looking and uniform energy strategies, the presence of various highly specialised and departmentalised laws and regulations for energy, and the lack of overall coordination among them have made the enactment of the *Energy Law* even more difficult. The third reason is about research support. The formulation of the *Energy Law* covers wide-ranging areas of interests and complex legal relations. Although relevant experts and authorities have made achievements through work practices and research over the past 20 years, the research on the legislation of energy law is still relatively inadequate as a whole. The fourth reason is that China's energy sector has not been completely free from the impact of the planned economy. There is still controversy over how to gain a thorough understanding of the differences and connections among different energy sectors and then proceed with the establishment of a relevant and specific promotion mechanism.

The timing for the introduction of the *Energy Law* is dependent upon when all sectors of the community will reach a full consensus on such major issues as the energy development strategies, development mode and management mechanism. In the future, as China's energy strategies are gradually clarified, the transformation of development model is accelerated, the management mechanism is continuously refined and the legislation research on the *Energy Law* is further intensified, then the time would gradually become ready for the introduction of the *Energy Law*.

Accelerating Introduction of Basic Laws for the Development of the Oil and Gas Industry and Atomic Energy

China has published some laws and regulations on the exploration, exploitation, storage, transportation, infrastructure protection and environmental protection for the oil and gas industry. However, due to the lack of coordination or the constraints of historical conditions, there is an absence of important legal systems or such systems remain to be improved for the paid use of oil and gas resources, price regulation, operating management, safety production and other aspects, resulting in the development of the oil and gas industry failing to secure the support of laws and regulations, and the lack of a legal basis for handling some relevant issues. Looking into the future, after drawing on the modern practices of such countries as the USA, Germany and Japan, reviewing the legislative experience in the oil and gas industry and giving consideration to the opinions from different sectors of society, China should formulate and release as soon as possible the basic law for the oil and gas industry which covers all relevant fields such as the exploration, exploitation, refining, storage, transportation, import/export, sale, production safety, quality control, supervision of the domestic oil market, and prescribe clear legal provisions for the strategic planning, regulatory model, market access system, pricing system, oil and gas storage, oil and gas finance and taxation policies for the oil and gas industry. This will not only benefit the refinement of the oil and gas legal system, it will also help push forward the formulation of the *Energy Law*, which may provide a reference and preparation for formulating the clauses on oil and gas development and supervision under the Energy Law, and serve as a legal norm for the exploration, exploitation and utilisation of unconventional oil and gas resources to facilitate the healthy development of unconventional oil and gas resources.

After becoming involved in atomic energy for the first time, many countries have promulgated their basic laws for atomic energy aimed at regulating the conduct of entities with different legal relationships in the atomic energy sector, and the atomic energy basic law has played an important role in the development and supervision of nuclear energy in these countries. For example, the *Special Law on Nuclear Energy Disaster Countermeasures*, the *Radioactive Materials and Radiation Hazards Prevention Law* and other supporting laws were adopted in Japan on the basis of its *Atomic Energy Basic Law*. The various contingency measures adopted in response to the nuclear leakage accident caused by the major earthquake in Japan in March 2011 were also launched on the basis of the *Atomic Energy Basic Law*. No progress has been made so far since the drafting of the *Atomic Energy Law* was contemplated in China in the 1980s. This is not commensurate with China's status as one of the world's major atomic energy powers. With the further development of atomic energy, it is time for the formulation of the *Atomic Energy Law*.

Amending the Electricity Law, Mineral Resources Law *and Other Existing Legislation in a Timely Manner*

With the development of the energy industry in China, the provisions of some existing laws and regulations, such as the *Electricity Law* and the *Mineral Resources Law*, can no longer cope with the current situation and must be amended without delay.

The main problems with the existing *Electricity Law* are: the lack of principled provisions and institutional arrangements for the sustainable development of the electricity industry; the lack of relevant provisions relating to the rules for electricity trading, tariff pricing system, the building of an electricity market and an electricity administrative system; some provisions are no longer consistent with the development status and future trend of the electricity industry; some provisions are too abstract to be implemented fully. As a result, the *Electricity Law* should be amended to enrich, modify and improve the aforesaid provisions and give full consideration to the future development trend of the electricity industry; encourage clean, efficient and environment-friendly electricity supply; further encourage and support the development of new energy, construction of Strong Smart Grids, optimisation of electricity resource allocation on a greater scale and the replacement of other resources by electricity; and clearly define the responsibilities of different entities to jointly ensure the security of the power system.

The main problems with the existing *Mineral Resources Law*: the failure of the specified mineral resource taxation system to reflect the scarcity of various mineral resources such as coal, and relevant over-fragmented provisions; and the lack of provisions on the transfer of exploration and mining rights. Hence, the *Mineral Resources Law* should be amended with a focus on adjusting the tax relating to mineral resources by fully incorporating the costs of energy resources into the prices of energy products to truly reflect the scarcity of energy resources, and on introducing new provisions governing the procedures for the transfer of exploitation and mining rights.

Actively Conducting Legislative Research on Major Issues of Energy Field

A number of major issues are encountered while energy for sustainable development is being carried out in China. Some of these issues have not received enough attention in terms of legislation, such as the energy reserve, energy early-warning as well as the development and utilisation of overseas resources. As the situation develops, legislative research on these issues has to be conducted in a timely manner so as to ensure the Chinese energy industry's continued move on the road to legalisation.

9.1.1.3 Strengthening Implementation of Laws and Regulations

Strengthening Publicity and Education on Energy Laws and Regulations

Energy legislation should be actively publicised through a wide variety of modern communication means and media platforms so that the major contents of energy legislation may become widely known and raise public awareness of the need to see energy legislation in a practical light. In particular, energy enterprises have to be built as a major platform for carrying out publicity and education on energy legislation, so as to closely integrate the implementation of energy legislation with corporate behaviour to run energy enterprises in accordance with the law.

Stepping up the Enforcement of Energy Legislation

Energy-related illegal and criminal acts such as theft and destruction of energy facilities which cause great harm to the safe supply of energy must be cracked down vigorously and earnestly through the strict enforcement of laws. The separation of government functions from enterprise management in the energy industry has invigorated the development of this industry, but at the same time this has also weakened the energy law enforcement power, especially the administrative law enforcement power, in China. This issue needs to be addressed. Through joint enforcement of laws or other means, the building of capacity for energy law enforcement must be strengthened and energy law enforcement stepped up so that the policy of ensuring legal compliance and law enforcement can be implemented fully in the energy sector.

9.1.2 Policy Guidance and Assurance

Policy is an important channel to deliver government opinions, and an important means for the government to apply macro-control initiatives. China is in the stage of rapid socioeconomic growth, facing a fast-changing environment for energy development. In these circumstances, China not only needs to step up the establishment of a legal regime for energy, but also roll out policies and measures in a timely manner in light of the new situation and new demand, so as to deal with the new circumstances and new issues arising from the process of development. Energy sustainability cannot be realistically pursued without policy support and assurance. To let policy play a good guiding role, China needs to set uniform and specific objective orientation, properly handle various significant relationships involved in policy formulation and implementation, strengthen overall coordination at the policy level and conduct post-implementation policy assessment.

9.1.2.1   Clearly Defining Orientation of Policy Objectives

The basic orientation of energy policies is to promote energy sustainability and accelerate the establishment of a safe, stable, economical and green modern energy system.

Enhancing the Capacity for Comprehensive Energy Development

China should continue to encourage the exploration and exploitation of domestic conventional energy resources. In particular, China must actively support the further exploitation and development of quality energy resources such as petroleum, natural gas and hydroelectricity. Policy guidance on the large-scale and comprehensive development of coal resources and on integrated development of coal and electricity should be strengthened. Support for the development of nuclear power should continue and more stringent policy requirements for nuclear power security should be imposed.

Support for the development of new and renewable energy resources needs to be strengthened. By such means as allocation of specialised fiscal funds, tax breaks and so on, China should support the exploration and exploitation of unconventional oil and gas resources. By comprehensively adopting various polices and measures, stronger support needs to be provided for promoting mature projects involving new and renewable energy resources. China should in a timely manner address signs that reflect unhealthy development of new and renewable energy resources.

China should actively encourage domestic energy enterprises to take part in international competition and cooperation in the energy field, provide support, standards and guidance to the energy enterprises that 'go global', moderately relax foreign exchange controls on investment in the exploration and exploitation of energy resources in overseas countries, and reduce the tariffs on imported energy resources and at the same time provide support at the diplomatic level for Chinese energy enterprises engaged in resources exploration and exploitation activities in overseas countries. China should also enhance the risk management and early-warning systems for overseas investments by Chinese energy enterprises and encourage those enterprises that 'go global' to strictly abide by the laws of the relevant country and to actively fulfil their social responsibilities so that the security of overseas investments can be safeguarded. A risk fund should be set up for overseas investments by energy enterprises to deal with any overseas investment risk these energy enterprises may be exposed to.

Encouraging Optimised Energy Distribution and Establishment of an Integrated Transport System

Energy distribution must be planned scientifically. Control must be exercised to guard against excessive growth of coal-fired power generation projects in the eastern and central regions of China. Development of large-scale coal and electricity bases should be encouraged in the western and northern regions with appropriate conditions to carry out in-situ conversion of coal to electricity for transmission to load centres over a long distance. China should optimise the distribution of nuclear power and oil refining projects.

China should push the establishment of a modern integrated energy transportation system and encourage the construction of such energy transportation channels and transportation systems as railways, highways, pipelines and power grids by simplifying approval procedures, optimising the energy pricing mechanism and structure, and stepping up financial and tax support. The development of Strong Smart Grids should be encouraged by building a grid network with UHV as the backbone, enhancing the intelligent level of power grids and carrying out long-distance power transmission on a greater scale. The development and expansion of the China's marine energy transport capacity should be supported.

Enhancing the Clean and Efficient Utilisation of Energy Resources

China should strengthen the policy incentives for energy conservation. The resources tax rates should be raised and differential tax rates imposed according to the different grades of resources by changing the tax base from one based on output quantity to one based on reserve quantity, and by changing the levying method from a specific basis to an ad valorem basis, with local governments allowed to take up an appropriately higher share of proceeds derived from resources, China should steadily implement tiered pricing for electricity and gas. It should encourage the development of new energy-saving industries such as contract energy management, step up support for investing in energy conservation, stimulate social investment by implementing such policies as subsidised loans, and boost energy conservation in various fields such as industry, building construction and transportation.

Stringent environmental policies in the process of energy development and utilisation must be implemented strictly. Pollution taxes should be levied when necessary and the external costs of environmental pollution and ecological damage internalised by tax charges. A system for voluntary energy conservation and emission reduction should be set up and those enterprises showing outstanding performance in honouring energy conservation and emission reduction commitments should be rewarded by the government. The mechanism for pollution credit trading, carbon emission trading and environmental capacity trading should be gradually improved.

Driving Progress in Energy Technology and Independent Innovation

China should actively create a policy environment that encourages technology innovation. It should coordinate efforts to establish a system for national energy technology innovation. Support should be increased for key technological breakthroughs in energy by various means such as the continuous implementation of key technology projects and the setting up of a fund for energy technology innovation. Enterprises should be encouraged to expand their R&D investment to address major technological difficulties focusing on energy sustainability. R&D spending should be made deductible for corporate income tax and high-tech enterprises in the energy industry should be offered financial incentive.

China should vigorously support the importation of advanced overseas technologies on the basis of independent innovation and encourage the introduction of complete production lines and the purchase of relevant patents instead of repeatedly importing foreign equipment and just foreign equipment. Funding should be increased for the assimilation and absorption of imported technology and equipment, and re-innovation encouraged on the basis of introduction, assimilation and absorption to continuously enhance the level of localisation.

China should increase support for the commercialisation of energy technology achievements. Domestic energy enterprises should be encouraged to purchase major domestic energy technology equipment by such means as government subsidies and tax concessions.

A specialised category for trading of the intellectual property of energy technology should be introduced for intellectual property trading agencies to push for speedy transformation of the results of energy technology research to drive productivity.

Driving the Establishment of Energy Market Operation Mechanism

China is committed to energy marketisation with Chinese characteristics and constantly modifies and optimises the system for energy management. It also actively and steadily deepens the energy pricing reform by accelerating the establishment of an energy pricing mechanism that reflects the supply and demand situation, the scarcity of energy resources and the cost of environmental damage and also the development of a price transmission mechanism that reflects the changing costs of various sectors such as the exploitation, production, transportation and sale of energy resource. It should fully consider the differences in energy development between different industries, between different regions and between urban and rural areas. It should steadily proceed with energy marketisation and continuously improve the rules for market operation. China should also open up the energy market in an orderly manner while at the same time strengthen the building of macro-control capacity to ensure the orderly and efficient operation of the energy market. The energy reserve system should be improved with fiscal subsidies or tax concessions for those enterprises which build energy reserves on a voluntary basis. The supervision of the energy market should be strengthened, energy market order and economies of scale of energy development maintained, the legitimate rights and interests of market players protected, and a subsidy system set up for basic energy supply directed at low-income groups to promote the popularity of energy services.

9.1.2.2 Handling Properly Several Major Relationships Concerning Policy Formulation and Implementation

Handling Properly the Relationship between Short-term Development and Long-term Planning

The implementation of energy strategies is a long-term and arduous task. Within a given period of time, energy policies are rolled out usually with a focus on the present by prioritising urgent matters, key sectors and important issues in the energy industry, which reflects the principle that important and urgent matters should take precedence. However, it should be noted that the formulation and implementation of energy policies must be based on a long term view by organically integrating short-term development programmes with long-term development planning. Guided by logical and clearly-defined energy strategies, efforts should be stepped up to overcome practical concerns and make policy more forward-looking and strategically oriented while reducing structural risks by unremittingly resolving some deep-rooted problems which may affect energy sustainability.

Handling the Relationship between Different Industries and Processes in the Energy Field

Judging by China's energy endowment and its development stage, in the coming two or three decades or even a longer period of time, conventional and unconventional energy resources as well as fossil and nonfossil energy resources need to be developed in a coordinated way, which must be reflected in policies. Meanwhile, China should highlight its policy orientation towards low carbon, environmental protection, high efficiency and energy conservation to promote the adjustment of energy structure and the transformation of energy development mode. China must concentrate its efforts on electricity development and coordinate development between secondary and primary energy resources. The focus should be on ensuring coordination and integration among different sectors by taking into account the development needs and balance of interests among different sectors on various issues such as planning, deployment, project arrangement, construction schedule, investment scale and price structure, for coordination of energy development, energy transportation and energy consumption.

Handling Properly the Interest Relationship between the State and Local Regions

National interest and local interest are inherently aligned. However, due to different focuses, there may be different demands in respect of particular issues and matters. In the energy industry, the differences are mainly reflected in such aspects as energy resources pricing mechanism, taxation system and distribution ratio, and control of energy resource exports. When promulgating policies relating to energy development, China has to fully consider the development of both the resource exporting regions and the resource importing regions (i.e. the overall needs of the national economic development) to ensure the continuous and stable supply of energy, take into account the reasonable demand of all parties, especially those of the energy resource exporting regions, and make efforts to maximise overall national interests. The resource exporting regions should fully benefit from and enhance sustainability when they are exporting energy and providing support for national development. China should give full support in respect of such problems as the treatment of ecological damage and environmental pollution arising from the exploitation and development of energy resources. China should also support efforts to address the issue of industry continuity after resources are exhausted. In the meantime, China insists that local interests must be subordinated to overall interests. Energy resource exporting regions should not keep on expanding their local interests without limit or improperly interfere in the energy market operation. They must not breach China's regulatory policy governing the energy market to the detriment of the healthy development of the energy industry.

Handling Properly the Relationship between Reform, Development and Stability

At present, the promotion of energy sustainability in China remains troubled by many deep-seated systematic and institutional problems. China must be firmly committed to reform, face up to the problems and actively explore effective solutions to them. In the meantime, China should take full consideration of the current situation in China and the support capacity of the economy and society. China must stay focused on boosting the scientific development of the energy industry to ensure priority is given to national energy security and social harmony. It shall seize the right moment for reform and maintain the pace of development while ensuring stability and strengthening the scientific orientation and effectiveness of energy policies. In particular on the issue regarding the building of the energy market, discussions should be conducted and decisions made in a scientific manner to actively and steadily push forward the building of the energy market. China must learn from and drawing on the international experience and be determined not to mechanically apply foreign practices not tailored for China's national conditions.

9.1.2.3 Strengthening Policy Coordination and Post-evaluation

Enhancing Convergence between Different Policies

As many policy instruments to drive energy development are available and many authorities may introduce energy-related policies, coupled with the fact that energy development requires the policy support and guidance in many aspects, China needs to earnestly strengthen the convergence between different policies during their formulation and implementation to make policies work and avoid policy misalignment or contradiction or the lack of support or concerted action. For example, the policy for raising the efficiency of energy resource allocation must converge and match with those policies for breaking the regional market barriers and building up a national electricity market, while the policy encouraging the construction of large-scale coal-fired power bases and new-energy power bases must be converged and matched with those policies encouraging the development of UHV power transmission and the construction of Strong Smart Grids. Current policies must be rationalised systematically to effectively correct problems, such as policy misalignment and contradiction, and make timely adjustment to or revoke any outdated policies.

Enhancing the Follow-up of Policy Implementation and Post-evaluation

A policy is only effective when implemented. While placing great emphasis on the formulation of energy policies and consistently making them more rational, China should substantially enhance the implementation of these policies, strengthen the follow-up analysis of policy implementation and resolve any identified problems in a timely manner. During the policy implementation, China must stay committed to established principles to maintain policy solemnity while avoiding over-rigidity and dogmatism to ensure solid and effective policy implementation. A post-evaluation system for energy policies implemented should be established to timely analyse and evaluate the effect of policy implementation for providing a basis for further policy improvement and adjustment.

9.2 Establishment of an Energy Standards System

Standards may be classified into national standards, industry standards, local standards, enterprise standards and so forth in China, depending on the nature of the subject publishers of the standards. These standards may be further categorised as technical standards, management standards, working standards and so forth, depending on the nature of the object entities. Developing and formulating a scientific energy standards system is essential for the healthy and scientific development of China's energy industry and energy enterprises, and for the enhancement of China's international competitiveness in the energy and related fields.

9.2.1 The Significance of Establishing an Energy Standards System

9.2.1.1 Strengthening the Establishment of an Energy Standards System is Essential for Healthy Development of the Energy Industry in China

Energy development in China is currently entering a new phase. New energy varieties, new energy products, new technologies for energy exploitation and utilisation, new energy offerings and service models are constantly emerging, giving rise to the need for compatible standards as guidance and norm for this process. As the existing energy standards system can no longer keep up with the fast-changing situation, it is necessary to further strengthen energy standardisation by revising the existing standards to accelerate the formulation of new energy standards and by constantly consolidating and refining the energy standards system to facilitate the healthy and orderly development of the energy industry.

9.2.1.2   Strengthening the Establishment of an Energy Standards System is Essential for Scientific Development of Energy Enterprises

Enterprise standards are those standards formulated for coordinating and unifying technology, management and work matters within an enterprise. Energy enterprise standards are an important part of an energy standards system. Standardisation is an important task at the fundamental and strategic levels for enterprises. Energy enterprises should strengthen the formulation of standards so that they can, on one hand, reduce its business operation and development costs, assure product and service quality, better satisfy the needs of consumers and customers and build excellent brand names and corporate image. On the other hand, they can make use of this opportunity to raise the market access threshold to nurture and build up unique competitive advantages that are difficult for competitors to imitate. Once the standards of an energy enterprise are upgraded to industry standards, national standards or even international standards, the enterprise would be able to secure a very favourable position in the competitive domestic and international markets, and seize much more development opportunities and wider market space. In recent years, many large Chinese firms have seen standardisation as a central task in the internationalisation process. 'Enterprise standardisation—international standardisation—enterprise globalisation' has become an important strategy for these enterprises to capture a dominant position in international competition.

9.2.1.3   Strengthening the Establishment of an Energy Standard System is Essential for Enhancing China's International Competitiveness

Standards constitute a basic element of China's core competitiveness. Since the dawn of the 21st century, many countries have formulated development strategies for standardisation, as marked by the successive introduction of such strategies in the European Union, the USA and Canada in 2000, and the launch of a general development strategy for international standardisation in 2006 in Japan. China published the *Standardisation Project Development under the 12th Five-Year Plan* in 2011, proposing that 'efforts be devoted to enhance the quality and benefit of standardisation; an innovative nation and a strong nation with emphasis on quality be built; and the contribution of standardisation to socioeconomic development greatly improved'. In the future, the international energy market for energy and related technologies and products will become increasingly competitive while emerging strategic industries such as new energy, Strong Smart Grids, new-energy vehicles and so on will become important arenas for competition. Strengthening the establishment of standards for the relevant sectors will help China take the initiative in international competition and encourage energy enterprises and energy equipment manufacturers to 'go global'.

9.2.2 Formulation of Energy Standards

The establishment of an energy standards system is a long-term regular undertaking. In line with the situation and the mission faced by the energy industry, efforts should be concentrated on the formulation of much-needed key standards in the near term to give China a relatively proactive position to deal with international energy reform, safeguard national energy security and promote energy sustainability.

9.2.2.1 Accelerating Improvement to Relevant Standards for New Energy Development

Lagging behind in the establishment of a standards system has become a key factor holding back the healthy development of new energy in China. Due to the lack of uniform standards and norms, the new energy industry in China is developing in a disorderly manner, with confusing standards and the lack of reliable assurance for equipment and engineering quality. This has posed numerous potential hazards to energy security, and has adversely impacted the development of the new energy industry in China. Internationally, many countries are vigorously exerting their clout over the formulation of standards for the global new energy industry, endeavouring to fight for a favourable position for themselves. China should focus on further stepping up the development of new energy standards, especially the introduction of a set of more systematic standards for wind and solar PV power generation to provide technical guidance and practices for the development of the relevant sectors.

9.2.2.2 Continued Efforts to Refine UHV and Smart Grid Standards

At present, Chinese domestic enterprises have already creatively carried out a considerable amount of work on developing standards for UHV and smart grids, which has enabled China to command a globally leading position in formulating standards for certain areas. State Grid Corporation of China has made achievements in this aspect. As at the end of 2011, the corporation had published 130 enterprise-level technical standards. It was also commissioned to draft and revise 16 national standards, with over 500 patents to its belt. Its UHV AC voltage standards have been recommended by the relevant international organisations as an international standard voltage. The corporation has also proposed a technical standards system framework for smart grids. It has published 166 enterprise-level technical standards, and was commissioned to draft or revise another 42 national and industry standards. In addition, the corporation has undertaken the formulation of 12 international standards for grid technology under authorisation by such international bodies as IEEE and IEC and by appointment of the Standardisation Administration of the People's Republic of China. In the next phase, China needs to continue to refine the technical standards system for UHV and smart grids in line with the development status of UHV and smart grid technologies, upgrade a set of more mature enterprise-level standards to national or industry standards as soon as possible, and strive to secure more rights to formulate international standards.

9.2.2.3 Organising the Formulation of Technical Standards for Electric Vehicles

To drive the establishment of a standards system for electric vehicles may help regulate the R&D and manufacture of electric vehicles as well as market order, reduce information inconsistency and also facilitate the effective integration of various processes in the electric vehicle industry chain to reduce the formulation of overlapping standards and increase investment returns. With respect to the establishment of a standards system for electric vehicles, the government and the industry associations should play an active coordinating role by relying on large corporations and business alliances to develop a technology roadmap and formulate relevant standards and practices to guide the healthy and orderly development of the electric vehicle industry after fully consulting various parties.

9.2.2.4 Constantly Improving and Refining the Compulsory Energy Efficiency Standards

The imposition of compulsory efficiency standards on major energy-consuming equipment and products is an important measure to implement the national strategic priority of energy conservation. China has already introduced compulsory efficiency standards for certain energy-consuming products such as television sets, air conditioners, home appliances, lighting and illumination equipment. The compulsory performance standards should be further expanded to cover more by launching compulsory energy performance standards for more energy-consuming products. In the meantime, China should gradually raise the bar for setting mandatory energy efficiency standards in China so as to fully fulfil the guiding role of these standards. For products without mandatory performance efficiency standards, the local governments should take the lead in establishing local standards in light of the local conditions by drawing on the experience of foreign countries.

9.2.2.5 Strengthening the Formulation of Standards for Energy Enterprises

China should actively guide and encourage energy enterprises to strengthen their standardisation work and continuously build up sound enterprise technical standards, management standards and various working standards. The standards in such areas as product quality, energy conservation and emission reduction must be made more stringent than the national, industry or local standards. Capable enterprises should actively adopt international standards and advanced overseas standards. Standardisation should be implemented as an important management mindset and management model. Enterprises should ensure that standards are always in place wherever possible in every aspect of business operation and management to achieve standardisation across the board. It is necessary to reinforce and review the enterprise-level standards at regular intervals to ensure the standards are relevant and advanced so as to enhance the enterprises' operational capabilities and competitiveness on an ongoing basis.

9.2.3   Bargaining Power over Development of International Energy Standards

At present, developed countries are in absolute control of the field of international energy standards. With increased independent innovation capabilities in recent years, China has improved its bargaining power over the formulation of international energy standards. Taking the electricity industry as an example, China has secured three chairmanships in the International Electrotechnical Commission and set up three international secretariats. China should continue to adopt specific measures for a more influential role and a greater say in the area of international energy standards.

9.2.3.1   Applying the International Standards as Important Reference for Formulating Domestic Energy Standards

With the close integration of the Chinese economy with the global economy, the mutual influence between the domestic and international energy markets is increasingly intensifying and international competition in the energy field has become increasingly complex and intense. Against such a backdrop, if the Chinese domestic energy industry and the related enterprises want to 'go global' better, China should play a proactive role in the competitive global energy industry and related fields, where factors relating to international competition and collaboration must be fully considered when formulating domestic energy standards and setting relevant benchmarks for energy technology, products, services, equipment and so on. Under the premise of safeguarding national economic and energy security, and without prejudice to the domestic industry's competitiveness, it is permissible to take the initiative to refer to or even directly adopt international standards. Where no generally accepted international standards are available, the world's strictest national or regional standards or internationally advanced enterprise-level standards may be referred to or adopted. Even if it is not possible to readily refer to international standards or the standards of developed countries (or advanced enterprises) after taking into account all practical considerations, it is nevertheless necessary to set relevant goals and plans for local standards to gradually fall in line with international standards or internationally advanced standards.

9.2.3.2   Strengthening International Exchange and Cooperation in the Field of International Energy Standards

Through different means, China should support and recommend domestic specialists in the energy and related sectors to join international bodies for energy standards, support and recommend domestic specialists and domestic enterprises to take part in drafting, formulating and revising international energy standards, and actively reflect domestic views and voices while contributing to the establishment of an international energy standards system. Moreover, it is necessary to encourage domestic enterprises, organisations and specialists to actively participate in exchange and cooperation in international energy standards in different ways, thereby gaining timely knowledge of the latest development in international energy standards and capitalising on all available opportunities to expand China's influence in the area of international energy standards. China should leverage the market advantages and technological edge brought about by the rapid development of its domestic energy industry to progressively enhance China's bargaining power over the formulation of international energy standards.

9.2.3.3 Taking a More Active Role in Formulating International Standards

A number of major energy corporations in China have made an important stride in 'going global' by establishing a rather close relationship of exchange and cooperation with their international counterparts and other relevant entities. Some of them have played a major role in formulating international standards. For example, some of State Grid Corporation of China's (SGCC) technical standards for UHV and smart grids have been or are being approved or accepted by relevant international standards bodies as international standards. In 2008, SGCC's 1000 kV UHV AC voltage was recommended by the International Electrotechnical Commission (IEC) and the Conference International des Grands Reseaux Electriques (CIGRE) as the international standard voltage. In 2010, SGCC submitted to IEC the proposals for 18 standards related to smart grids, covering user side interface, smart dispatching and other areas. Of these standards, including grid access on the user side, 12 have since been approved. In addition, SGCC has undertaken part of the work for IEC's strategic working group on smart grid standards system (IEC-SG3) and the working group of Institute of Electrical and Electronics Engineers (IEEE) on smart grid standards system (P2030). In February 2011, the Standard for Test Procedures for Electric Energy Storage Equipment and Systems for Power System Applications as drafted by SGCC was formally catalogued by IEEE as IEEE P2030.3 standard.[1] In June 2011, IEEE specifically convened a working group meeting in Beijing on the 4 international grid standards drafted by SGCC,[2] granting formal approval for the establishment of a project group. In the next phase, it is necessary to further expand the role played by the large energy corporations in participating in the formulation of international energy standards, support large energy corporations to take part in the relevant work, and encourage large energy corporations to gradually transform their role in international energy standards from one of an introducer and recipient to one of a developer and leader.

9.3 Large Energy Groups

The development of the energy industry is linked to a nation's energy security and economic lifeline. With industrialisation, information technology development, urbanisation, marketisation and globalisation, especially with globalisation and China's increasing dependence on foreign energy, the internal and external circumstances of China's energy

---

[1] IEEE P2030.3 Standard was jointly proposed by the Institute of Electrical and Electronics Engineers and the National Institute of Standards and Technology ('NIST') with the goal of formulating a set of standards and communication principles for smart grids that can be adopted on a global scale. The standard document covers mainly the standards and principles for power engineering, information technology and communications.

[2] The 4 standards are: 'Overvoltage and Insulation Coordination for 1000 kV and Higher Voltage UHV AC Systems'. 'On-site Testing Standards and System Commissioning Procedures for 1000 kV and Higher Voltage UHV AC Equipment', 'Voltage and Reactive Power Standard for 1000 kV and Higher Voltage UHV AC Systems' and 'Standard for Test Procedures for Electric Energy Storage Equipment and Systems for Power System Applications'.

development have become increasingly complex, and the country's energy companies are facing increasingly fierce competition from their international counterparts. As flagships of the energy industry, large energy groups are important players in the implementation of China's energy strategies. They assume an important role of ensuring the energy security of the country. Not only do they represent the competitiveness of China's energy industry, but they also play a role as the leaders and vanguards in the globalisation of China's energy industry. The development of large energy groups is the objective requirement for the transformation of the development model of China's energy industry and the key to protecting national energy security.

China should attach great importance to the role of large energy groups and create a favourable policy environment for their development. We should also actively support the horizontal and vertical expansion of energy companies, promote innovation in management and the institutional systems, and speed up the pace of 'going global' by taking advantage of the favourable conditions at home and abroad, so as to foster a number of energy flagship enterprises that meet the requirements of modern enterprise institutional systems, and boast strong international competitiveness and good brand images.

9.3.1 Significance of Developing Large Energy Groups

9.3.1.1 Large Energy Groups are Important Players in the Implementation of China's Energy Strategies and a Guarantee for the Nation's Energy Security

Internationally, oil, electricity and other large energy groups have always been the main agents of implementation of their national energy strategies at different stages of their countries' economic development. Using their advantages as large enterprises, they have effectively protected their countries' energy security. For example, Électricité de France (EDF) actively implemented the energy development strategy of France, effectively promoting the development of the power industry and ensuring supply security to meet the demand for electric power from the country's rapid economic growth. After the French economy entered a phase of stable development, France effectively enhanced its energy security level and achieved the strategic objective of self-sufficiency in electricity supply by innovation and R&D on nuclear power technology. The country has achieved the strategic target of electricity self-sufficiency.

China's energy enterprises have played a significant role in supporting the growth of the national economy and social development. The country's energy self-sufficiency has always been maintained at quite a high level of 90% or more. Unique capabilities in innovation, organisation and resource allocation ensure that large energy groups will be the main engines of implementation of the national energy strategy. Large energy groups also need to perform their pillar roles in accelerating a series of strategic tasks like developing alternative energy and renewable energy, adjusting and optimising the energy structure, promoting the transformation of the mode of energy development, as well as intensifying the protection of the ecological environment and responding to global climate change. Therefore, it is necessary to expedite the development of large energy groups to provide strong energy assurance for China's industrialisation and modernisation.

9.3.1.2 Large Energy Groups are China's Main Champions in Global Energy Competition

In a globalised world, competition between different countries is often manifested as competition between their corporations. The international competitiveness of enterprises is not only related to the development of the enterprises themselves, but also linked to the nation's power; it is an important manifestation of national strength and prosperity. In order to occupy a dominant position in the field of international energy competition, China must accelerate the development of large energy groups that are internationally competitive.

Currently, new trends are emerging in the international competition for energy: at the national level, there is simultaneous competition and cooperation between the energy exporting and energy importing countries, between energy exporting countries, and between energy importing countries. At the corporate level, some large global energy groups are not only increasing their scale, but also diversifying their businesses, extending up and down the industry chain and expanding into other related fields. Major international energy groups have not only become the main focus of the countries involved in energy competition, but they are also the main force in the global development of energy.

The M&A of overseas energy resources and the capture of overseas energy markets by 'going global' are of important strategic significance to China. They ensure the supply security of important energy resources and the implementation of China's energy diplomacy, which will in turn bring about the upgrade of industrial infrastructure for the domestic energy industry, promote the transformation of the mode of economic development, and advance sound and rapid social development. Hence, accelerating the implementation of 'going global' is a major strategic mission of China's large energy groups. At the same time, this is also the objective requirement of their own transformation and development as it enhances the value of their corporate brand names and competitiveness in the international market. Large energy groups are able to enter new markets, adjust their business directions, develop new businesses, and seek new profit growth centres by 'going global'. They can optimise their asset structures, promote the enhancement of overall business performance by managing and developing their overseas assets, and realise their globalisation goals in terms of assets, business operations, management and personnel. They can also establish strategic alliances with international partners to combine their strengths, share risks, enjoy mutual benefits and build win-win relationships. In recent years, China's large energy groups have been actively 'going global' and have achieved positive results (see Table 9.1). As early as 2003, Huaneng Group had already acquired a 50% equity interest in Australia's Ozgen Retail Pty Ltd. This was the first time a Chinese power corporation acquired electric power assets in a developed country. After a successful bid for the franchise to operate a national grid in the Philippines in 2008, State Grid Corporation successfully purchased a number of franchised transmission projects in Brazil in 2010. In February 2012, State Grid Corporation successfully tendered for a stake in National Energy Grid Company in Portugal to become the largest shareholder of this company. This was the first successful acquisition of a national-level transmission grid and natural gas transmission network, and it is of strategic importance to China's foray into the European energy market.

Table 9.1 Examples of 'Go Global' by large energy companies in China.

No.	Enterprise	Period	Description
1	Sinohydro	June 2003	It obtained the Merowe Dam Project Civil Engineering Works Contract in Sudan, the largest hydropower project in Africa
2	Huaneng Group	December 2003	It acquired 50% of the shares of Ozgen Retail Pty Ltd in Australia with an investment of USD227 million
3	Yankuang Group	December 2004	It acquired the Southland Coal Mine in Australia with an investment of AUD32 million
4	PetroChina	August 2005	It purchased PetroKazakhstan in Kazakhstan with an investment of USD4.18 billion
5	China Power Investment Corporation	December 2006	It acquired the right to develop 7 cascade hydropower stations in Myanmar
6	CNOOC	January 2008	It bought all the equities of Awilco Offshore ASA with an investment of USD2.5 billion
7	China Datang Corporation	January 2008	It invested in the construction of the Dapein Hydropower Project in Myanmar
8	Shenhua Group	March 2008	It invested in the GH EMM INONEDIA Project of Indonesia to build a $2 \times 150MW$ coal-fired plant
9	Huaneng Group	March 2008	It purchased all the shares of Tuas Power in Singapore with an investment of SGD4.235 billion
10	Sinopec	September 2008	It purchased all the shares of Tanganyika Oil Company in Canada with an investment of USD1.93 billion
11	Shenhua Group	November 2008	It gained the permit to explore the Watermark Exploration Area in Australia
12	PetroChina	November 2008	It was awarded the development and service contract of Al-Ahdab oil field in Iraq
13	State Grid	December 2008	It won a 25-year franchise to operate a national power grid in the Philippines with an investment of USD3.95 billion
14	CGNPC	April 2009	It established a company with Kazatomprom in Kazakhstan to develop its uranium resources
15	Sinopec	June 2009	It acquired all the shares of Addax Petroleum in Switzerland with an investment of USD7.24 billion
16	PetroChina	June 2009	It acquired a 45.51% stake in Singapore Petroleum Co Ltd with an investment of SGD1.47 billion
17	Sinopec CNOOC	July 2009	Jointly acquired a 20% interest in an oil-bearing block in Angola held by Marathon Oil Company in the US with an investment of USD1.3 billion

Table 9.1 (*continued*)

No.	Enterprise	Period	Description
18	Yankuang Group	August 2009	It purchased Felix Resources Co Ltd in Australia with an investment of AUD3.333 billion
19	PetroChina	November 2009	It was awarded the development and service contract of Rumaila oil field in Iraq
20	PetroChina	December 2009	It was awarded the development contract of Turkmenistan's South Iolotan natural gas field worth USD3 billion
21	PetroChina	December 2009	It bought Oil Sands in MacKay River and Dover in Canada worth about USD1.8 billion
22	PetroChina	January 2010	It was awarded the development, production and service contract of Halfaya oilfield in Iraq
23	Sinopec	January 2010	It acquired 40% of the shares of the Brazilian Branch of Spain's Repsol with an investment of USD7.1 billion
24	CNOOC	February 2010	It bought the shares of Uganda oil field from Tullow Oil plc in UK with an investment of USD2.5 billion
25	CGNPC	February 2010	It bought 66% of the shares of Energy Metals Limited in Australia with about USD100 million in investment
26	PetroChina	March 2010	It purchased Arrow Energy Limited in Australia in partnership with Royal Dutch Shell with an investment of USD3 billion
27	CNOOC	March 2010	It acquired 60% of the shares of Pan American Energy in Argentina from BP p.l.c. with an investment of USD7.06 billion
28	Sinopec	June 2010	It bought 9.03% of the shares of Syncrude Canada Ltd with an investment of USD4.65 billion
29	China Huadian Corporation	August 2010	It invested in the 2×65MW coal-fired plant in Indonesia's Batam Island
30	China Coal Group	September 2010	It was granted approval to develop the Surat Basin of Columboola, Australia
31	China Huadian Corporation	September 2010	It invested in the class I 2×90MW hydropower plant in Asahan, Indonesia
32	China Guodian	November 2010	It obtained the right to develop 2 hydro stations in Stungcheayareng and Sambor of Cambodia
33	State Grid	December 2010	It acquired 7 franchise operators of power transmission networks in Brazil with an investment of USD989 million
34	Sinopec	February 2011	It bought all the shares of Occidental Petroleum's Argentinean subsidiary with an investment of USD2.45 billion

(*continued overleaf*)

Table 9.1 (*continued*)

No.	Enterprise	Period	Description
35	China Huaneng	April 2011	It purchased 50% of the shares of InterGen with an investment of USD1.23 billion
36	China Investment Corporation	August 2011	It acquired 30% of the shares of GDF Suez SA with an investment of USD3.24 billion
37	Sinopec	October 2011	It acquired all the shares of Daylight Energy Ltd in Canada with an investment of USD2.13 billion
38	Sinopec	November 2011	It bought 30% of the shares of the Brazilian subsidiary of Portugal's Galp Energia SGPS SA with an investment of USD3.54 billion
39	CNOOC	November 2011	It purchased all the shares of OPTI Canada Inc. with an investment of USD2.1 billion
40	China Three Gorges Corporation	December 2011	It bought 21.35% of the shares of Energias de Portugal with an investment of EUR2.69 billion
41	Yankuang Group	December 2011	It acquired 77% of the shares of Gloucester Coal Ltd in Australia with an investment of USD2.09 billion
42	Sinopec	December 2011	It purchased 15% of the shares of Australia Pacific LNG with an investment of USD1.77 billion
43	State Grid	February 2012	It bought 25% of the shares of Portuguese energy group REN—Redes Energéticas Nacionais, SGPS, S.A. with an investment of EUR387 million

9.3.1.3 Large Energy Groups are Important Players in the Promotion of Innovation in Energy Technology

The energy industry is a technology-intensive industry, in which technology innovation can be an enormous driving force for both energy development and utilisation. Advancing energy technology innovation is an objective need for energy companies to seize the initiative in their own future economic and technological development and to accelerate their own development. It is also an important requirement to promote the technical upgrade of China's energy industry. As energy technology is characterised by massive investment, multiple interrelationships, long cycles and strong inertia, large energy groups tend to have stronger capabilities to embark on investment and bigger technological advantage than the smaller energy companies. Throughout the world, large energy groups are the leaders in technology innovation in the relevant sectors in their home countries, and a number of large-scale energy groups have become the leaders of research and technological development of the global energy industry.

Building an innovative nation is a major strategic plan for the overall development of the socialist modernisation programme. As an important force in building an innovative nation, large energy groups are the main advocates of China's energy technology innovation. Large energy groups are the main force of energy technology innovation, which have irreplaceable advantages in such areas as investing in science and technology, building up scientific and technological expertise, training of industry personnel, and integrating production and research. In the overall strategy of building an innovative nation, China must give full play to the principal role of innovation. It needs to increase support for technology innovation in energy companies to promote China's scientific and technological progress in the energy field and safeguard China's future energy security. We must tap the potential for energy supply by technical innovation and look for alternative energy, so as to improve energy utilisation efficiency. In recent years, large-scale energy groups have invested more science and technology R&D, and have achieved great progress in scientific research. The large energy companies won 3 out of 10 National Science and Technology Progress Awards (special class and first class) in 2011.

9.3.2 Supporting the Development of Large Energy Groups

Corporate development depends not only on the corporation's own efforts, but also on a favourable policy environment. The scarcity of resources and the special strategic nature of the energy industry determine that in the development and strengthening of China's energy enterprises, it is both necessary to pay attention to the basic role of the market in resources allocation and to fully rely on the government's macro-control. Every historic spike in the growth of large energy groups in the USA, the UK, France and Japan was a direct result of the support of their governments. The development and strengthening of China's energy enterprises also need strong support from the state.

9.3.2.1 Supporting Large Energy Groups' Horizontal and Vertical Expansion of Operations

The development and expansion of an enterprise is driven by economies of scale, synergies and an enhanced ability to withstand risks, and it is required for the enterprise's sustainable development and effective participation in global competition. Judging by their growth history, multinational energy companies invariably underwent the process of expansion, strengthening and globalisation despite their differences in the nature of assets held and development background. China's large energy groups are mainly state-owned holding companies, with a relatively narrow range of businesses of a highly specialised nature. In order to promote the development of its large energy groups, China should draw on the development experience of international energy companies, promoting horizontal merger and vertical integration as well as developing beyond the energy field.

Firstly, support the integration and mergers of large energy groups. Larger scale of operation does not only bring about economies of scale, it also greatly enhances an enterprise's ability to withstand risks and stand up to large global energy groups. Having integrated more than 1400 small local power companies in the course of its development, Électricité de France has maintained a dominant position in development and operation in the French power industry for the past 40 years. Exxon Mobil Corporation also gradually grew up through continuous mergers of weak oil companies in the mid- and late 19th century. The enterprises of the Chinese energy industry and energy equipment manufacturing sector are small, numerous and not competitive. China should speed up the corporate M&A of the coal and new energy industries, break down geographical barriers, develop large corporations, improve industry concentration, and achieve economies of scale in production, so as to increase the competitiveness of the Chinese energy industry.

Secondly, support large energy groups to integrate upstream and downstream along the industry chain. The higher level of economic development will give rise to a higher degree of integration between the upstream and downstream operations and stronger competitiveness. Integration of the upstream and downstream industry sectors is a common trend in the development of the world's largest energy companies. Through merger and integration, energy companies are able to develop with corporatisation and economies of scale, while reducing costs, improving operational efficiency of the whole industry chain, and improving their enterprise management standards and competitive advantages. China should support its large energy groups to capture the high-end segments of the industry chain and core technologies, suitably extend the industry chain to cover the upstream and downstream sectors, optimise and integrate the business value chains, enhance the integration capacity and the value-added potential of the industry chain, make efforts to expand services, and improve corporate profitability. It should support large energy groups in accelerating the expansion of such operations as exploration, exploitation, trading, stockpiling and management of energy resources, and in speeding up the merger and reorganisation of coal and electricity groups for joint development of the coal and power industries. China should develop the energy transport and logistics industry and promote the collaborative development of coal, power, road transport and port facilities; and accelerate the development of the downstream sectors like chemical, refining, fertiliser, professional and technical services.

Thirdly, support energy groups to expand into related fields. Energy companies may diversify into related fields for the reason of ensuring business stability and security, by leveraging their own management and technical advantages as well as related resources advantages through synergies. They can provide users with a wide range of energy and related services, expand their market shares, and improve business performance. Électricité de France, E.ON Group and other power companies have continued to expand into natural gas and other new business areas. The business interests of the world's top three oil companies, Exxon Mobil, Royal Dutch Shell and BP Plc, cover not only the upstream and downstream sectors of oil and gas, but also extend to port facilities, transportation, coal mining, power generation and other energy-related fields. In addition, as large energy companies generally have stronger cash flows and a larger asset base, appropriately developing into the financial service industry and actively exploring the ways and modes of integration of production and financing operations are also important options for the expansion of the space for enterprise development.

9.3.2.2   Supporting Large Energy Groups to Speed up the Pace of 'Going Global'

After 30 years of reform and opening up, and with the rapid development of the energy industry, the overall strength of China's large energy groups has been significantly enhanced and they enjoy clear comparative advantages in some areas. In 2011 Sinopec, PetroChina and State Grid Corporation became three of the top 10 Fortune Global 500 companies, ranking 5th, 6th and 7th respectively. They are among the world's most influential corporations. Over the past 10 years, many domestic energy enterprises made bold attempts at 'going global', accumulating abundant experience, both positive and negative. Compared to large overseas multinational energy companies, Chinese energy companies generally operate along a single business line and have weak international capabilities, but they have laid the basis and fulfilled the conditions for speeding up the pace of 'going global'.

Large Chinese energy groups are given an important historic opportunity to accelerate the pace of 'going global'. On one hand, as the developed countries in Europe and the USA are slowly recovering after the financial crisis, with declining investment and financing capacity, the developing countries with rich resource endowments are badly in need of capital for economic development. As overseas asset prices have fallen, accompanied by a reduction of certain restrictions, large energy groups are seeing greater interest in cooperation for investment in overseas countries. On the other hand, since the beginning of the 21st century, with the acceleration of economic globalisation, especially since the global financial crisis, the world's economy and the international division of labour have experienced significant changes. China's international status and overall strength have significantly increased, as the continued appreciation of the RMB has boosted purchasing power, and strong foreign exchange liquidity has been accumulated. Now it is a hard to come by opportunity to speed up the pace of 'going global' for China's large energy groups.

The state may support large energy groups in 'going global' in the following five ways.

Firstly, encourage large energy groups to enter the international market from the policy and administration levels. Reinforce macro policy guidance for the energy enterprises to 'go global', support resource-based investment, technology acquisition investment and market development investment of strategic significance, achieving the strategic synergy between the micro objectives of foreign investment by the energy companies with the country's macroeconomic objectives. Simplify the approval procedures for foreign investment by energy companies, put in place sound financial support measures for the energy enterprises to 'go global', formulate and implement such support policies in such areas as fiscal, taxation, foreign exchange, and capital finance, provide support and guidance for the energy companies in the area of introduction, exchange, training, and remuneration systems for international talents, creating a favourable policy environment for energy enterprises to intensify their 'going global' efforts and conduct international operations.

Secondly, give full play to the important supportive and cooperative role of diplomatic resources. China should give full play to the role and advantages of the overseas agencies and institutions for foreign trade and economic cooperation, intensify information support and guidance for 'going global' by the domestic energy companies, and guide the domestic energy companies to pay attention to foreign market opportunities in a timely manner and adopt appropriate strategies for 'going global' on the basis of the result of scientific risk-benefit assessment. At the same time, China should effectively integrate diplomatic resources, fully leveraging the country's influence in the political, economic, military, and diplomatic spheres, alleviating any interference in our energy enterprises' 'going global' undertakings by political and noncommercial factors from the target regions, and boosting the smooth implementation of the 'going global' undertakings.

Thirdly, lead the energy companies to realise strategic synergy in the process of 'going global'. The energy enterprises in China started multinational operations late. On the whole, most energy companies lack experience or understanding in developing and implementing transnational business strategy. They need to increase information exchange and collaboration in the process of 'going global' to avoid vicious competition in overseas investment and M&A. China should strengthen guidance and coordination of the energy enterprises' 'going global' undertakings, reinforce information sharing and give full play to the unique characteristics and advantages of enterprises, so as to make concerted efforts and form a united front, avoiding the situation where everyone fights their own battles.

Fourthly, support large energy groups to 'go global' using the model of 'joint acquisition and joint development'. Combining strengths is the main trend of the current global investment activities, and this helps to spread out the market risks for China's large energy groups in the process of 'going global'. China should actively help large domestic energy groups establish long-term stable cooperative relationships with well-known and influential international enterprises in the target regions to carry out overseas investment.

Fifthly, strengthen risk warning, prevention and control for the energy enterprises in the process of 'going global'. Due to a variety of uncertainties in the international operations, large energy groups will also face political, economic, legal, security and cultural risks in the process of 'going global'. Therefore, the large energy groups should be both active and prudent when 'going global'. They should place risk prevention as an important priority and carry out risk control and strategic security measures to advance the steady and sound development of their overseas businesses. The state should also make full use of its advantage in information resources, put in place relevant risk assessment mechanisms for the international market and the investment target regions, guide the enterprises to set up internal risk control systems, so as to effectively prevent and respond to any risk in 'going global'. In case of the occurrence of a significant risk, take precaution in a timely manner to ensure the security of overseas assets and personnel.

9.3.2.3 Supporting Large Energy Groups to Promote Innovation of Systems and Mechanisms

Most of the large energy groups are holding companies under state control, and they typically had highly centralised production and management systems combining government functions and enterprise management before China's reform and opening up. With China's gradual transition from a planned economy to a socialist market economy, the energy industry has now separated government administration from enterprise management and undertaken market-oriented reforms, and the management systems and operational mechanisms of the large energy groups are being continuously adjusted. Due to a complex history and the gradual nature of the market-oriented reforms of the energy industry, China's energy industry faces some deep-seated conflicts of systems and mechanisms in the process of building up large modern energy groups with international competitiveness. The state needs to offer support and guidance in finding the solution to these deep-seated problems.

Firstly, we should accelerate the establishment of a modern enterprise system for large energy groups. Most of the large energy groups are state-owned or state-controlled enterprises formed after the reforms undertaken by government departments to separate government administration from enterprise management, and they are plagued with such typical problems as imperfect governance structures and inflexible management mechanisms. Although in recent years we have gradually improved the governance model and the supervision and management mechanisms of the large energy enterprises through establishing boards of directors, introducing external directors, implementing cross appointments of Party members and administrators, and establishing and improving the performance assessment and incentive mechanisms for the state-owned enterprises. Overall, they have yet to develop a genuine modern enterprise system. In order to develop large energy groups, we need to further improve their corporate governance structures, and establish a scientific and standardised governance mechanism based on previous experience. At the same time, we need to continue to improve the methods of performance assessment and incentive mechanism for the large energy groups. We should actively promote listing by eligible large energy groups, especially public listing as conglomerates, to encourage the enterprises to improve their governance structure through a public listing, and change their operational mechanisms and reinforce the development of modern enterprise systems.

Secondly, we need to help large energy groups to effectively solve a variety of problems bequeathed by history. There are widespread historical problems such as enterprises running social services and nonconformant employment practices in China's large energy groups, which are bearing rather heavy burdens. These problems have been created under special historical conditions, involving the national regulations and policies and the immediate interests of the workers, and cannot be effectively solved by simply relying on the efforts of the enterprises. In this regard, on one hand the state should strengthen guidance to encourage enterprises to actively and steadily address these historical issues under the framework of existing regulations and policies. As for those problems that cannot be solved in the short term, we should solve it step-by-step in accordance with the principle of respecting history and seeking the truth from facts. On the other hand, we should give policy support to enterprises to solve the problems left over by history and alleviate the cost pressure and security and stability pressures on them.

Thirdly, we should support large energy groups in carrying out internal management innovation. The development model of many large Chinese energy groups are typically rather unsophisticated, with such problems as multiple organisational hierarchies, long management command chains, a lack of integration in resources, weak management control of group activities and unscientific management. State Grid Corporation of China is a very typical example. Due to historical reasons, in the early days of its establishment, the internal management structure had grown to eight levels at the highest point, and the organisation of labour relations in its power supply subsidiaries was also a fixed system unchanged for decades, which meant it lagged far behind the development of the productive workforce. With the advance of reforms, the State-owned Assets Supervision and Administration Commission of the State Council has put forward a clear policy requirement for central government-controlled enterprises to cut down the complexity of their management hierarchies, strengthen the operational management and control of the group activities as well as implementing scientific management systems. With support from the state, the large energy groups should be encouraged to follow the conventions of enterprise development, implement modern management concepts, models and methodology and adjust the organisational structures. It should be in their interest to cut down on the complexity of management hierarchies, create new management and control models, optimise the business processes and continue to build new internal management systems in accordance with the principle of adopting a flat, specialised, standardised and intensive management structure. This will help promote resource integration of the core elements, enhance the group control and execution capabilities, fully leverage the advantages of size and strategic synergy of the combined group operations, and enhance the operational efficiency of assets and economic benefits.

9.3.3   Market Position of Large Energy Groups

Whether it is market or social impact or the key factors related to competitiveness such as scientific and technical R&D and modernisation level of equipment, large Chinese energy groups will always occupy a dominant position in China's energy industry. In recent years, accompanied with public debates on issues such as 'the state-owned enterprises advance while the private enterprises retreat', anti-monopoly and energy price adjustments, the dominance of large energy groups has attracted much attention and they are often dogged by allegations of being 'monopolistic'. In this regard, it is necessary to conduct a comprehensive and objective analysis to gain a good understanding of the issue.

Firstly, the uniqueness of the energy industry in terms of economies of scale has determined that monopoly is a factual phenomenon in the development of energy industries around the world, and only a monopolistic operation is suitable for a business with natural monopoly characteristics. The energy industry needs huge investment in fixed assets and possesses the obvious characteristics of economies of scale. Such characteristics determine that the huge energy companies possessing more resources with stronger market positions will be the winners in the competitive energy sector. For this reason, the energy markets of various countries are generally dominated by a handful of large energy groups and are characterised by oligopoly. For example, 76% of the countries in the world have only one oil company each, and the rest have three oil companies at most. Both EON Group and Électricité De France are power enterprise groups with apparent competitive advantages in the world today and they have maintained strong market influence and control for a long time.

There are typical natural monopolies in some segments of the energy industry, such as the electricity transmission and distribution grids, long-distance transmission pipelines of oil and natural gas, and the municipal pipelines and power grids. No matter how these entities are divided up, the characteristics of natural monopoly in their business areas will not change. Therefore, for power grids, oil and gas pipelines and town gas pipeline networks, the implementation of exclusive monopoly operation is the most efficient institutional arrangement.

Secondly, in the context of economic globalisation, we should not focus solely on the domestic market when analysing the monopoly problem, but should have an international outlook and take full account of the need for participating in the international competition for energy. With greater economic globalisation, the competition in the energy market is becoming increasingly global. For China, large energy groups are the main participants in the international energy competition. Perhaps from the perspective of the domestic market, a particular energy group has a monopoly position and strong market influence, but from the perspective of the international energy market, it is just one of many market players participating in the international competition for energy. In order to enhance our competitiveness in the international energy market, and better protect our national energy security, we must accelerate the development of China's large energy groups, and never ignore international competitive factors and never limit them on account of their domestic monopoly positions. Not only should we refrain from restricting them, but we should also actively support China's large energy groups in actively participating in international competition. Through this competition, they can continue to grow and develop, as well as build up decisive influence in the international market as soon as possible.

Thirdly, international experience has shown that effective supervision can make up for the deficiency in monopoly, and that anti-monopoly measures target monopolistic behaviour rather than the monopolistic market position of the enterprises. In the energy sector, through innovative regulatory systems and improvements in the assessment mechanism, we can make the best of an enterprise's economies of scale while exercising effective regulation of monopolies. It is a universal consensus and common practice in many developed market economies. Therefore, from the perspective of the state, we should improve the energy market rules and the regulation of energy markets, as well as further standardise the behaviour of energy companies to encourage them to improve management, reduce costs and improve efficiency. We should maintain fair market practices and protect the interests of society. We should also provide large energy groups with stable policy expectations and a standardised institutional environment for their healthy development and help them better fulfil their responsibilities and missions.

Over the past 20 years, Western developed countries have also gradually adjusted the 'monopoly standard', changing the legislative principle that a 'dominant market position' determined exclusively on the basis of market share and industry concentration shall be the criteria for judging monopoly cases and imposing sanctions. They are beginning to shift to a new criterion based on judgment of whether a enterprise's 'behaviour has abused its dominant position in the market'. China has already promulgated the *Anti-monopoly Law*, with the major objective of preventing enterprises from harming consumer interests and the general public by abusing their monopoly positions.

Fourthly, China's basic economic system requires the development of large state-owned energy groups to maintain absolute control over the energy industry. China's basic economic system where public ownership plays the dominant role while diverse forms of ownership develop side by side with state ownership is the fundamental characteristic of the socialist market economy with Chinese characteristics. The energy industry is an important basic industry and key sector affecting national security and the nation's economic lifeline, over which the state-owned capital must maintain absolute control. Therefore, it becomes an inevitable choice to foster and develop large state-owned energy groups. Large state-owned energy groups have stronger intrinsic motivation and bear more important social responsibilities in such aspects as implementing the national macro-control measures, protecting the security of energy supply, fulfilling corporate responsibilities, and promoting a harmonious society.

9.3.4   Social Responsibilities of Large Energy Groups

Large energy groups are an important pillar of the national economy and the development of the energy industry, affecting all aspects of social production and life with their business scope covering urban and rural areas. Therefore, we must enhance their sense of responsibility and mission, so that they will of their own accord implement the scientific concept of development, protect national energy security and serve the country's modernisation efforts. They should be committed to a spirit of openness and strict self-discipline, and continue to improve the quality of service. They must accept supervision and regulation, as well as fulfil their social responsibilities. All these are important for achieving sound and rapid development.

9.3.4.1   Committing to Openness and Progress while Continuing to Make Improvements

Large energy groups must be fully aware of their important responsibilities, and take the initiative in planning development in the context of socioeconomic growth, and consciously implement and support the national development strategy to serve the socialist modernisation programme and development of various local undertakings. We should reform and innovate, strengthen scientific and technological progress, and change the pattern of development to continuously improve the operational efficiency of corporate assets when providing the state with a stable, safe and reliable energy supply. Starting from the realities of our country, we should focus on the domestic and international markets, establish a global vision and regard enhancing national strength and international competitiveness as our mission. We should never be satisfied with ruling the roost in the domestic market, but should fight for the initiative in the complex and intense international competition for energy.

9.3.4.2   Enhancing Self-discipline and Standardising Operations According to the Law

Due to the impact of unhealthy practices deriving from the long history of a planned economy and the stage of social transformation, there are many weak links in the enterprise management and ethics in the energy sector. We need to attach great importance to and enhance our self-discipline. Those large energy groups having a strong influence on the market should strengthen self-discipline, and constantly improve their management styles to aim to run the enterprises according to the law, properly manage the enterprises and run the enterprises diligently and thriftily.

9.3.4.3   Innovating Services and Continuing to Improve Service Quality

The quality and reliability of energy services relate directly to the people's working and everyday lives, development of the national economy and improvement of people's living standards. The large energy groups should do a good job in the construction of energy facilities and management of energy risks and emergencies, as well as maintain the operational security of the energy systems. They should adapt to the development of energy technologies and user needs, while actively creating new energy services and improving the delivery of energy services. We should also change our mode of energy service from one of offering passive services to one of offering active services and from one of singular service of purchase and sale to one of integrated intelligent services, which will form a new service structure marked by two-way interaction, diversity and high quality.

9.3.4.4 Handling the Relationship between Economic and Social Benefits

The scientific concept of corporate social responsibility requires large energy groups to improve efficiency and effectiveness while requiring them not to solely pursue economic self-interests. Large energy groups should pay attention to the needs of the stakeholders in the process of implementation of energy strategies and accelerating the development of enterprises, and should set an example in assuming responsibilities and taking the lead in implementing the relevant national guidelines and policies to implement the requirements of energy conservation and environmental protection through every stage of the process in enterprise operation. They should support the work of 'agriculture, countryside, and farmers' and do a good job in providing integrated energy services; they should safeguard the legitimate rights and interests of workers, building harmonious labour relations, and support the development of social welfare undertakings to strive to maximise the comprehensive value of the economy, society and environment.

9.3.4.5 Strengthening Communication with Communities, and Consciously Accepting Social Supervision and Government Regulation

As the main developers, operators and maintainers of national energy facilities and the main provider of energy products and energy services, large energy groups should correctly handle their relations with stakeholders, and improve the information disclosure mechanism in accordance with the basic principles of openness and transparency. They should maintain open communication channels with the relevant parties and improve the communication methods, focusing on various demands, and properly articulate their values and development concepts to improve the efficiency and effectiveness of communication. At the same time, they must consciously accept the supervision of the government, users and media, and actively assimilate a variety of views. They must continue to improve their work accordingly and strive for understanding and support.

References

[1] Zemin Jiang (2008) *Research on China's Energy Problem*. Shanghai: Shanghai Jiao Tong University Press.

[2] Peng Li (2005) *Electric Power Must Come First, Peng Li's Power Diary*. Beijing: China Electric Power Press.

[3] Guobao Zhang (2010) *Report on China's Energy Development*. Beijing: Economic Science Press.

[4] Tienan Liu (2011) *Report on China's Energy Development*. Beijing: Economic Science Press.

[5] National Bureau of Statistics of China (2011) *China Statistical Yearbook 2011*. Beijing: China Statistics Press.

[6] Department of Energy Statistics and National Bureau of Statistics of China (2012) *China Energy Statistical Yearbook 2011*. Beijing: China Statistics Press.

[7] China Electricity Council and Environmental Defense Fund (2009) *The Current Status of Air Pollution Control for Coal-fired Power Plants in China: 2009*. Beijing: China Electric Power Press.

[8] Guihui Wu (2011) The energy situation and the ways of development in China. *Engineering Sciences*, April: 10–11.

[9] Jiagui Chen (2007) *The Report on Chinese Industrialisation—Review and Study of China's Provincial-Level Industrialisation 1995–2005*. Beijing: Social Science Academic Press (China).

[10] Yuan Chen (2007) *Energy Security and Energy Development Strategy Research*. Beijing: China Financial & Economic Publishing House.

[11] Qingyi Wang (2007) Fundamental change facing China's coal industry, *China Coal*, February: 19–23.

[12] Yue Chen and Qinhua Xu (2010) *China Energy International Cooperation Report (2009)*. Beijing: Current Affairs Press.

[13] Yishan Xia (2009) *China's Perspective on International Energy Development Strategy*. Beijing: World Knowledge Publishing House.

[14] Zheng Lyu, Kesha Guo and Qizi Zhang (2003) Experience and cost of traditional industrialisation strategy in China, *China Industrial Economy* 1: 48–55.

[15] Jiu Yan, Xinzhu Zhang, Yongsheng Feng and Yuan Ma (2010) *Study of China's Grid Management System Reform*. Nanchang: Jiangxi People's Publishing House.

[16] Dan Shi (2006) *Research Report on the Market Reforms of China's Energy Industry*. Beijing: Economic Management Press.

[17] Haiyun Wang (2011) *Review of China's Energy Diplomacy 2010*. Beijing: Energy Diplomacy Research Center, March.

[18] Carbon Politics: New International Politics and China's Strategic Option, Economy of China, September 2009.

[19] Zhiznin Z. Stanislav (2005) *International Energy Politics and Diplomacy*, transl. Xiaoyun Qiang. Shanghai: East China Normal University Press.

[20] Boqiang Lin (2009) *Study of China's Energy Policy*. Beijing: China Financial & Economic Publishing House.

[21] Boqiang Lin and Guangxiao Huang (2011) *Energy Finance*. Beijing: Qinghua University Press.

[22] China Energy Report (2006): Strategy and Policy Research, Beijing: Science Press, 2006.

[23] Yiming Wei, Ying Fan, and Zhiyong Han (2008) Opportunities and challenges for CBM development in China, *Natural Gas Industry*, March: 1–4.

[24] Zhou Yi (2011) The breakthroughs in R&D and manufacture of large oil and gas tankers, *China Petrochemical News*, 5 May 2011.

[25] Bogeng Chen (2005) Holding well the key point to make cities and countryside develop harmoniously, *Shanghai Rural Economics*, October: 13–15.

[26] Jianjun Guo (2007) Status, issues and policy proposals on China's urban–rural development, *Review of Economic Research*, January: 24–44.

[27] Angang Hu (2005) A look at the rise of modern China, *Research on Development*, March: 1–5.

[28] Yage You, Wei Li, Weimin Liu, *et al.* (2010) Development status and perspective of marine energy conversion systems, *Automation of Electric Power Systems*, July: 1–12.

[29] Kai Ming Feng (2009) Controlled nuclear fusion and ITER project, *China Nuclear Power*, September: 212–19.

[30] Yu Ouyang (2009) Research progress in advanced nuclear energy technology, *China Nuclear Power*, June.

[31] Weiping Yan (2008) *Clean Coal Power Generation Technology*, *2nd Edition*. Beijing: China Electric Power Press.

[32] Lihong Yu (2008) Research of Strategic Substitution of Energy Resources. Beijing: China Economic Times.

[33] Beijing Energy Club (2011) New pricing trends in international natural gas market, *International Petroleum Economics*, March: 48–52.

[34] Xiaogang Long (2009) Price formation system for petroleum products in Korea and Japan: A comparative study and its findings, *Prices Monthly*, November: 63–5, 75.

[35] Ying Fan and Jianling Jiao (2008) *Oil Prices: Theoretical and Empirical Study*. Beijing: Science Press.

[36] Gang Gu (2004) Technology roadmap of hydrogen energy and its enlightenment to China, *Studies in International Technology and Economy* **6**: 34–7.

[37] Research Team, Center for Environment, Natural Resources and Energy Law, Tsinghua University (2008) *Expert Proposal with Narration on China Energy Law (Draft)*. Beijing: Tsinghua University Press.

[38] Shuang Tan (2010) Administration system for R&D funding should be reformed, *People's Daily*, 11 November.

[39] Rongsi Ye and Zhonghu Wu (2006) *A Study of China's Energy Legal System: Strategic Security and Sustainability of Energy Legislation*. Beijing: China Electric Power Press.

[40] Zhenzhong Huang, Qiuyan Zhao and Baiping Tan (2009) *China Energy Law*. Beijing: Law Press.

[41] Drafting Group for Amending the Energy Conservation Law (2008) *Energy Conservation Law of the People's Republic of China*. Beijing: Peking University Press.

[42] Guangrong Zhang (2010) *Research on Several Basic Issues of China's Resources and Energy Oversea Investment*. Beijing: China Economic Publishing House.

[43] Jigang Xie (2010) *Report on Mergers and Acquisitions of Central State-owned Enterprises*. Beijing: China Economic Publishing House.

[44] Huiyao Wong (2011) *China Overseas Development: Views from Overseas on China Going Global*. Beijing: Dongfang Publishing Press.

[45] Williamson E. Oliver and Winter G. Sidney (2007) *The Nature of the Firm: Origins, Evolution, and Development*, transl. Yao Hai-xin and Xing Yuan-yuan. Beijing: Commercial Press.

[46] Tianning Wang and Wei Ding (2009) The discussion of energy-saving and emission-reduction effects of electric locomotive application, *Shanghai Energy Conservation*, March: 21–3.

[47] Jizun Li (2008) *On Chinese Energy Early Warning Model*. Beijing: Science Press.

[48] Siqiang Wang, Yufeng Yang, Zhiyong Tian, *et al.* (2008) Energy data and experience of energy forecasting system in IEA and other international organizations, *Energy of China* **22** (January): 28–30.

[49] Editorial Committee on Fundamental Knowledge of Emergency Relief (2010) *Fundamental Knowledge of Emergency Relief*. Beijing: China Petrochemical Press.

[50] Hui Hou, Jianzhong Zhou, Yongchuan Zhang, *et al.* (2010) Analysis of power emergency drills system at home and abroad and its inspiration for China, *Power System Protection and Control*, December 236–41.

[51] Minxuan Cui (2011) *Annual Report on China's Energy Development*. Beijing: Social Sciences Academic Press.

[52] Shengli Ma and Fei Han (2010) Natural gas reserves situation and experiences of foreign countries, *Natural Gas Industry*, August 62–6.

[53] Yaoqi Wu (2010) *Innovative Model of New-Energy Development: Study and Application of Laws of Energy Industry*. Beijing: Science Press.

[54] Kneese V. Allen, Sweeney L. James, *et al.* (2010) *Handbook of Natural Resources and Energy Economics (Volume 3)*, transl. Xiaoxi Li, Peijun Shi, *et al.* Beijing: Economic Science Press.

[55] Xiaoming Wu (2009) *The Road to a Great Nation: China's Energy Development Strategy*. Beijing: People's Daily Press.

[56] Jun Zhang and Xiaochun Li (2007) *Progress in International Energy Strategy and Innovative Technology*. Beijing: Science Press.

[57] Editorial Committee on the Report on China's Economic Situation and Energy Development (2010) *The Report on China's Economic Situation and Energy Development 2010*. Beijing: China Electric Power Press.

[58] Energy of China (2011) *Annual Report on China's Energy Development 2010*. Beijing: China Science Technology Press.

[59] United Nations Intergovernmental Panel on Climate Change (IPCC) (2007) *Climate Change 2007,* Synthesis Report. Geneva: IPCC.

[60] IEA (2011) *Energy Balance of Non-OECD Countries*. IEA.

[61] IEA (2011) Energy *Balance of OECD Countries 2011*. IEA.

[62] IEA (2011) *Electricity Information 2011*. IEA.

[63] IEA (2011) *World Energy Outlook 2011*. IEA.

[64] IEA (2011) CO_2 *Emissions from Fuel Combustion 2011*. IEA.

[65] IEA (2010) *Energy Price & Taxes, 2nd Quarter 2010*. IEA.

[66] WNA (2011) *World Nuclear Power Reactors & Uranium Requirement*. London, UK: WNA.

[67] GWEC (2011) *Global Wind Report 2010*. Beijing: GWEC.

[68] REN21 (2011) *Renewables 2011 Global Status Report*. Paris, France: REN21.

[69] NREL (2010) *Eastern Wind Integration and Transmission Study*. Denver, Colorado, USA: NREL.

[70] IMF (2011) *World Economic Outlook Database*. Washington, DC, USA: IMF.

[71] US Energy Information Administration (2011) *Annual Energy Outlook 2011 with Projections to 2035*. Washington, DC, USA: EIA.

[72] British Petroleum (2011) Statistical Review of World Energy 2011. BP.

Postscript

All who have worked in the energy sector will feel very proud when they look back at the development of China's energy industry in the six decades since 1949. The generations' hard work has brought about the achievements in China's energy sector, which have garnered global acclaim and which have provided a strong foundation for socialism's modern development. I am a veteran in the energy sector. Forty years have gone by since I began devoting myself to China's electric power industry in the 1970s. In the period of time, I was a part of many important electrical engineering projects and witnessed the growth and progress of China's power industry. Before every obstacle in their paths, the older generation of energy pioneers demonstrated dogged resolve, keen enterprise, the desire to serve the people, and the courage to take responsibility. These qualities compelled me to surmount all difficulties and soldier on.

In the 21st century, with the rapid development of China's economy and the profound changes in the global energy situation, China's energy sector faces a string of tests and challenges. Frequent shortages in energy supply, as well as resources, the environment, technology and frameworks have become important factors in holding back the sustainable development of China's energy sector. As a backbone enterprise in China's energy industry, State Grid Corporation is also facing the problem of finding the right development path to better safeguard China's electricity supply. Strategic thinking with a global vision is required to ensure the right direction of State Grid Corporation's development. Grid development should be considered and planned within the context of the overall development of China's energy sector or even within the wider context of the nation's modernisation. And so, I began to think analytically about the strategic issues of China's energy sector.

I have always believed that to solve China's energy problems, the piecemeal approach of 'treating the head when it aches and treating the foot when it hurts' does not work. A grand energy vision is needed. All the thoughts and proposals set forth in this book are based on this vision, whether it is the planning and promotion of basic direction in transforming the model of economic development, transforming the model of energy development, and transforming the international competitive framework; or the strategic implementation of measures such as the simultaneous development of energy exploration and energy conservation, the simultaneous exploration of traditional energy and new

Electric Power and Energy in China, First Edition. Zhenya Liu.
© 2013 China Electric Power Press. All rights reserved. Published 2013 by John Wiley & Sons Singapore Pte. Ltd.

energy, the simultaneous exploitation of domestic resources and foreign resources, the simultaneous development of improving the energy framework and improving the transmission of energy, the simultaneous development of technological innovation and systemic creativity; or the core beliefs of electricity as the mainstay and the 'One Ultra Four Large' (1U4L) strategy.

A clear strategic basis is necessary to expedite the transformation of the model of energy development. I am of the opinion that this strategic basis is the 1U4L. Bringing about the change in the development of electricity through 1U4L, resulting in electricity's scientific development and ultimately, the transformation of the model of energy development and realignment of the energy strategy, is an approach that is in line with practical circumstances and energy development trends. The key to implementing 1U4L lies in ultra high voltage (UHV). After I began working in Beijing in 2000, I have been able to consider the issues of electricity and energy from a higher plane. Coupled with my practical work experience, I continue to acquire a deeper understanding of the development of UHV power grids. An academic dissertation that I completed some years back provides a more systematic exposition of this issue.

Like any other important and innovative project, the development of UHV in China in recent years has garnered not only applause and praise, but doubts as well. There have also been different perceptions on the construction of the synchronous UHV grid in Northern, Central and Eastern China. I believe that the development of anything new would always go through this stage. As long as all parties concerned are committed to the common good and are subjective and fair, both supporters and detractors will eventually reach a consensus that will serve the national interest.

It is not easy to sort out the strategic thinking on China's energy and harder still to implement the strategies effectively. The structure and regulation of the market system, the standardisation of and support for laws and policies, the support from and leadership of scientific innovation, and the nurturing and growth of modern energy corporations are all important issues concerning the implementation of China's energy strategies. To resolve these issues, the first thing to do is to liberate our thinking, change our perceptions and free ourselves from outmoded ideas. Then, we need to properly manage the various interested parties, with the overall strategic situation of China's energy and the nation's fundamental interest as the prerequisites. To properly manage the interested parties, however, requires us to first change our mode of thinking.

The situation of energy in China is grim. Despite my deep misgivings, I have always remained confident about the development of China's energy and its future. If we can build up the relatively complete and large-scale modern energy industry that we have today from practically nothing in the early years of the People's Republic, we can definitely surmount all challenges to provide safe, reliable and sustainable energy security to fuel the nation's development and renaissance, and improve our people's lives. This is achievable if we abide by the Scientific Outlook on Development, remain united and seek innovation.

I am very aware that there is much room for improvement in this book due to my inadequate levels of knowledge and physical capacity. For example, I am less familiar with oil, gas and coal than I am with electricity. Many of the expositions in this book are made from the standpoint of electricity, and the discussions on oil and gas are much weaker than those on electricity. The book has many other flaws, and I sincerely hope readers will be generous with their criticisms and corrections.

I received assistance and support from various people during the writing of this book. Many leaders and experts of the electricity and energy industries gave their valuable opinions and suggestions, and the colleagues in State Grid Corporation of China were very meticulous in the collection and organisation of a large volume of data. My most sincere thanks to all of them!

Liu Zhenya
May 2013

Index

Electric Power and Energy in China, First Edition. Zhenya Liu.
© 2013 China Electric Power Press. All rights reserved. Published 2013 by John Wiley & Sons Singapore Pte. Ltd.